T0340183

Clinical Ethics at the Crossroads of Genetic and Reproductive Technologies

Clinical Ethics at the Crossroads of Genetic and Reproductive Technologies

Edited by

Sorin Hostiuc

ACADEMIC PRESS

An imprint of Elsevier

Academic Press is an imprint of Elsevier
125 London Wall, London EC2Y 5AS, United Kingdom
525 B Street, Suite 1650, San Diego, CA 92101, United States
50 Hampshire Street, 5th Floor, Cambridge, MA 02139, United States
The Boulevard, Langford Lane, Kidlington, Oxford OX5 1GB, United Kingdom

Notices
Knowledge and best practice in this field are constantly changing. As new research and experience
broaden our understanding, changes in research methods, professional practices, or medical
treatment may become necessary.

Practitioners and researchers must always rely on their own experience and knowledge in
evaluating and using any information, methods, compounds, or experiments described herein.
In using such information or methods they should be mindful of their own safety and the safety
of others, including parties for whom they have a professional responsibility.

To the fullest extent of the law, neither the Publisher nor the authors, contributors, or editors,
assume any liability for any injury and/or damage to persons or property as a matter of products
liability, negligence or otherwise, or from any use or operation of any methods, products,
instructions, or ideas contained in the material herein.

Library of Congress Cataloging-in-Publication Data
A catalog record for this book is available from the Library of Congress

British Library Cataloguing-in-Publication Data
A catalogue record for this book is available from the British Library

ISBN 978-0-12-813764-2

For information on all Academic Press publications
visit our website at https://www.elsevier.com/books-and-journals

Working together
to grow libraries in
developing countries

www.elsevier.com • www.bookaid.org

Publisher: John Fedor
Acquisition Editor: Peter B. Linsley
Editorial Project Manager: Leticia Lima
Production Project Manager: Punithavathy Govindaradjane
Cover Designer: Victoria Pearson

Typeset by SPi Global, India

Contents

Contributors

Numbers in parentheses indicate the pages on which the authors' contributions begin.

Maria Aluas (81), Iuliu Haţieganu University of Medicine and Pharmacy; Babes-Bolyai University, Cluj-Napoca, Romania

Joana Araújo (195), Instituto de Bioética, Universidade Católica Portuguesa, Porto, Portugal

Diana Badiu (55), Ovidius University, Constanta, Romania

Ursela Barteczko (335), Center for the Study of Bioethics, University of Belgrade, Belgrade, Serbia

Ana S. Carvalho (195), Instituto de Bioética, Universidade Católica Portuguesa, Porto, Portugal

Sandrine de Montgolfier (99), Institut de Recherche Interdisciplinaire sur les enjeux Sociaux (IRIS), Bobigny; Université Paris Est Créteil (UPEC), Créteil, France

Jelena Dimitrijevic (335), Center for the Study of Bioethics, University of Belgrade, Belgrade, Serbia

Ina Dimitrova (367), Faculty of Philosophy and History, Plovdiv University "Paisii Hilendarski", Plovdiv, Bulgaria

Rebecca Dimond (31), Cardiff University, Cardiff, Wales, United Kingdom

Lauren Diskin (313), Obstetrics and Gynecology, WellSpan York Hospital, York, PA, United States

Maureen Durnin (385), University of Guelph, Guelph, ON, Canada

Jazmine L. Gabriel (313), WellSpan York Cancer Center, York, PA, United States

Vladimir Gasic (335), Institute of Molecular Genetics and Genetic Engineering, University of Belgrade, Belgrade, Serbia

Fermín J. González-Melado (263), High Centre for Theological Studies, Badajoz, Spain

Sorin Hostiuc (205, 229, 293), Department of Legal Medicine and Bioethics, Faculty of Dental Medicine, "Carol Davila" University of Medicine and Pharmacy, Bucharest, Romania

Mihaela Hostiuc (205), Department of Internal Medicine and Gastroenterology, Faculty of Medicine, "Carol Davila" University of Medicine and Pharmacy, Bucharest, Romania

Michael Hoy (385), University of Guelph, Guelph, ON, Canada

Daniela Iancu (1), Centre for Nephrology, University College London, London, United Kingdom

Valentina Kaneva (367), Faculty of Philosophy, Sofia University "St. Kliment Ohridski", Sofia, Bulgaria

Susana Magalhães (195), Instituto de Bioética, Universidade Católica Portuguesa; Universidade Fernando Pessoa; CEGE, Research Centre in Management and Economics, Católica Porto Business School, Universidade Católica Portuguesa, Porto, Portugal

Stefan Micic (335), Center for the Study of Bioethics, University of Belgrade, Belgrade, Serbia

Amir Muzur (113), Faculty of Medicine and Faculty of Health Studies, University of Rijeka, Rijeka, Croatia

Valentina Nastasel (55), Kantonsspital Graubünden, Chur, Switzerland

Ionut Negoi (205), Department of Surgery, Faculty of Medicine, "Carol Davila" University of Medicine and Pharmacy, Bucharest, Romania

Rafael Pardo (149), BBVA Foundation, Spain

Sonja Pavlovic (335), Institute of Molecular Genetics and Genetic Engineering, University of Belgrade, Belgrade, Serbia

Iva Rinčić (113), Faculty of Medicine and Faculty of Health Studies, University of Rijeka, Rijeka, Croatia

Mugurel Constantin Rusu (205), Department of Anatomy, Faculty of Dental Medicine, "Carol Davila" University of Medicine and Pharmacy, Bucharest, Romania

Dario Sacchini (131), Institute of Bioethics and Medical Humanities, Fondazione Policlinico Universitario "A. Gemelli" – Università Cattolica del Sacro Cuore, Rome, Italy

Margarida Silvestre (195), Instituto de Bioética, Universidade Católica Portuguesa, Porto, Portugal; Faculty of Medicine, University of Coimbra, Coimbra, Portugal; UNESCO Chair in Bioethics, Institute of Bioethics, Universidade Católica Portuguesa, Porto, Portugal

Stephen O. Sodeke (113), Tuskegee University National Center for Bioethics in Research and Health Care, Tuskegee, AL, United States

Antonio G. Spagnolo (131), Institute of Bioethics and Medical Humanities, Fondazione Policlinico Universitario "A. Gemelli" – Università Cattolica del Sacro Cuore, Rome, Italy

Milena Ugrin (335), Institute of Molecular Genetics and Genetic Engineering, University of Belgrade, Belgrade, Serbia

Chapter 1

Genomic Editing—From Human Health to the "Perfect Child"

Daniela Iancu

Centre for Nephrology, University College London, London, United Kingdom

1. INTRODUCTION

In August 2015, the first successful application of genomic editing on a human embryo was announced. This was not only a milestone on the long road of curing genetic diseases, but also a start for an intense debate about the ethics, morality, and legality of deliberately modifying the human genome.

The discovery of recombinant DNA in the 1970s proved that genomic DNA could be modified to incorporate a DNA fragment originated from another organism. Two years later, the scientists organized a conference at Asilomar, in California, United States, to discuss the ethical and moral implications and decide how and under what conditions to use the new technology (Berg et al., 1975a). Forty years later, CRISPR/Cas9 genomic editing brings us not only close to solving the therapy for genetic diseases but also to the redesigning of human beings and many other organisms around us. To be able to extract the most beneficial aspects of technology, we need to understand how it works, what are its strengths and limitations, and what can be done to make it safer.

1.1 Definitions and Context

Genome editing signifies the intentional change of the DNA sequence in a genome by replacing, inserting, or deleting one or more nucleotides. The term is related and sometimes used interchangeably with "**gene editing**," which means acting on one gene to modify its sequence. Related to these is the term "**gene therapy**" meaning the correction of a genetic mutation by replacing or adding a functional copy of the mutated gene to the genome (Naldini, 2015). Gene therapy has a broader meaning than genome editing, and it has been used for decades related to etiologic genetic treatments.

A **genome** comprises the whole genetic information included in the DNA molecules of a cell, an organelle, or organism (Strachan and Read, 2011).

Clinical Ethics at the Crossroads of Genetic and Reproductive Technologies.
https://doi.org/10.1016/B978-0-12-813764-2.00001-5

Therefore we can talk about the nuclear genome, the mitochondrial genome, or the human genome as a whole. The human nuclear genome contains more than 3 billion letters or bases distributed in 23 pairs of DNA molecules. The mitochondrial genome is much smaller (about 16,500 base pairs in humans), with fewer genes and much more variability than the nuclear genome (Strachan and Read, 2011). When we refer to the genome of a whole organism, like a human or mouse genome, we generally have in mind the nuclear genome. The 23 pairs of DNA molecules in the nucleus of a human cell are associated with proteins and packed in as many pairs of **chromosomes**. The first 22 pairs of chromosomes are named autosomal, and they are identical, pairwise, in both genders while the other two chromosomes, designated X and Y, are present in different combinations in men and women: XY and respectively XX (Strachan and Read, 2011). The **genes** are discrete units found along the DNA molecules, containing the information needed to synthesize a protein or a functional RNA molecule. Even if regulatory RNA molecules are also encoded by genes, it is widely accepted that whenever we refer to the coding part of the genome, we consider only the protein-coding genes. Most of DNA is noncoding (about 98%) and only about 2% encodes proteins. The protein-coding genes have a discontinuous structure, represented by exons (the coding part) and introns, noncoding spacers of variable length between the exons, where most frequently are located regulatory elements and sometimes other genes. Each gene depends on close (*cis*) and further (*trans*) located elements, which can be accessed by protein complexes to regulate the onset, rate and time of their expression (Strachan and Read, 2011). The sequence in the coding part of the genes is translated into amino acid sequence in the proteins, with a set of three bases corresponding to one amino acid. Any mistake (**mutation**) in the DNA structure is likely to be translated differently if it alters the way the code is read. Every DNA molecule is copied only once per cell division and the new copy is distributed in the new cell. This is a very precise process due to both polymerase's proofreading mechanism and other repair and correction mechanisms (Miyabe et al., 2011). However, at times, a letter is misplaced and a variation is generated. Occasionally, physical, chemical, or biological factors can also alter the genome, requiring the intervention of repairing mechanisms. The consequences vary from none to the complete alteration of a protein, depending on where the mutation is located (Strachan and Read, 2011). Evolution itself is the result of an accumulation of genetic changes leading to survival, development, or extinction. When a mutation is present in the **germinal cells**, meaning the sperm, the egg, and their precursors, it can be transmitted to the offspring and can determine a genetic disease. Because all the cells of the new organism are derived from a single, initial cell, resulted from the fecundation, this mutation will also be present in all nucleated cells. Every individual inherits half of the genome (one autosomal chromosome of each pair plus one sex chromosome) from each parent. In some diseases, the presence of a single mutation, on one copy of the gene, is enough for pathogenicity (**dominant diseases**) while in others both copies need to be altered (**recessive diseases**); the X chromosome can also be affected and generate a different gender-based distribution of the disease in the family. Occasionally, a mutation occurs in one of the cells of an organ, after the organism has been formed. If the cell is actively proliferating, it will generate a clone with the same genome, but the change will remain without consequences if the cell is quiescent. This variant is named **somatic mutation** and will not be inherited by the offspring, as opposed to **germline mutations**, which are present in germ cells and therefore can be transmitted. Cancers, for example, start with a somatic mutation activating one or more genes involved in cell proliferation. Not all genetic changes are detrimental; some are innocent, or silent, as they do not alter any protein or other regulatory molecules. It is usual to refer to pathogenic genetic changes as mutations and the nonpathogenic ones as polymorphisms (Strachan and Read, 2011). Some DNA variants can either provide interindividual variation (e.g., hair or eye color) or support a better adaptation to the environment.

Given the severe consequences of pathogenic genetic changes, scientists looked for means to correct them. DNA modification of higher organisms is inspired by natural protection mechanisms present in bacteria or less evolved eukaryotes like yeast (Fernandez et al., 2017). The story started decades ago and followed the accumulation of knowledge about enzymes able to recognize and process DNA, in parallel with finding out how a genome is capable of maintaining its sequence and repairing itself after injury or polymerase errors.

1.2 Recombinant DNA Technology—The Basis for DNA Modification

The history of genomic editing started in 1972, when Paul Berg and his team obtained the first **recombinant DNA** molecule ex vivo (Jackson et al., 1972), from two viral parts: the SV40 simian virus and a lambda phage. Shortly, he was followed by Stanley Cohen and Herbert Boyer, who demonstrated that a DNA molecule resulted from the combination of DNA from two different sources can be functional and able to be replicated in a host organism (Cohen et al., 1973). This technology made use of **restriction enzymes**, which are part of bacteria's immune system, to recognize and eliminate the DNA of infecting viruses (phages) from their genomes. A characteristic of restriction enzymes is that they can specifically recognize a short DNA sequence, irrespective of its origin, and then precisely cut both DNA strands, always in the same location, producing two ends where another DNA molecule treated in the same way can be inserted. The resultant DNA molecule can be introduced into a host organism where it can be replicated independently of the host genome and provide new characteristics to the cell (Cohen et al., 1973). The DNA providing the new feature may contain a gene or just the coding part of the gene, and it is called the **insert** while the other component, containing the elements required for self-replication and control of gene expression called **vector**. The vector can be a modified plasmid (bacterial DNA, usually responsible for resistance to chemicals or other functions and located outside bacteria's main genome), a virus, or a complex, engineered DNA molecule (Strachan and Read, 2011). This system allows for a limited size of the DNA insert, variable between thousands and tens of thousands of nucleotides, depending on the vector used (Strachan and Read, 2011). Recombinant DNA obtained from viral vectors modified to contain human genes were later used to integrate new DNA molecules into the mammalian genome, either in vitro, to perform functional studies, or in vivo, to test gene therapy or to introduce markers that could identify modified organisms. Restriction enzymes require a specific sequence that can be found anywhere in the genome, within genes, or in the intergenic space this being a disadvantage for the applications of genomic editing in eukaryotes, particularly mammalians, because it does not allow targeting of a single gene/region.

1.3   Genome Editing

The capacity to induce targeted modifications in the genome of higher eukaryotes evolved progressively. According to the tools used to identify and open the target sequence, there are two approaches:

- protein-based genome editing: the target sequence is recognized by a protein, which can be a naturally occurring restriction enzyme (meganuclease) or an engineered enzyme (zinc finger nuclease or TALENs); in this case the protein is also the "cutter" (Chandrasegaran and Carroll, 2016);
- RNA-based genome editing: the target sequence is recognized by the complementarity between an RNA molecule and the DNA; a protein is required to recognize the RNA-DNA complex and cut the DNA strands (Doudna and Charpentier, 2014). CRISPR/Cas9 is the classic example in this case and it will be described in detail further down.

The process involves cutting out the mutated sequence and replacing it with a corrected version. Alternatively, one might choose to introduce a mutation in a gene to study the changes induced by its loss of function. In both cases, an enzyme must recognize and cut a specific sequence in the 3 billion base pairs of the human genome and this is not an easy task. The longer the sequence recognized by an enzyme, the more specific the process is. The second part in the process is delivered by the cell's own repair mechanisms. There are two types of repairs involved in genome editing: nonhomologous end-joining (NHEJ) and homology-directed repair (HDR). The first consists of simple reattachment of the two ends of the break and can result in new mutations like insertions, deletions, or inversions. The second needs a template and produces targeted, more precise changes. The disadvantage of the system is that NHEJ is the most

TABLE 1 DNA Repair Mechanisms

Direct repair

This mechanism is dedicated to the repair of nicks (rupture of a single dsDNA phosphodiester bond), or chemical modification of bases (introduced by physical or chemical agents).

Excision repair

- Base excision repair (BER): the wrong base is removed, followed by removal of a few adjacent bases on either sides, on the same strand and resynthesis of the DNA stretch.
- Nucleotide excision repair (NER): the DNA segment containing the mutation is removed and replaced by synthesis; this mechanism can repair more complex mutations.

In humans more than 40 genes are involved in this type of repair.

Mismatch repair

A complex of proteins recognize the mismatch, select the strand that needs to be repaired, and finally perform the correction by synthesizing a new strand.

Nonhomologous end-joining

The double-strand brakes are repaired by simply linking the two ends together, disregarding the precision; this mechanism can, by itself, generate a mutation.

(Based on Strachan, T., Read, A., 2011. Human Molecular Genetics, Garland Science, New York; Brown, T.A., 2006. Genomes, Garland Science, New York.)

frequent choice. Other types of DNA repair mechanisms available in living cells and that can be used in editing technologies are presented in Table 1.

1.3.1 Meganucleases

In the second half of the 1990s, Choulika et al. used a restriction nuclease with a very large recognition sequence, a meganuclease, to cut the host DNA prior to ligating the insert DNA molecule. The first enzyme to be used was a yeast restriction enzyme, *I-SceI*, whose restriction site of 18 nucleotides is very rare in the yeast's genome (Choulika et al , 1995). For comparison, an average restriction enzyme recognizes a stretch of 4–6 nucleotides. The meganucleases' requirement for a specific restriction site has limited the practical applications, so they have been engineered to recognize different target sites and to integrate into the genomes of a larger variety of organisms. For example, this technology was used to edit the adenosine deaminase (*ADA*) gene in human cells (Grizot et al., 2009).

1.4 Zinc Finger Nucleases (ZFNs)

By combining the nuclease domain of the restriction enzyme *FokI* with the DNA recognition domain of a **zinc-finger transcription factor**, via a recombinant DNA technique, resulted in the first programmable nuclease, zinc finger nuclease (ZFN) [reviewed in Chandrasegaran and Carroll (2016)]. The **transcription factors (TFs)** are proteins able to recognize the regulatory sequence of a gene and to promote gene expression by attaching to it and attracting other proteins of the transcriptional complex. There are several types of TFs, classified based on their secondary structure, the zinc finger TFs being the most abundant. The zinc finger motif contains a zinc atom bound to two conserved cysteine and histidine residues and forms a secondary structure similar to a finger. Each motif recognizes a stretch of 3–4 DNA bases and by engineering the structure of these motifs, they can be made to recognize a precise target sequence (Chandrasegaran and Carroll, 2016). Once the zinc finger motif recognizes the target DNA, the nuclease domain is activated to cut it.

This technology was used successfully to create the first transgenic rat model (Geurts et al., 2009) and spread rapidly in research labs as a tool to study disease mechanisms. It was also used to enhance and ameliorate the genetic profile of livestock productivity. Zinc fingers nucleases were also used in a clinical trial for a genome-editing therapy against HIV infection (Maier et al., 2013). The ZNFs have the advantage on being programmable and more effective than meganucleases but they are also expensive and imprecise, having a large number of off-target effects, some with particularly toxic consequences on cells and animals (summary in Shim et al. (2017)).

1.4.1 Transcription-Activator Like Nucleases (TALENs)

The next step in genome-editing history was represented by another type of chimeric nucleases incorporating *FokI* nuclease domain: transcription-activator like nucleases or TALENs. Transcription activator-like effector proteins act in plants to overcome the host's immune protection and redirect resources to support the pathogen. Even more, TALENs' molecule can contain a code to direct recognition toward a specific DNA sequence, based on a simple correspondence between the amino acid structure of the DNA binding domain and the nucleotide sequence of the target (Carroll, 2011). TALEN's started to be used in functional studies, basic research, and also in gene therapy using iPSCs (Hatada and Horii, 2016). TALENs have been successfully used to correct a genetic defect responsible for macular degeneration in mice (Low et al., 2014).

TALEN as a genomic-editing tool is based on the discovery of a TALE (transcription activator-like effector) module in the plant virulence factor of *Xanthamonas* bacteria. The domain contains almost identical repeated units of 33–35 amino acids: the only variation is located at positions 12 and 13, where the amino acids are highly variable and thus provide specificity to nucleotide recognition. This variable part is named repeat variable di-residues (RVDs) (Chandrasegaran and Carroll, 2016). Moreover, the DNA recognition by TALE modules is context-independent. TALEs have been used in combination with *FokI* cleavage domain in order to develop a new set of genome-editing nucleases with high specificity for a specific locus. TALENs are easier to generate and less toxic than ZNFs but more difficult to multiply and administer into host organisms due to the much bigger size of these genes. Its use has been limited by the very high price and the competition from the newly developed CRISPR/Cas9 technology (Kim, 2016).

1.5   CRISPR/Cas9 Technology

Zinc finger nucleases and transcription activator-like effector nucleases recognize the DNA sequence based on specially designed protein motifs, which makes them more difficult to program and control. These problems have been solved by the introduction of a different genomic-editing system, where the recognition of the target DNA region is realized via an RNA molecule, complementary to the target sequence. This solved three problems: the size (a short RNA molecule being much smaller than a protein), the precision (RNA to DNA complementarity ensures a more precise recognition of the target site), and the cost.

CRISPR history started in 1987, when Yoshizumi Ishino had discovered a set of strange repeated sequences in the 3' region of *E. coli iap* gene and he called them REP sequences, without being able to explain their role (Ishino et al., 1987). Later, Francisco Mojica found 30 base-pair repeats separated by 36 base-pair spacers in archaea (Mojica et al., 1993) and correlated these structures with the ones reported by Ishino in *E. coli*.

Following his studies, he showed that these repetitive sequences are present in 40% of bacteria and 90% of archaea and he initially named them Short Regularly Spaced Repeats (SRSR) which was later on changed to Clustered Regularly Interspaced Palindromic Repeats (CRISPR) and the characteristics of these loci, including the presence of *cas* genes, have been described (Jansen et al., 2002). The biological role of this system in the immune protection of bacteria against viruses was not found for another 10 years (Barrangou et al., 2007). In 2005, it was recognized the extrachromosomal origin and the role of the spacers (sequences between repeats) (Bolotin et al., 2005). The first use of Cas9-crRNA system as a genomic editing tool was reported in 2012 (Peng et al., 2016) and the system became shortly the golden standard for performing functional studies to understand gene functions or disease mechanisms. In 2013, the first application of CRISPR/Cas9 technique in mammals was recorded (Cong et al., 2013). The ease of use and the precision recommended the technology as a major gene therapy tool in humans.

Three types of CRISPR systems have been described (Peng et al., 2016; Makarova et al., 2011). Of these, type I and type III require two or more Cas proteins while type II only needs one Cas protein.

Genome-editing experiments use the type II system, as this requires only one Cas protein, respectively Cas9 or an alternative (Doudna and Charpentier, 2014). The mechanism is very simple: the guide RNA molecule contains a part complementary to the target sequence (crRNA) and a part required for Cas9 recognition and activation (tracrRNA). These two components are synthesized separately in bacteria and they were used separately in the beginning but nowadays, crRNA and tracrRNA are synthesized together, in a single molecule named sgRNA (single guide RNA), containing a stretch of 20 nucleotides complementary to the target at the 5′ end and a 3′ double-stranded tracrRNA-like end. One or more regions can be targeted at the same time, provided that the adequate sgRNAs are provided. crRNA and Cas9 act when a NGG sequence called *Protospacer Adjacent Motif* or PAM is present adjacent to the complementary genomic sequence (Doudna and Charpentier, 2014). Cas9 protein has a HNH, nucleasic domain, and a RuvC domain. The nuclease domain cleaves the strand, which is complementary to the crRNA, while the RuvC domain cleaves the opposite strand (Doudna and Charpentier, 2014). Mutations in the two essential domains of Cas9 can impact on its function by either eliminating the capacity to cut the opposite strand or completely the capacity to create DSBs (Doudna and Charpentier, 2014). tracrRNA is encoded by a sequence located upstream of Cas9 in the CRISPR/Cas operon in bacteria and it has been shown that this is essential for crRNA maturation (Doudna and Charpentier, 2014).

CRISPR/Cas generates DSBs repaired mostly by NHEJ and with a much lower frequency, below 10%, by HDR. An alternative is microhomology-mediated end-joining (MMEJ) and this technique has been successfully applied in mice to edit genes in vivo (Yao et al., 2017). The efficiency of these repair mechanisms is dependent on the cell cycle: HDR happens in cells found in late S and G2 phases while MMEJ in cells in G1 and early S phases (Yao et al., 2017). According to some authors this is more efficient, particularly in slow or less dividing cells as it is the case for adult tissues and therefore it would be more efficient for somatic editing. In addition, this technique would require a shorter homology arm acting as donor template, a fact that can be very useful in vivo, provided that the specificity can be ensured (Yao et al., 2017). It was shown that using NHEJ for integrating genes into the genome can generate random insertions, deletions, and can reverse the orientation of the genes, thus removing the functional genomic context and making the effects more difficult to predict. In addition, NHEJ is not able to replace a point mutation or a small deletion or insertion, but is only able to introduce a new sequence between the two sides of the cut.

As for other programmable nucleases, CRISPR/Cas9 system may have off-target defects, variable according to the sgRNA design, cell types, delivery methods, cell cycle stage, persistence in the cell before inactivation (Kim, 2016).

1.6   Base Editing Technology

In an attempt to circumvent CRIPSR disadvantages and limitations, the team lead by David Liu at Harvard University derived an editing enzyme by merging elements of Cas9 and a cytidine deaminase, which can change a base in DNA without introducing double-strand breaks (Komor et al., 2016). In this variant of genome-editing technology, the base is no longer removed but converted into another one by a chemical modification. For example, a cytidine (C) would be replaced by a uridine (U), and subsequently, the other strand will have an A instead of a G, after DNA replication. The system is still guided by an RNA molecule but no longer requires a template to perform homology-directed mutagenesis (Komor et al., 2016). Two mutations, p.Asp10Ala and p.His840Ala neutralize the catalytic activity of Cas9, rendering it unable to cut the DNA strands and the modified enzyme is called "catalytically-dead Cas9" (dCas9). Genome-editing efficiency can be increased above 50%, while the indels form with a frequency below 0.1% and there are no off-target effects (Komor et al., 2016; Gaudelli et al., 2017). Liang et al. applied a high fidelity base editing system to perform a C to T transition in mouse embryos and obtained 100% efficiency on both alleles, meaning that this technique could be used to correct recessive mutations (Liang et al., 2017b). The same team edited a mutation causing beta thalassemia in human embryos, with an efficiency of 23% (Liang et al., 2017a). While CRISPR/Cas9 can only be applied to double-strand DNA, base editing can also be used to modify RNA molecules, an option that can prove very useful for a better controlled genetic treatment, at least for diseases with a limited evolution in time (Kim et al., 2017).

1.7   Principles of Using Genome Editing in Research and Clinical Practice

The DNA modification techniques are essential tools for basic and translational research (Lau and Davie, 2017) but they are much less used in clinical practice. This is about to change in the near future as new methods of gene therapy become more precise and more affordable.

Classical gene therapy implies addition of a correct, functional copy of the affected gene to the patient's genome, using a carrier DNA molecule (vector). The vector and all its elements are delivered with variable efficiency to the nucleus of target cells, where it functions separately or is integrated into the genome at the first cell division. Most vectors used in clinical applications are modified viruses lacking the infectious capacity: retroviruses, lentiviruses, or adenoviruses. Gene therapy has been tested in clinical trials for diseases like hemophilia (Manno et al., 2006) or X-linked severe combined immune deficiency (SCID) (Touzot et al., 2015).

Genomic editing involves the use of a complex system to correct the mutation in one or both copies of a gene. Therapies based on genomic editing can be developed for the treatment of genetic diseases, cancer, infectious diseases, and even complex diseases when one or more variants are proven to be significantly associated. The presence of a vector is not required but another vehicle might be needed to deliver the complex to the place of action.

About 20 gene-editing therapeutic trials are currently in process worldwide (Shim et al., 2017).

There are two ways to address the targeted treatment of a genetic disease: induce corrective changes into the specific tissue(s) affected by the disease (**somatic editing**) or act very early during development (on the germ cells or the zygote before four-cell stage), also named **germline editing** (Evitt et al., 2015).

Most efforts have been made toward the development of curative treatments for rare diseases but some cancers and infectious diseases are also potential targets for genome-editing therapy.

Genome editing can be applied to all developmental stages, but any genetic change affecting germline cells is still banned by the international community as this can have permanent consequences on future generations.

2.   ETHICAL ISSUES IN CLINICAL GENOME EDITING

The moral and ethical values of medical practice transcend the physical, legal, economic, and social boundaries and follow the same fundamental principles: respecting the autonomy of the moral agents, beneficence, nonmaleficence, and justice (Beauchamp and Childress, 2001). The rapid development of gene editing, accessibility, the many "unknowns" related to the long-term effects, the risk of accidental changes in other genes than those intended, or intentional misuse, raised concerns among scientist, ethicists, politicians, and the general public.

2.1   Nonmaleficence and Risk/Benefit Assessment

"Do no harm" was a fundamental conundrum of the medical profession since Hippocrates. Each time we introduce a new therapy, we should perform every reasonable test to demonstrate that the procedure does not generate a significant medical harm compared to the expected/known benefits (therefore there is a positive benefit/risk ratio). Gene therapy, with its latest development, namely genomic editing, makes no exception. One can say that finding a treatment for severe diseases, characterized by a rapid evolution toward incapacity and death, can only be beneficial for the patient or at least it cannot be worse than the disease. However, because the medium and long-term effects of introducing precise changes in the genome cannot be predicted based on current knowledge, added to the fact that these effects can extend to more than one generation, genomic editing raises concerns and intense debates (Bosley et al., 2015; Baltimore et al., 2015).

Genetic diseases rarely affect a single member of a family, even if the patient is the only one showing the phenotype and, therefore, any gene therapy should take into account potential effects on other family members. Some authors recommended a moratorium on the applications of genome-editing technologies in humans (Carroll and Charo, 2015), while others considered that the knowledge potentially offered by the technology justifies the risks of moving further (Savulescu et al., 2015; Gyngell et al., 2017). The moratorium supporters are inspired by a similar decision that followed the discovery of recombinant DNA (Berg et al., 1975b) and recognize the insufficient information about potential long-term effects, the limitations and the risk of misuse of genome-editing techniques, particularly CRISPR (Lanphier et al., 2015). While somatic editing is accepted as a potential solution for severe genetic diseases and cancer, it is generally agreed that edited germline cells should not be used for reproductive purpose (Bosley et al., 2015).

2.1.1 Risk to Benefit Analysis

The risk to benefit analysis is a multifactorial concept, including parameters such as the age of the patient, the severity of the disease, the availability of alternatives, the social, psychological, and economic costs of both performing and not performing the procedure, and the yet unpredictable factors. While for the traditional therapies, the risks are weighed against the benefits for the patients or group of patients, the assessment of risks and benefits in gene editing therapy should be extended beyond the physician-patient relationship, as it can alter the well-being of future generations, and even of the human race as a whole (Committee on Science and NASM, 2016). The subjective perception of the risk varies between patient, his family, health-care professionals and general public, and no stakeholder alone should assess it, being needed a combined approach to maximize the acceptability of the analysis.

Risk assessment should take into account the nature, severity, probability and imminence of the risk. Different levels of risk are tolerated at different development stages and ages, with procedures performed on prenatal or pediatric patients having the lowest acceptable risk to benefit ratio (NASM, 2017).

To exemplify this, let's consider a few cases from our clinical experience, each presenting particular ethical aspects with regards to gene editing-based therapy.

Case 1

A 5-year-old boy, with Duchenne muscular dystrophy (DMD) in its early stages, is the first one with this disease in the family and his mother is pregnant with a second boy. Genetic testing identifies a mutation in the *DMD* gene in the mother, a result that means that the second child is at risk of being affected. The mother has a sister who has no children but who is also at reproductive age. This is a model for a severe disease, with very early onset, in a family with multiple (potentially) affected individuals. There is no cure and very few therapeutic options are available. There are several aspects to consider in this case and these will be detailed further.

Even if the child is only mildly affected at this stage, there is no doubt about the fatal evolution of the disease. Somatic gene editing, able to maintain a certain amount of functional muscular fibers, extend the ambulatory period and delay cardiovascular and respiratory complications by several years or more, can be seen as beneficial, compared to the "no treatment" option. The risks of immune reactions, off-target effects, and later onset cancers or other equally severe diseases cannot be estimated because there is limited information available about this type of therapy. Previous information from gene therapy trials performed for SCID, showed that immune reactions or cancers might occur within 2 or 3 years from treatment (Branca, 2005), an interval that is shorter than the natural evolution of the disease. This is true both in the case of the diseases treated in the trial, and it could also be true for DMD, in the case presented above. The parents have to decide whether their child should risk the new therapy, or continue with the conventional treatment (e.g., Prednisone), waiting for better knowledge to be accumulated in the field. Gene therapy has the potential not only to alleviate or even cure the muscular dystrophy but also to initiate another condition (e.g., severe immune reaction or cancer). By choosing to wait and obtain additional data about the risks associated with somatic gene editing from other studies, the parents face the risk of limiting the beneficial effect from the genetic treatment because the number of actionable muscular cells decreases as the child grows older.

In genetic diseases with early onset and severe evolution the benefits of a clinical trial, based on gene editing or any other therapy, can only be obtained if the relevant population is involved (Jaffe et al., 2006). For example, the gene therapy trials for cystic fibrosis showed limited benefit on adults, because of the differences in lung morphology and physiology, accumulation of chronic changes, reactions to the vectors used for delivery, and level of inflammation, compared to children (Jaffe et al., 2006). In our case, this means that a clinical trial performed on older patients would not provide the same information and benefit. Given the evolution of the disease, the risks associated with such a study could be higher in older patients, but they should be the initial subjects nonetheless as they can autonomously choose to enter in the trial, unlike children who, even if could benefit more, are unable to do so on their own.

Regulations in individual countries may vary, but in all cases, it is required to characterize the risks to the greatest possible extent and to present them to the participants in an open manner. Trials with a higher risk can be pursued for those patients with very limited or no alternatives (Evitt et al., 2015). Only after an initial assessment of the risks is performed, we should evaluate the risk to benefit ratio. When the trial is beneficial for the child, and the estimated risk is low, one parent can be enough to decide to include the child in the study. When the risk is high, it is preferable to ask both parents for consent, if available (NASM, 2017).

The social, educational, and economic context, as well as the support available for the families who care for a child with a severe disease can influence the way parents perceive the risks. Media and social networks, close friends and other members of the extended family may also play a role in parents' perception of the risk.

The male fetus in a mother's ongoing pregnancy can be tested for the presence of the mutation and the result can influence medical and family decisions depending on the age of the pregnancy; personal, cultural, and social values; as well as economic status of the parents and existing support in the society. The fetus could receive somatic gene therapy after birth or in utero, assuming that this would be available. The risks involved would depend on the methods developed for the treatment in utero, but we can consider the risk of intrauterine death or spontaneous abortion, deformation due to accidental damaging of an organ or segment, hemorrhage, or delayed effects associated with the vehicle used to deliver the gene editing complex or with the off-target effects of the complex. However, the child is unlikely to transmit potentially damaging genome changes to his offspring. It is now difficult to estimate the frequency of adverse reactions following gene-editing therapy as there is not enough experience, yet, but this compares to the 100% risk of developing a severe disease early in childhood.

Different levels of risk are tolerated at different development stages and ages, with procedures performed on prenatal or pediatric patients having the lowest acceptable risk allowed. In the United States, the consent for an experimental intervention on fetuses is needed from one or both parents, as follows: (1) if both the fetus and the mother benefit, the mother's consent is enough; (2) if the treatment/research is potentially beneficial for the fetus only, the consent of the father is also needed; (3) if neither the mother nor the fetus benefit, just bringing developmental information, only a minimal risk is allowed with regards to the fetus and only if this is the only way to obtain this information (NASM, 2017). In the United Kingdom, the court of law can reverse the decision of the parents if this is not considered to be in the child's best interest, as recently could be seen in the case of Charlie Gard (BBC, 2017).

The mother's risk to have more children with DMD depends on the gender: every boy has half a chance of inheriting the mutation and developing the disease. The probability of having an affected boy would be ½ (probability to have a boy) × ½ (probability that the boy is affected) = ¼ or 0.25. Being tested for the disease and having the mutation identified, this risk can be altered by applying selection methods: genetic testing on the embryo, followed by gene therapy, pregnancy termination, in vitro fertilization, and either selection of healthy or female embryos or using the egg from a donor. Each of these associates a certain amount of physical, medical, and emotional risk, which is difficult to predict and can have a different impact on the family. It is known that IVF technologies have a success rate dependent on the age of the woman, the quality of the germline cells used, and the presence of other health-related and technical factors (Vaegter et al., 2017). In the world, the average success rate is 40%. The procedure is emotionally stressful but can also involve health risks for the mother, related to hormonal stimulation and egg collection. PGD is shown to decrease the pregnancy success rate and can also be affected by diagnosis errors, particularly in the mosaic cases or when more than one genetic change is present in the genome (Klitzman, 2017). The parents and particularly the woman will need to have all this information in mind and to balance the expected benefit (a healthy child) against all immediate and long-term risks. Unfortunately, a significant number of couples with affected children choose nowadays not to have additional offspring, as a means of protecting them from such a severe disease. Media can have an influence on how the risk is perceived in such families but can also bring the language closer to the patient and can facilitate understanding and stimulate questions.

About one-third of the mutations generating DMD are new mutations, meaning that an affected child would appear unexpectedly in a family. Because it is impossible to predict when and where this is going to happen, the alternative preventive methods like PGD are not available, and therefore postnatal gene therapy following detection of the first sign of disease remains the only option in such cases.

Because many genetic diseases run in the family, the sister of the mother presented here is also at risk of being a carrier and giving birth to an affected child. Sharing genetic information within the family is recommended and encouraged as this can provide the opportunity for early diagnosis in other members. This issue challenges the principle of confidentiality, and it is a matter still under debate if the doctor should consider the benefit to the family and communicate genetic information (Lucassen and Parker, 2003, Offit et al., 2004). Should the child presented above receive any genetic therapy, this is covered by the rules of confidentiality and because there is no implication on other members of the family, this information should remain confidential. However, because a curative therapy is available, this could constitute a reason for sharing the genetic information to the mother's sister who is at risk of having an affected child who could benefit from early treatment (Dheensa et al., 2017).

The risks associated with gene therapy can be only partially assessed, based on the current knowledge. An immediate risk can be related to a potential inflammatory reaction to the components of the editing system. A similar reaction but to the vector used in a classical gene therapy trial was seen in the famous case of Jesse Gelsinger who was treated for an inborn ornithine transcarbamylase deficiency (Branca, 2005). Gene editing can be done without the intervention of a viral vector but can still initiate an immune reaction to the Cas protein, the DNA template, and the guide RNA.

Failure of the gene editing system to produce results can be observed at short or medium intervals after treatment and oncogenetic or off-target effect after a medium or long-term evolution. The patient needs to be aware of these and attend regular follow-up appointments.

Case summary: The risk assessment in a case of a severe genetic disease, with early onset, must take into consideration: the familial aspect of the disease, natural evolution, the age and health status of the patient, type of therapy (somatic or germline), known or potential toxic effects, alternative treatment and associated risks, involving the family in making the decision. Gene-editing therapy is tailored for a single mutation, which can be present in just one family. This can bring the costs at higher levels and affect the patient's right to benefit from the treatment. Sharing genetic information is seen as very sensitive by many patients and requires a particular attention to autonomy and confidentiality issues.

Case 2

A 50-year-old man presents with early signs of Huntington's chorea. He is the first in the family with the disease, but he has children who are at risk of inheriting the mutation. Disease onset can vary even among people who belong to the same family but the evolution is fatal in 100% of the cases. Given the onset of the disease at a mature age, the patient has the chance to live a fulfilling life before this and have a family. The type of mutation is different compared to the first case: Huntington disease is caused by a dynamic mutation (expansion of a trinucleotide encoding for glutamine in the 5′ part of the gene encoding *huntingtin*). Editing this type of mutation is more difficult and requires a different approach than the usual gene-editing techniques (Haussecker, 2016; McMahon and Cleveland, 2016). The risk assessment in this case must consider both the option of somatic gene therapy for the patient and preventive gene editing for his children.

The risk of a long and severe physical and mental deterioration followed by premature death is 100% in chorea Huntington, and there is no curable treatment. Because it is a degenerative disease, the gene therapy must be initiated at early stages, when there are still enough viable neurons to be effective, and the editing system must reach the whole mass of the brain, an objective that is difficult to achieve.

The risks for the patient would be: (1) ineffective therapy and consequently the necessity go through several rounds of treatment, thus multiplying the risks and allowing even more nerve cells to degrade; (2) immediate or delayed adverse reactions; (3) wearing-off of the beneficial effects and the requirement to repeat the administration. In this case, there is a difficult decision to make: if we wait for the disease to advance before initiating the treatment, this can be rendered ineffective by the natural evolution of the disease. On the contrary, if we act early, there is a potential risk of toxic reactions.

In the case of the patient's children, the risk presents different aspects. Initially, the presence of the mutation must be ascertained in them. However, this would represent predictive testing in a person who cannot consent for herself, because the children do not have any sign of disease. Predictive testing is only supported by the availability of a curative therapy and can be accompanied by significant emotional stress in the family. Assuming they are younger than 16, even if they inherited the mutation, they can be asymptomatic and have a good life for several decades before showing any sign of disease, while new, safer, and more effective therapies are developed. The risk of having early gene-editing treatment, in an attempt to prevent degeneration must be weighed against the possibility of more adverse reactions that can initiate either immune or oncogenic effect and severely affect their health, much before Huntington disease would do it. More experience accumulated about the frequency of adverse reactions would help to establish the probability of these risks.

Should the patient decide to have more children, there is a 50% chance that they would inherit the mutated allele of the gene and therefore, IVF combined with PGD or gene editing would be available, with the associated risks of germline editing.

Case summary: In a late-onset severe degenerative genetic disease, the risk assessment should include the disease stage, the probability of adverse reactions, efficiency related to the possibility to deliver the editing complex to all the target cells. Regarding technology, this should be adapted to the type of mutation and the targeted tissues.

Case 3

A 30-year-old woman with a recent diagnosis of autosomal dominant polycystic kidney disease (ADPKD), is a member of a family with several affected individuals. ADPKD is known for the phenotypic variations among members of the same family, and thus the evolution of this patient's disease would be difficult to predict. End-stage renal failure can be reached earlier or later in life or can be absent. The patient has a 50% chance of transmitting the disease to her offspring and could benefit from prenatal testing and/or IVF—PGD prevention.

In this case, the risk of adverse reactions from gene-editing therapy might outweigh the benefits and precede by many years the deterioration of patient's kidney function. In additions, the presence of multiple cysts in patient's kidney might limit the target tissue available for gene therapy, and therefore the beneficial effect might not be reached. Other tissues (liver, pancreas) potentially involved in the disease should also be targeted. Would this treatment be more beneficial to patient's children and is this a reason to approve predictive genetic testing? Finding out the mutation responsible for ADPKD is not an easy task because there are multiple pseudogenes in the genome and the result can be confusing, meaning that there is a risk of mistargeting and possibly deleterious effects.

Case summary: When a disease has late onset and variable severity, one must consider the fact that the potential risks from gene editing can be higher than the expected benefit and limit this option to the situations where other alternatives are not available. When the exact mutation responsible cannot be determined, gene editing should be avoided. The recent whole genome sequencing projects could help identify the potential modifying genes and select the patients who are more likely to suffer from a severe form of the disease.

Genetic diseases are essentially familial. The targeted mutation should be the same in all related individuals as it is very likely that they all inherited the same mutation. However, there might be a variation in the side effects, given by different off-target effects and different background between the individuals. This means that a very good outcome in one patient is not necessarily reproduced in a relative. Treating a known mutation does not exclude the presence of another unknown mutation in another gene. An unwanted mutation resulted from the therapy could affect more than one generation. The case of multifactorial diseases is even more complicated. We do not know today if a polymorphism creating a higher risk for a chronic condition is not, in fact, protecting for another (Camporesi and Cavaliere, 2016; Bosley et al., 2015), like, for example, sickle cell trait protects against malaria in countries where this is very common. Recently, a modification in the genome of T cells aiming to induce resistance to HIV virus, was found to increase susceptibility to West Nile virus (Baker, 2016).

There is almost no information about how potential errors introduced by the editing mechanism can influence epigenetic regulation or the recently discovered genomic interactions and this is a subject that should be addressed before initiating clinical trials.

When this type of intervention becomes available, it will need to be accompanied by reliable mechanisms to prevent or remove these unwanted effects. A list of suggested elements to consider when assessing benefit-to-risk ration is presented in Table 2.

TABLE 2 Elements to Consider When Assessing the Risk to Benefit Ratio

Disease: Mendelian OR multifactorial OR cancer

Mutation: Point mutation OR dynamic mutation OR chromosomal rearrangements OR epigenetic

Penetrance (proportion of people who manifest the disease of those who present a genetic change)

Variable expressivity (presence of a particular combination of signs and symptoms in people having the same disease and mutations in the same gene)

De novo mutation OR inherited?

Presence or postimplantation occurrence of other mutations in one or more different genes

Autosomal (dominant OR recessive) OR X-linked disease?

Age of the patient: Adult OR child OR fetus OR embryo

Familial OR sporadic disease

How many members of the family are at risk of being affected?

Natural evolution of the disease, including life expectancy and the quality of life

What dosage provides the best and safest outcome?

Are there any safer alternative treatment and preventive means?

Knowledge about the beneficial and adverse effects

Potential impact on other family members, the community, and the environment

The medical team: How well the team masters the technology?

Previous experience

How comfortable the team is with performing the treatment for the specific case of the patient?

Is there a protocol in place to deal with immediate and long term unexpected adverse reactions?

Is there a follow-up protocol set up to early detect and treat complications?

2.2   Beneficence in Gene-Editing Therapies

By applying the principle of beneficence, as defined by Beauchamp and Childress, the physician has the duty to care and provide benefit to the patient while keeping the risks to the minimum.

In genetic diseases, beneficence can have multiple layers: (1) a cure for the patient; (2) cure or prevention for other family members; (3) alleviation or delay of severe symptoms where no cure is available; (4) minimizing the risks; (5) a favorable risk/benefit ratio.

It is very likely that a well-designed gene-editing therapy would be considered as beneficial and preferable to the conventional therapies or the natural evolution of the disease when it is severe (Cases 1 and 2 above). On the other hand, applying gene therapy for an isolated defect, like deafness, which has limited effects on the quality of life and allows a relatively normal lifespan, can prove to have a much higher risk to benefit ratio, making the procedure not recommendable, unless the risks are not more than minimal. Some diseases have a wide range of phenotypic variation, from mild to severe and it is often difficult to assess the long-term evolution at the moment of diagnosis. This is why a higher risk associated with gene-editing therapy in Case 3 must be balanced against the probability of the disease to have a very mild course. Previous experience from gene-therapy trials supports the idea that any intervention at the genetic level in human should be done only after thorough research into the possible effects (Branca, 2005).

In order to provide the best options for the patient and the family, and the best care available it is essential to work in a multidisciplinary team if the genetic disease affects several organs and systems. This is also valid for the long-term follow-up after gene therapy, when besides the evolution of the main condition, the physician is looking for evidence of off-target or secondary effects. Both before and after therapy, it is recommended a full transparency regarding available treatment or trials, new knowledge and reinterpretation available. Encouraging the patient to provide feedback contributes to a prompt adjustment of the therapeutic protocol (Table 3).

Given the higher risk of these therapies, they should be chosen voluntarily and particularly recommended to patients where validated therapies are not available (Kimmelman, 2005).

Children and adolescents can be involved and benefit from gene therapy trials when there is no safer treatment for their condition and the data support a low risk of toxicity. The same applies for adults with limited capacity, who cannot comprehend and appreciate the implications of therapy.

When used on embryos, an immediate benefit would be that the family would have a healthy child, while it is spared of repeated IVF cycles, emotional and psychological stress; in parallel, this could contribute to the decrease in the number of embryos destroyed. The supporters of germline editing argue that editing mutations in embryos inheriting a severe genetic disease will limit the pool of mutations

TABLE 3 Elements of Beneficence in Gene-Editing Therapies

I. For the patient:
  - being offered an efficient and safe cure for his condition;
  - being informed about his options and professionally assisted to make an informed decision;
  - avoiding potentially harmful solutions offered by commercial agents acting outside regulatory framework of the medical profession;
II. For the family:
  - improvement of the overall health status;
  - information regarding options for other affected members;
  - prevention of disease for the future generations or, at least, better management;
III. For the community and society:
  - Increase of general well-being;
  - Reducing the health-care costs;
  - Public involvement in the decision about controversial issues: germline editing, use of gene editing for human enhancement;
  - International collaboration to improve knowledge and therapeutical protocols.

and this will contribute to general beneficence for humans (Cavaliere, 2017). This is true for many monogenic diseases but other diseases are sporadic (they occur in one member of a family) or are caused by large chromosomal rearrangements or de novo mutations, which cannot be anticipated and therefore corrected.

In countries like Germany, Italy, or Switzerland, where the law limits the number of embryos to be produced and the manipulations that can be done on them, a successful gene-editing therapy can help the family to have a healthy child (Bock von Wülfingen, 2016; Bayefsky, 2016). In the United States, PGD can be freely performed, not only for severe diseases like Tay Sachs or cystic fibrosis but also to select child's gender in as high as 9% of the cases or other comparably more benign diseases like deafness (Bayefsky, 2016). The benefit of having gene-editing done on an embryo at risk of inheriting a mutation might be seen differently in PGD-permissive countries, like the United States, compared to other countries where the restriction is severe and family's options are more limited. In both situations, availability of the means to obtain a healthy child can be source for reproductive tourism. In research, the limitations imposed on using human embryos determined researchers to move to countries with more permissive regulations. The lack of therapies would push the patients or their parents to look for less conventional and validated therapies (Cook, 2004). This is why, when considering ethical and legal regulations it is recommended to have in mind the international context in the field. Educating the general public in the domain of genetics, gene therapy, and genome editing can provide a general basis for autonomy in selecting and participating in gene therapy and clinical trials plus disarm potential harmful influences from pseudoscientific sources.

2.3   Respect for Autonomy

It is a central part of the medical care to involve the patient in the decision-making process regarding his/her diagnosis/therapeutic management, which is able to take into account not only the health-related benefits, but also his beliefs, cultural values, and needs. A practical approach is to provide the patient with a complete and clear explanation of the problems, available solutions, and recommendations, which would allow him to make an informed choice. The patient's consent can be considered **informed** if the information is given by a trained person to a patient who is able to understand and retain it, make a decision and communicate it to the research or medical staff. This raises the problem of capacity, linked to the age of the person giving consent and/or intellectual ability. The consent is documented in writing, by filling in a form containing all the relevant information about every aspect of the project/therapy, ended by patient's signature. In some situations, where the risk is minimal, the consent can be verbal or implied, however this is not an adequate approach in such a complex case like genome editing. The information should be conveyed to the patient in a way that allows him to make his own decision and nondirectional, even if the patient would ask for a personal opinion and this would relieve his anxiety (Wertz et al., 2003).

When confronted with reproductive choices and prenatal gene therapy the family and ultimately the woman must be the one who chooses (Wertz et al., 2003).

All competent adults, meaning able to understand, retain, and act upon, have the right to know about their own genetic changes and about how these can be transmitted to their offspring (Wertz et al., 2003). **Genetic information** is, at times, more difficult to understand for the general public or even for a part of the health-care staff, and this is why it should be presented by a geneticist, a genetic counselor, or another person trained in genetics, to ensure the correct understanding of the therapeutic intervention. Clear and detailed information should include: prognosis, diagnosis, foreseeable risks and benefits, details about the envisaged procedure/s, possible adverse effects, alternative treatment, confidentiality, and solutions in case of adverse effects (NASM, 2017). For example, the genome can be presented as an instruction manual, where some words can be misspelled or misplaced and this is making the text difficult to understand; the purpose of the therapy is to correct these mistakes and restore the significance of the "text." To achieve this, we can either include a correct sentence in addition to the incorrect one (as in the classical gene therapy) or to remove precisely the wrong letters and replace them with the right ones (as in CRISPR-based genome editing). This example minimizes the complexity of the procedure but can help the patient understand the pathological changes that lead to his condition and the way therapeutical intervention works. Additional details about the means and the protocols used in the therapy can be added.

The risks are not always considered carefully by a patient with an untreatable severe disease who might see any therapy as preferable to his condition. However, in this context, the risks should be correctly underlined, as well as potential solutions, reversibility, or alternatives. For example, in the case of somatic gene-editing therapy for a cancer, traditional methods (chemo- or radiotherapy) can be presented as an option; in embryo editing for the prevention of a genetic disease running in the family, can be offered in vitro fertilization and preimplantation genetic diagnosis to select the healthy embryos. The parents should be aware in this latter case that any genetic change induced into their child's genome will be inherited by his offspring and this is true both for the wild-type sequence that was restored and for any accidental off-target changes that might have occurred.

A patient who receives a genetic treatment consisting of somatic gene editing would need to know about the means used to deliver the treatment: cells collected from him, transformed in vitro with a gene-editing system, and returned to his body. He would also need to know about the risk of off-target effects, or what other side effects might be and what attitude would follow such an effect.

As every other treatment, gene therapy can have side effects and the efficiency can vary from patient to patient, without a well-understood and documented reason, even in the case of the new and more efficient techniques like CRISPR/Cas. The genome editing itself or the delivery means can be the source of unwanted changes, which can cause either malfunctioning of another gene or initiating an oncogenetic process. It is important to present the knowns and unknowns with a neutral attitude so that the patient can decide mostly based on the facts.

A potential influence on patient's decision comes from media invaded of partial or biased information, or Internet sources advertising for a "miracle" treatment in severe diseases available in remote places where no validation and controls are present. The more severe the disease, the more likely it is that the parents or patient choses to believe what is favorable, and this will affect the autonomy. Even more, recently DIY kits have been made available for people willing to try CRISPR at home (Sneed, 2017). Most people purchasing these kits are more likely to use them for brewing a glowing pint of beer than solving a difficult medical problem, but the patients are also looking into them as a potential solution, misled by the commercial presentation. Therefore it is useful if these elements are known and discussed, together with relevant explanations to what might work or not in the specific case of each patient.

2.4 Confidentiality

Confidentiality is a central principle of medical care, preventing the physician from disclosing any information about the patient to others without consent. With regards to genetic information, the duty of confidentiality conflicts with the duty to warn other persons at risk and therefore promote their beneficence and autonomy. Before the age of gene therapy, by "genetic information" one could understand exclusively elements related to clinical and genetic diagnosis for a genetic disease. The guidelines and recommendations mostly cover these aspects, without referring to the attitude in the case of gene therapy results. Some authors support the idea that the obligation of confidentiality prevails (Hallowell et al., 2003) while others support a more flexible approach, influenced by the severity and curability of the disease (Clarke et al., 2005; Dheensa et al., 2016, 2017). We rely on the patient to inform the relatives at risk but for this to happen the physician should explain the familial nature of the disease and the potential harms and benefits in all cases. All authors agree that the disclosure of genetic information should be encouraged but not coerced (Clarke et al., 2005; Dheensa et al., 2016; Forrest et al., 2010). Genetic therapy, though, is restricted to the patient and it is unlikely to have an effect on other family members when only somatic editing is performed. In the case of germline editing though, any unwanted genomic changes will be present in a child who did not consent to this treatment, and they can be passed on to their offspring. In this case, more than one generation will be affected, and even if the individual has the right to choose his treatment, the effect will be borne by people who were not part of the consent process. This situation can be treated similarly to the naturally occurring mutation in a family member. Depending on the potential effect, the physician could discuss a partial breach of confidentiality with the purpose of protecting the relatives from a severe disease and support their autonomy. The American Society of Human Genetics recommends professional disclosure in exceptional situations (severe disease, treatable or preventable) when the "harm resulted from nondisclosure outweighs the harm induced by disclosure" (ASHG, 1998) but never for diseases that cannot be treated or prevented. The British Society for Human Genetics Guide enforces the need to respect confidentiality but acknowledges the exceptional need to act differently in order to prevent "serious harm" (Physicians et al., 2011). In Norway, any disclosure without specific consent from the patient is prohibited by the law (Dheensa et al., 2016). The law prevents clinicians from discussing health-related issues with third parties but there are no provisions when other people or the community can be affected by the results of a therapy or enhancement based on genomic editing.

In the context of genome editing, potentially leading to inheritable mutations, it becomes important to establish what should be the procedure of data sharing regarding the genetic treatment.

2.4.1 Germline Editing

When gene-editing is applied on germline cells, any change that is introduced into the genome is inherited by the offspring and can be spread in the general population. These changes can be made into the sperm or egg and their precursors or the very early zygote. Germ cells and their precursors are not covered by the current regulations (e.g., Human Tissue Act in the United Kingdom). However, the way they are used to obtain an embryo and how this is used are regulated variably between different countries. A common ruling is that an embryo cannot be created for research purpose and only spare embryos resulted from IVF can be used where the legislation allows. One of the main subjects is the moral status of the embryo, which vary from a cellular mass to full human rights (Deleidi and Yu, 2016; Ormond et al., 2017). See also the corresponding chapter from this book.

In vitro gametogenesis means that germline cells are generated in vitro from stem cells (either iPSCs or embryonic stem cells). This technique is seen as a source of germline cells for research (to study early phases of development) or potentially source of gametes for in vitro fertilization (Bourne et al., 2012). The state can enforce restrictive regulations but these can be weakened by private, commercial activities, which do not require specific legal permission but only lack of legal punitive measures against it.

The risks associated to germline editing are not limited to the patient or the family but can reflect on the whole community. Let's assume that an embryo receives a gene-editing therapy affecting genes involved in the immune response. Should the effect be an increased resistance to germs, the child can harbor infectious agents that are damaging for other members of the community. The law of confidentiality prevents any medical professional from disclosing the special features of the child. These characteristics can be further transmitted to the next generation.

2.5 Applying the Principle of Justice in Clinical Genome Editing

CRISPR-based genome editing can be performed with limited resources, a fact that can support a larger access to this type of treatment and therefore improve the fairness in health care. The current cost of CRISPR/Cas9 system for research is as low as $30 (Ledford, 2015a). It is expected that the clinical cost will be much higher due to the extensive research work needed to make the system safe and efficient enough for use in humans; to this will be added the patenting costs. Conversely, the cost of existent drugs used to treat some rare diseases is prohibitive for patients with less economic means (Rhee, 2015). Would the cost of new therapies allow fairness in the distribution of benefits and risks among patients belonging to different populations? The elements accumulated to date support a positive answer to this question. In addition to the economic factors, large differences in regulations and politics may introduce inequalities or exposure to misuse (Carroll and Charo, 2015) and this is another field where public involvement is needed. The use of CRISPR is international and therefore the principle of justice would require international oversight of clinical use and particularly when germline editing is involved (ACMG, 2017; Bosley et al., 2015; Carroll and Charo, 2015).

The most sensitive field is that of germline editing. The postconceptional age from which the baby is considered a person and therefore protected by the laws applying to any human being is different in each country. Some prohibit or severely limit the research on any human embryos, while others permit it explicitly or implicitly (NASM, 2017; Ledford, 2015b). Once the clinical use of genome editing on germline cells or embryos will be permitted in one or more countries, we can face a new wave of health-related tourism from those who can afford the expenses, thus deepening the imbalance.

2.6 Eugenics, Enhancement, and "Designer Babies"

One of the major concerns associated with the use of genome editing in therapy is linked to its potential to be used in nonmedical purposes, for enhancement or eugenic practices (Ormond et al., 2017; Skinner, 2014). **Eugenics** can have different definitions but in general it represents a way of introducing nonrandom changes in the genetic makeup of a population to improve its phenotype. In a more restrictive way, it is defined as "a coercive policy intended to further a reproductive goal, against the rights, freedoms, and choices of the individual" (Wertz et al., 2003). Human **enhancement** mostly refers to genetic engineering of human beings, by modifying the genetic material to provide a selective advantage (e.g., immune protection, height, muscle mass, intelligence). Advanced knowledge in the field of genomics and genome editing reopened the discussion about **human enhancement** and possibility to choose the features of our children the same way we choose clothes in a shop. "**Designer babies**" is the term most used to designate engineered human beings, whose genes are selectively modified to provide enhanced features and/or cosmetic traits, eyes, hair, or skin color, according to the choices made by his parents.

Classically, eugenics was achieved by preventing the "nonsuitable" individuals from reproducing and/or selecting for parenting only those presenting the desired features or genetic inheritance (Krishan et al., 2016). It has been used in livestock for a long time, to select the most productive animals for reproduction but it is not a new idea for human beings, either. In humans, eugenics is mostly associated with Nazism but in other countries (e.g., Sweden, Austria, Switzerland, the United States, and Canada), individuals with mental or physical disabilities were sterilized according to laws that lasted for more than 50 years, until about 1975 (Economist, 1997). In some cultures, there is a restrictive approach to eugenics (e.g., Germany or Switzerland) while others use a broader meaning (the United States) and this is reflected in the regulations with regards to genetic testing and gene therapies (Bayefsky, 2016). The reasons for these differences can be found in cultural, historical, and economical references (Bayefsky, 2016). It is still under debate if choosing an embryo for IVF, out of a set of other embryos with different genetic characteristics, or interrupting a pregnancy with an affected embryo can be considered as eugenic or just the right of the family to having a healthy child and a manifestation of reproductive autonomy. Some authors support the idea that a free, informed decision of a family to select a healthy embryo and remove one with a damaging genetic mutation is not eugenics as this is limited to an individual or a family and will not be reflected on a larger population (Wertz et al., 2003). However, availability of medical care and social constraints can influence the choice of the patient or his family (Epstein, 2003).

Human enhancement has been dreamed of and imagined in science fiction books and films. Translating this into practice would mean overcoming knowledge and technical gaps, but most importantly moral and ethical issues. In terms of knowledge, we know very little about what sequence in our genes makes us more intelligent, taller, or stronger; all these are multifactorial characters, resulted from interaction between our genetic makeup and environment. Should we try to modify our genome to get improved features, we would need to perform multiple changes at once, without the possibility to accurately predict the outcome at individual, family, and society level. There is a general consensus in the scientific world that germline editing and particularly the nonmedical applications should be banned and a large public consultation should take place before any such application be initiated. A proper oversight system should be in place when genome editing is used for therapy, to prevent its deviations toward misuse (Baker, 2016; Wang et al., 2017; Darnovsky, 2017).

3. CONCLUSIONS

Rapid development of genome-editing technology promises to bring the most effective therapeutic solution for genetic diseases and cancer. The fact that genome editing can modify human germline determined ethical concerns both among the scientists and the public. As in any other medical intervention, genetic therapy must be governed by the core ethical principles of benevolence, nonmaleficence, respect for autonomy and justice. Before recommending a gene-editing therapy to a patient affected from a genetic disease, health professionals must take into account the type of disease, inheritance, natural evolution, penetrance and expressivity, familial character, potential immediate and long-term adverse reactions or off-target effects, alternative treatment, and associated risks. All information must be openly presented to the patient prior to treatment and updated as the knowledge advances.

Patients' associations and global society should be involved in the decision about germline editing and how the use of gene editing should be regulated internationally to avoid potential harmful effects. Scientists, medical professionals, patients, and general public must decide when, in what context, for what diseases and ages can we apply this technology in the clinical context, to provide the maximum benefit and the minimal risk.

We do not know yet to what extent changes in the genetic pool will influence our evolution as species and occurrence of new mutations or changes in our environmental fitness, and this is not a result that can be obtained through a simple laboratory experiment. However, the few results of targeted genetic changes performed so far in human beings, recommend caution and a deep understanding of the mechanisms before proceeding to any clinical application of genome-editing techniques.

REFERENCES

ACMG, 2017. Genome editing in clinical genetics: points to consider-a statement of the American College of Medical Genetics and Genomics. Genet. Med. 19, 723–724.

ASHG, 1998. Professional disclosure of familial genetic information. Am. J. Hum. Genet. 62, 474–483.

Baker, B., 2016. The ethics of changing the human genome. Bioscience 66, 267–273.

Baltimore, D., Berg, P., Botchan, M., Carroll, D., Charo, R.A., Church, G., Corn, J.E., Daley, G.Q., Doudna, J.A., Fenner, M., Greely, H.T., Jinek, M., Martin, G.S., Penhoet, E., Puck, J., Sternberg, S.H., Weissman, J.S., Yamamoto, K.R., 2015. Biotechnology. A prudent path forward for genomic engineering and germline gene modification. Science 348, 36–38.

Barrangou, R., Fremaux, C., Deveau, H., Richards, M., Boyaval, P., Moineau, S., Romero, D.A., Horvath, P., 2007. Crispr provides acquired resistance against viruses in prokaryotes. Science 315, 1709–1712.

Bayefsky, M.J., 2016. Comparative preimplantation genetic diagnosis policy in Europe and the Usa and its implications for reproductive tourism. Reprod. Biomed. Soc. Online 3, 41–47.

BBC, 2017. Charlie Gard: The Story of his parents' Legal Fight. (BBC News).

Beauchamp, T.L., Childress, J.F., 2001. Principles of Biomedical Ethics. Oxford University Press, New York.

Berg, P., Baltimore, D., Brenner, S., Roblin III, R.O., Singer, M.F., 1975a. Asilomar conference on recombinant DNA molecules. Science 188, 991–994.

Berg, P., Baltimore, D., Brenner, S., Roblin, R.O., Singer, M.F., 1975b. Summary statement of the Asilomar conference on recombinant DNA molecules. Proc. Natl. Acad. Sci. U. S. A. 72, 1981–1984.

Bock Von WUlfingen, B., 2016. Contested change: how Germany came to allow PGD. Reprod. Biomed. Soc. Online 3, 60–67.

Bolotin, A., Quinquis, B., Sorokin, A., Ehrlich, S.D., 2005. Clustered regularly interspaced short palindrome repeats (CRISPRS) have spacers of extrachromosomal origin. Microbiology 151, 2551–2561.

Bosley, K.S., Botchan, M., Bredenoord, A.L., Carroll, D., Charo, R.A., Charpentier, E., Cohen, R., Corn, J., Doudna, J., Feng, G., Greely, H.T., Isasi, R., Ji, W., Kim, J.S., Knoppers, B., Lanphier, E., Li, J., Lovell-Badge, R., Martin, G.S., Moreno, J., Naldini, L., Pera, M., Perry, A.C., Venter, J.C., Zhang, F., Zhou, Q., 2015. Crispr germline engineering—the community speaks. Nat. Biotechnol. 33, 478–486.

Bourne, H., Douglas, T., Savulescu, J., 2012. Procreative beneficence and in vitro gametogenesis. Monash Bioeth. Rev. 30, 29–48.

Branca, M.A., 2005. Gene therapy: cursed or inching towards credibility? Nat. Biotechnol. 23, 519–521.

Camporesi, S., Cavaliere, G., 2016. Emerging ethical perspectives in the clustered regularly inter-spaced short palindromic repeats genome-editing debate. Pers. Med. 13 (6), 575–586.

Carroll, D., 2011. Genome engineering with zinc-finger nucleases. Genetics 188, 773–782.

Carroll, D., Charo, R.A., 2015. The societal opportunities and challenges of genome editing. Genome Biol. 16, 242.

Cavaliere, G., 2017. Genome editing and assisted reproduction: curing embryos, society or prospective parents? Med. Health Care Philos. (Epub ahead of print).

Chandrasegaran, S., Carroll, D., 2016. Origins of programmable nucleases for genome engineering. J. Mol. Biol. 428, 963–989.

Choulika, A., Perrin, A., Dujon, B., Nicolas, J.F., 1995. Induction of homologous recombination in mammalian chromosomes by using the I-SceI system of Saccharomyces cerevisiae. Mol. Cell Biol. 15, 1968–1973.

Clarke, A., Richards, M., Kerzin-Storrar, L., Halliday, J., Young, M.A., Simpson, S.A., Featherstone, K., Forrest, K., Lucassen, A., Morrison, P.J., Quarrell, O.W., Stewart, H., 2005. Genetic professionals' reports of nondisclosure of genetic risk information within families. Eur. J. Hum. Genet. 13, 556–562.

Cohen, S.N., Chang, A.C., Boyer, H.W., Helling, R.B., 1973. Construction of biologically functional bacterial plasmids in vitro. Proc. Natl. Acad. Sci. U. S. A. 70, 3240–3244.

Committee on Science, Technology and Law; Policy and Global Affairs, NASM, 2016. International Summit on Human Gene Editing: A Global Discussion. National Academies Press, Washington, DC.

Cong, L., Ran, F.A., Cox, D., Lin, S., Barretto, R., Habib, N., Hsu, P.D., Wu, X., Jiang, W., Marraffini, L.A., Zhang, F., 2013. Multiplex genome engineering using CRISPR/Cas systems. Science 339, 819–823.

Cook, G., 2004. Desperate Parents Chase a Stem-Cell Miracle. Boston Globe. Available at: http://garethcook.net/desperate-parents-chase-a-stem-cell-miracle/. Accessed 30 October 2017.

Darnovsky, M., 2017. Editing humans. Biosci. Technol.

Deleidi, M., Yu, C., 2016. Genome editing in pluripotent stem cells: research and therapeutic applications. Biochem. Biophys. Res. Commun. 473, 665–674.

Dheensa, S., Fenwick, A., Shkedi-Rafid, S., Crawford, G., Lucassen, A., 2016. Health-care professionals' responsibility to patients' relatives in genetic medicine: a systematic review and synthesis of empirical research. Genet. Med. 18, 290–301.

Dheensa, S., Fenwick, A., Lucassen, A., 2017. Approaching confidentiality at a familial level in genomic medicine: a focus group study with healthcare professionals. BMJ Open 7, 1–10.

Doudna, J.A., Charpentier, E., 2014. Genome editing. The new frontier of genome engineering with CRISPR-Cas9. Science 346, 1258096.

Economist, T., 1997. Here, of all places. In: The Economist. The Economist Group Limited, London.

Epstein, C.J., 2003. Is modern genetics the new eugenics? Genet. Med. 5, 469.

Evitt, N.H., Mascharak, S., Altman, R.B., 2015. Human germline Crispr-Cas modification: toward a regulatory framework. Am. J. Bioeth. 15, 25–29.

Fernandez, A., Josa, S., Montoliu, L., 2017. A history of genome editing in mammals. Mamm. Genome 28, 237–246.

Forrest, L.E., Delatycki, M.B., Curnow, L., Skene, L., Aitken, M., 2010. Genetic health professionals and the communication of genetic information in families: practice during and after a genetic consultation. Am. J. Med. Genet. A 152A, 1458–1466.

Gaudelli, N.M., Komor, A.C., Rees, H.A., Packer, M.S., Badran, A.H., Bryson, D.I., Liu, D.R., 2017. Programmable base editing of A•T to G•C in genomic DNA without DNA cleavage. Nature 551, 464–471.

Geurts, A.M., Cost, G.J., Freyvert, Y., Zeitler, B., Miller, J.C., Choi, V.M., Jenkins, S.S., Wood, A., Cui, X., Meng, X., Vincent, A., Lam, S., Michalkiewicz, M., Schilling, R., Foeckler, J., Kalloway, S., Weiler, H., Menoret, S., Anegon, I., Davis, G.D., Zhang, L., Rebar, E.J., Gregory, P.D., Urnov, F.D., Jacob, H.J., Buelow, R., 2009. Knockout rats via embryo microinjection of zinc-finger nucleases. Science 325, 433.

Grizot, S., Smith, J., Daboussi, F., Prieto, J., Redondo, P., Merino, N., Villate, M., Thomas, S., Lemaire, L., Montoya, G., Blanco, F.J., Paques, F., Duchateau, P., 2009. Efficient targeting of a SCID gene by an engineered single-chain homing endonuclease. Nucleic Acids Res. 37, 5405–5419.

Gyngell, C., Douglas, T., Savulescu, J., 2017. The ethics of germline gene-editing. J. Appl. Philos. 34, 498–513.

Hallowell, N., Foster, C., Eeles, R., Ardern-Jones, A., Murday, V., Watson, M., 2003. Balancing autonomy and responsibility: the ethics of generating and disclosing genetic information. J. Med. Ethics 29, 11.

Hatada, I., Horii, T., 2016. Genome editing: A breakthrough in life science and medicine. Endocr. J. 63, 105–110.

Haussecker, D., 2016. Stacking up CRISPR against RNAI for therapeutic gene inhibition. FEBS J. 283, 3249–3260.

Ishino, Y., Shinagawa, H., Makino, K., Amemura, M., Nakata, A., 1987. Nucleotide sequence of the IAP gene, responsible for alkaline phosphatase isozyme conversion in Escherichia coli, and identification of the gene product. J. Bacteriol. 169, 5429–5433.

Jackson, D.A., Symons, R.H., Berg, P., 1972. Biochemical method for inserting new genetic information into DNA of simian virus 40: Circular Sv40 DNA molecules containing lambda phage genes and the galactose operon of Escherichia coli. Proc. Natl. Acad. Sci. U. S. A. 69, 2904–2909.

Jaffe, A., Prasad, S.A., Larcher, V., Hart, S., 2006. Gene therapy for children with cystic fibrosis—who has the right to choose? J. Med. Ethics 32, 361–364.

Jansen, R., Embden, J.D., Gaastra, W., Schouls, L.M., 2002. Identification of genes that are associated with DNA repeats in prokaryotes. Mol. Microbiol. 43, 1565–1575.

Kim, J.S., 2016. Genome editing comes of age. Nat. Protoc. 11, 1573–1578.

Kim, K., Ryu, S.-M., Kim, S.-T., Baek, G., Kim, D., Lim, K., Chung, E., Kim, S., Kim, J.-S., 2017. Highly efficient RNA-guided base editing in mouse embryos. Nat. Biotechnol. 35, 435–437.

Kimmelman, J., 2005. Recent developments in gene transfer: risk and ethics. BMJ 330, 79–82.

Klitzman, R., 2017. Challenges, dilemmas and factors involved in PGD decision-making: providers' and patients' views, experiences and decisions. J. Genet. Couns.

Komor, A.C., Kim, Y.B., Packer, M.S., Zuris, J.A., Liu, D.R., 2016. Programmable editing of a target base in genomic DNA without double-stranded DNA cleavage. Nature 533, 420–424.

Krishan, K., Kanchan, T., Singh, B., 2016. Human genome editing and ethical considerations. Sci. Eng. Ethics 22, 597–599.

Lanphier, E., Urnov, F., Haecker, S.E., Werner, M., Smolenski, J., 2015. Don't edit the human germ line. Nature 519, 410–411.

Lau, V., Davie, J.R., 2017. The discovery and development of the CRISPR system in applications in genome manipulation. Biochem. Cell Biol. 95, 203–210.

Ledford, H., 2015a. CRISPR, the disruptor. Nature 522, 20–24.

Ledford, H., 2015b. The landscape for human genome editing. Nature 526, 2.

Liang, P., Ding, C., Sun, H., Xie, X., Xu, Y., Zhang, X., Sun, Y., Xiong, Y., Ma, W., Liu, Y., Wang, Y., Fang, J., Liu, D., Songyang, Z., Zhou, C., Huang, J., 2017a. Correction of β-thalassemia mutant by base editor in human embryos. Protein Cell 8, 811–822.

Liang, P., Sun, H., Sun, Y., Zhang, X., Xie, X., Zhang, J., Zhang, Z., Chen, Y., Ding, C., Xiong, Y., Ma, W., Liu, D., Huang, J., Songyang, Z., 2017b. Effective gene-editing by high-fidelity base editor 2 in mouse zygotes. Protein Cell 8, 601–611.

Low, B.E., Krebs, M.P., Joung, J.K., Tsai, S.Q., Nishina, P.M., Wiles, M.V., 2014. Correction of the Crb1rd8 allele and retinal phenotype in C57BL/6N mice via Talen-mediated homology-directed repair. Invest. Ophthalmol. Vis. Sci. 55, 387–395.

Lucassen, A., Parker, M., 2003. Confidentiality and serious harm in genetics—preserving the confidentiality of one patient and preventing harm to relatives. Eur. J. Hum. Genet. 12, 93.

Maier, D.A., Brennan, A.L., Jiang, S., Binder-Scholl, G.K., Lee, G., Plesa, G., Zheng, Z., Cotte, J., Carpenito, C., Wood, T., Spratt, S.K., Ando, D., Gregory, P., Holmes, M.C., Perez, E.E., Riley, J.L., Carroll, R.G., June, C.H., Levine, B.L., 2013. Efficient clinical scale gene modification via zinc finger nuclease-targeted disruption of the HIV co-receptor CCR5. Hum. Gene Ther. 24, 245–258.

Makarova, K.S., Haft, D.H., Barrangou, R., Brouns, S.J., Charpentier, E., Horvath, P., Moineau, S., Mojica, F.J., Wolf, Y.I., Yakunin, A.F., Van Der Oost, J., Koonin, E.V., 2011. Evolution and classification of the CRISPR-Cas systems. Nat. Rev. Microbiol. 9, 467–477.

Manno, C.S., Pierce, G.F., Arruda, V.R., Glader, B., Ragni, M., Rasko, J.J., Ozelo, M.C., Hoots, K., Blatt, P., Konkle, B., Dake, M., Kaye, R., Razavi, M., Zajko, A., Zehnder, J., Rustagi, P.K., Nakai, H., Chew, A., Leonard, D., Wright, J.F., Lessard, R.R., Sommer, J.M., Tigges, M., Sabatino, D., Luk, A., Jiang, H., Mingozzi, F., Couto, L., Ertl, H.C., High, K.A., Kay, M.A., 2006. Successful transduction of liver in hemophilia by AAV-factor IX and limitations imposed by the host immune response. Nat. Med. 12, 342–347.

McMahon, M.A., Cleveland, D.W., 2016. Gene-editing therapy for neurological disease. Nat. Rev. Neurol. 13, 7.

Miyabe, I., Kunkel, T.A., Carr, A.M., 2011. The major roles of DNA polymerases epsilon and delta at the eukaryotic replication fork are evolutionarily conserved. PLoS Genet. 7, e1002407.

Mojica, F.J., Juez, G., Rodriguez-Valera, F., 1993. Transcription at different salinities of Haloferax mediterranei sequences adjacent to partially modified PstI sites. Mol. Microbiol. 9, 613–621.

Naldini, L., 2015. Gene therapy returns to centre stage. Nature 526, 351–360.

NASM, 2017. Human Genome Editing: Science, Ethics and Governance. NASM, Washington, DC.

Offit, K., Groeger, E., Turner, S., Wadsworth, E.A., Weiser, M.A., 2004. The "duty to warn" a patient's family members about hereditary disease risks. JAMA 292, 1469–1473.

Ormond, K.E., Mortlock, D.P., Scholes, D.T., Bombard, Y., Brody, L.C., Faucett, W.A., Garrison, N.A., Hercher, L., Isasi, R., Middleton, A., Musunuru, K., Shriner, D., Virani, A., Young, C.E., 2017. Human germline genome editing. Am. J. Hum. Genet. 101, 167–176.

Peng, R., Lin, G., Li, J., 2016. Potential pitfalls of CRISPR/Cas9-mediated genome editing. FEBS J. 283, 1218–1231.

Physicians, R.C.O., Pathologists, R.C.O., Genetics, B.S.F.H., 2011. Consent and confidentiality in clinical genetic practice: guidance on genetic testing and sharing genetic information. In: Report of the Joint Committee on Medical Genetics, second ed. RCP, London.

Rhee, T.G., 2015. Policymaking for orphan drugs and its challenges. Am. J. Ethics 17, 776–779.

Savulescu, J., Pugh, J., Douglas, T., Gyngell, C., 2015. The moral imperative to continue gene-editing research on human embryos. Protein Cell 6, 476–479.

Shim, G., Kim, D., Park, G.T., Jin, H., Suh, S.K., Oh, Y.K., 2017. Therapeutic gene-editing: delivery and regulatory perspectives. Acta Pharmacol. Sin. 38, 738–753.

Skinner, D., 2014. Beyond Whac-a-mole? Rethinking "race" in social studies of genetics. New Genet. Soc. 33, 450–457.

Sneed, A., 2017. Mail-order CRISPR kits allow absolutely anyone to hack DNA. In: Scientific American. Springer Nature, New York.

Strachan, T., Read, A., 2011. Human Molecular Genetics. Garland Science, New York.

Touzot, F., Moshous, D., Creidy, R., Neven, B., Frange, P., Cros, G., Caccavelli, L., Blondeau, J., Magnani, A., Luby, J.M., Ternaux, B., Picard, C., Blanche, S., Fischer, A., Hacein-Bey-Abina, S., Cavazzana, M., 2015. Faster T-cell development following gene therapy compared with haploidentical HSCT in the treatment of SCID-X1. Blood 125, 3563–3569.

Vaegter, K.K., Lakic, T.G., Olovsson, M., Berglund, L., Brodin, T., Holte, J., 2017. Which factors are most predictive for live birth after in vitro fertilization and intracytoplasmic sperm injection (IVF/ICSI) treatments? Analysis of 100 prospectively recorded variables in 8,400 IVF/ICSI single-embryo transfers. Fertil. Steril. 107, 641–648 e2.

Wang, J.H., Wang, R., Lee, J.H., Iao, T.W.U., Hu, X., Wang, Y.M., Tu, L.L., Mou, Y., Zhu, W.L., He, A.Y., Zhu, S.Y., Cao, D., Yang, L., Tan, X.B., Zhang, Q., Liang, G.L., Tang, S.M., Zhou, Y.D., Feng, L.J., Zhan, L.J., Tian, N.N., Tang, M.J., Yang, Y.P., Riaz, M., Van Wijngaarden, P., Dusting, G.J., Liu, G.S., He, Y., 2017. Public attitudes toward gene therapy in China. Mol Ther Methods Clin Dev 6, 40–42.

Wertz, D., Fletcher, J., Berg, K., 2003. Review of Ethical Issues in Medical Genetics. World Health Organization, Geneva.

Yao, X., Wang, X., Liu, J., Hu, X., Shi, L., Shen, X., Ying, W., Sun, X., Wang, X., Huang, P., Yang, H., 2017. CRISPR/Cas9—mediated precise targeted integration in vivo using a double cut donor with short homology arms. EBioMedicine 20, 19–26.

FURTHER READING

Brown, T.A., 2006. Genomes. Garland Science, New York.

Chapter 2

Ethics of Mitochondrial Gene Replacement Therapy

Rebecca Dimond
Cardiff University, Cardiff, Wales, United Kingdom

1. INTRODUCTION

In February 2015, the United Kingdom Parliament voted in favor, by a large majority, of legalizing controversial in vitro fertilization (IVF) procedures known as mitochondrial donation. The decision followed an extensive review and consultation process to explore safety and efficacy, address ethical questions, and examine whether there would be broad public support for changing the law. The techniques were developed for women with maternally inherited mitochondrial disease to have healthy genetically related children. But they are controversial because they use the cytoplasm from a donated egg, which contains genetic material in the form of mitochondrial DNA (mtDNA). In the United Kingdom and elsewhere, the techniques have attracted intense media interest, with "three-parent babies" dominating the headlines and raising concerns about a "slippery slope" to designer babies and further human modification. Many of the terms used to describe mitochondrial donation have been contested. Mitochondrial donation and mitochondrial donor are both used throughout this chapter, in alignment with current United Kingdom policy, although it is noted that there are alternatives. This chapter explores the situation in United Kingdom and other countries and then considers questions of safety, ethics, and the implications for genetic counseling.

2. WHAT IS MITOCHONDRIAL DISEASE AND MITOCHONDRIAL DONATION?

Mitochondria are small structures contained in the cytoplasm of a cell, producing energy in the form of adenosine triphosphate. Each cell contains hundreds to thousands of mitochondria, depending on the energy requirements of particular tissues. mtDNA is made up of 37 genes, which are primarily responsible for maintaining the function of the mitochondria, making up $<0.1\%$ of our body's

Clinical Ethics at the Crossroads of Genetic and Reproductive Technologies.
https://doi.org/10.1016/B978-0-12-813764-2.00002-7

total DNA. Although the genetic contribution of mitochondria is small, the impact when they fail to function is considerable. Mitochondria dysfunction can be due to mutations in either nuclear or mtDNA sequences, this chapter focuses only on diseases caused by mutations in mtDNA sequences. As mitochondria are derived through the oocyte (only one case of paternal inheritance of mitochondria DNA has been identified, see Schwartz and Vissing, 2002) disease caused by mutations of mtDNA display a maternal inheritance pattern. Both sexes can inherit the disease but it is only women who are at risk of transmitting it to their children.

Mitochondrial disease is extremely variable according to which organs are affected and to what extent, and patients can be mildly, severely, or fatally affected. Symptoms can include diabetes, epilepsy, digestive disorders, fatigue, cardiomyopathy, deafness, restricted sight, and difficulties with mobility and balance. The term "mitochondrial disease" was introduced in the late 1980s, but it encompasses a range of distinct disorders, including mitochondrial encephalomyopathy, lactic acidosis, and stroke-like episodes, myoclonic epilepsy with ragged red fibers, Leber's hereditary optic neuropathy, and Leigh syndrome. The distinctions between these classifications and their implications for clinical management are beyond the scope of this chapter and are described elsewhere (see e.g., Craven et al., 2017a,b; DiMauro, 2011; Kisler et al., 2010; Chinnery and Hudson, 2013).

It is estimated that one in 400 people carries a disease-causing mitochondrial mutation (Craven et al., 2017a,b; Manwaring et al., 2007; Schaefer et al., 2008) although identifying disease prevalence can be complicated because of the diverse spectrum of symptoms, as well as differences in national diagnostic criteria and clinical practice (see, e.g., the consensus statement by Parikh et al., 2017). Although mitochondrial disease might not be well known at the local level, patients within the United Kingdom are relatively well provided for, and research is supported. The National Health Service (NHS) supports a rare mitochondrial disorders service, which has three centers of expertise in Newcastle, London, and Oxford. It is a highly specialized service for research and clinical care, and hosts a national patient database through which patients can participate in research and clinical trials.

Mitochondrial disease cannot yet be cured, and treatment is generally limited to strong vitamins (Hargreaves, 2014), although advances are being made. In the context of a serious disease with limited treatment options, attention has turned to reproductive options, which could prevent the transmission of disease to offspring. Mitochondrial donation is one such option. The Wellcome Trust Centre for Mitochondrial Research in Newcastle, UK, has developed several related techniques known collectively as mitochondrial donation. Maternal spindle transfer (MST) involves removing the maternal spindle (a formation in the nucleus of an egg) from the egg of a woman with mitochondrial disease for transfer into a donated egg, which has already had its maternal spindle removed. The egg would then be fertilized and implanted. Pronuclear transfer

(PNT), which is at a more advanced stage of development within the United Kingdom, occurs after fertilization. During PNT, the pronuclei in the newly fertilized donor egg is replaced with the pronuclei from the newly fertilized egg of the intended mother. The result of MST and PNT are the same—the resulting child would inherit nuclear DNA (nDNA) from the intending mother (and father) and mtDNA from the donor.

Although the United Kingdom was the first to legalize, many countries have also initiated discussion about the benefits and risks of facilitating the clinical introduction of mitochondrial donation. For some, the United Kingdom presents a "gold standard" of regulation and a guide for clinical adoption of emerging technologies. For others, it presents an example of a liberal technocracy. But the United Kingdom also presents a particular context that facilitated legalization, including a history of close but permissive regulation, central record keeping of fertility clinics and treatments, and specifically in the case of mitochondrial disease, clinical provision for patients, and a centralized research cohort database. The route from bench to bedside within the United Kingdom is now considered.

3. THE UK TIMELINE

The United Kingdom has a long history of supporting, regulating, and licensing embryo research and IVF technologies. Following the birth of Louise Brown, the first "test tube" baby in 1978, an inquiry was set up to report on human fertilization and embryology amid concerns about the rapidly increasing capacity of assisted reproduction technologies. Almost 30 years ago the Human Fertilisation and Embryology Authority (HFEA) was established, and it continues to be the UK Government's independent regulator overseeing the use of gametes and embryos in assisted reproduction and research within the United Kingdom. Through licensing, inspection, and setting standards, the HFEA ensure that all fertility centers and human embryo research complies with relevant legislation, particularly the Human Fertilisation and Embryology (HFE) Act 1990 (as amended) and the HFE Act 2008.

The legalization of mitochondrial donation within the United Kingdom was possible because parliament had previously voted for a number of amendments to the 1990 Act. The HFE Act 2008 permitted the creation of human-animal embryos for research purposes, prevented the use of sex selection for nonmedical purposes, and importantly for this chapter, allowed for the future possibility of the clinical use of mitochondrial donation to prevent the inheritance of mitochondrial disease, depending on further parliamentary approval (Jones and Holme, 2013). Following the first proof of concept study on human zygotes (Craven et al., 2010), which demonstrated the potential to prevent disease, the government was invited to use the power granted in the HFE Act 2008 to support mitochondrial donation as a viable clinical option.

Between 2011 and 2015, the UK government and several influential UK institutions initiated reviews to explore safety, ethical concerns, and whether a change in law would have broad public support. HFEA convened an independent expert panel to consider the safety and efficacy of the techniques, reporting their findings in 2011. Further updates were published in 2013, 2014, and 2016, each broadly concluding that the techniques were "not unsafe." A process of public consultation and engagement took place alongside these reviews. In 2012, the Nuffield Council on Bioethics (NCoB) called for evidence on the ethical questions raised by mitochondrial donation. It reported on several issues including identity and genetic parentage, the status of the mitochondrial donor, and implications for society. It concluded that if proven safe and effective, then mitochondrial donation would be ethical for families to use, "if they wish to do so and have been offered an appropriate level of information and support" (NCoB, 2012, p. xvi). In 2013, the HFEA conducted its own public consultation, aiming to understand public attitudes toward legalization through questionnaires, focus groups, and open meetings. It reported that despite strong objection, "the overall view is that ethical concerns are outweighed by the arguments in favour…" (HFEA, 2013, p. 4). The Department of Health (DoH) also conducted their own public consultation, although this was slightly different because it focused on the detail of how legislation would work in practice, for example, how the relationship between child and donor should legally be defined (DoH, 2014).

The nature of UK regulation is that changes to the HFEA Acts required parliamentary scrutiny and approval. Mitochondrial donation had been discussed several times in parliament, and following extended debates, parliamentarians were asked to vote for or against the change in law in February 2015. On February 3, 2015 the House of Commons voted by a majority of 382 to 128 in favor of legalizing mitochondrial donation. On February 24, 2015, the House of Lords also voted in favor, with a majority of 280 to 48. As a consequence of parliamentary approval, the HFE (Mitochondrial Donation) Regulations 2015 came into force in October 2015 (HFEA, 2015). The regulations specified that the law change only related to the clinical introduction of MST and PNT, they could only be offered to those at risk of transmitting severe mitochondrial disease, and they could only be offered through a licensed clinic. The regulations also specified that the mitochondrial donor had no parental rights although the child and donor could access nonidentifying information about each other (HFEA, 2015; also see Castro, 2016; Varvaštian, 2015). Although the law itself had been changed, licenses could not be allocated to clinics until the HFEA were satisfied that they had the specialist skills in place, and that the techniques were ready to be introduced into the clinic. In November 2016, the final report on efficacy and safety was published (HFEA, 2016), and in December 2016, the HFEA declared that clinics could apply for a license. The Wellcome Trust Centre for Mitochondrial Research at Newcastle was granted the first UK clinical license in March 2017.

4. THE INTERNATIONAL POSITION

During the mitochondrial debates within the United Kingdom, it was clear that other countries were waiting to see what direction the United Kingdom would take. One letter, published in a prominent newspaper was authored by 40 scientists from 14 different countries, explained why the techniques should be legalized:

> *A positive vote would not only allow affected families to choose to use this new procedure under the care of the globally respected Newcastle team, with proper advice and safeguards; it would also be an international demonstration of how good regulation helps medical science to advance in step with wider society. (Carroll et al., 2015)*

A contrasting view was represented in a letter to the Times, when 55 Italian Members of Parliament urged members of the House of Lords to vote against legalization, drawing attention to the pace of development and the consequences for "the whole of humanity":

> *The creation of such embryos could have uncontrollable and unforeseeable consequences, affecting future generations, and modifying genetic heritage in an irreversible way, inevitably affecting the human species as a whole. It is a dangerous intervention involving genetic engineering, which affects the whole of humanity, and cannot possibly be contained within the confines of the United Kingdom. (Roccella et al., 2015)*

The different perspectives about mitochondrial donation and its legalization highlight both global connectedness and local boundaries, and the challenges for national governments in regulating emerging medical technologies. The United Kingdom stance mattered to scientists and regulators in other countries, but what technologies are seen as a viable solution in one country might not be possible in another, particularly with differences in patient populations and diagnostic capabilities. For some countries seeking to introduce mitochondrial donation, lack of information about the patient population can compound problems caused by gaps in clinical management (see, e.g., Meldau et al., 2016 exploring the current situation in South Africa).

Decisions to legalize mitochondrial donation or initiate discussions will also vary according to general attitudes to embryo research and assisted reproduction and what kinds of regulatory bodies exist. Araki and Ishii (2014) have highlighted the variability of the international regulatory landscape in relation to gene editing by conducting an initial survey of 14 countries, where the vast majority prohibited gene modifications. They then expanded to 39 countries, 29 of which banned germline modification, whereas others were more "ambiguous" in their position. They found that some of those who banned modification (such as China, India, Ireland, and Japan) did so on the basis of flexible guidelines rather than enshrined in law. Those suggesting that the United

Kingdom would be breaking the law by legalizing mitochondrial donation did so through appeals to the UNESCO Universal Declaration on the Human Genome and Human Rights, the Council of Europe's Convention of Human Rights and Biomedicine, and the EU Clinical Trials Directive (also see Varvaštian, 2015). Ishii (2015) found that countries such as Belgium, Bulgaria, Canada, Denmark, Sweden, and the Czech Republic prevent genetic modification because of the potential to inherit modified genes or because gene modification may harm embryo development. Araki and Ishii (2014) also point to vast differences in wealth between and within countries, and inequality of access to reproductive technologies, and urge further research to understand the different positions of each country and the local contexts in which reproductive technologies might be introduced or contested.

The United States regulatory landscape is very different from that of the United Kingdom, and is restrictive rather than specifically banning mitochondrial donation. The technologies are currently impermissible, partly because of the Dickey-Wicker Amendment, which prohibits public funding for embryo research. The United States does not have a specific institutional body tasked with regulating reproductive technologies and the US Food and Drug Administration (FDA) have been prevented from evaluating mitochondrial donation as a reproductive technology (Adashi and Cohen, 2017; Lyon, 2017). The United States also has a particular history of mitochondrial donation, first in supporting but then affectively banning cytoplasm injection at the turn of the 21st century (see below, and Castro, 2016). The United States holds a much stronger position around embryo politics than the United Kingdom, and debates about mitochondrial donation have been conflated with concern about embryo research and abortion (Lyon, 2017). Debates have been initiated in the United States, exploring the clinical utility of mitochondrial donation through expert discussion (in contrast to the United Kingdom, which draws on a strong model of public participation). The consensus report of the National Academies of Sciences, Engineering, and Medicine (2016) identified benefits, but suggested a cautious approach, particularly recommending only transferring male embryos to avoid future generational transmission. Sex selection for this purpose would not currently be legal within the United Kingdom.

One of the concerns about legalizing mitochondrial donation is that it could lead to the legalization of other forms of genetic modification, a move toward designer babies, and undesirable uses and abuses. There have been calls for a moratorium before collective decisions are made about the risks and benefits of technologies such as mitochondrial donation, while noting a difference between research and clinical use. For Ishii (2015), the United Kingdom becoming the first country to legalize mitochondrial donation significantly changed the global regulatory landscape. For Varvaštian (2015, p. 418), as a germline technology, the United Kingdom's decision to legalize mitochondrial donation was akin to opening Pandora's box, which could lead to "selective genetic engineering of human beings" and "would shatter the existing consensus of nonmeddling with human genome."

5.   ARE THE TECHNIQUES SAFE?

Safety and efficacy are key issues for any new genetic technologies, particularly those that involve human reproduction. Because mitochondrial donation involves modifying the germline, the changes introduced would be irrevocable, which means they pose a risk not just to the health of the child, but also to the health of future generations if proven unsafe. Many commentators continue to express concern about introducing the techniques into the clinic (Baylis, 2017; Haimes and Taylor, 2017; Gómez-Tatay et al., 2017; Ishii, 2015). This section focuses on current understanding of the technologies, the safety and efficacy reviews that were conducted within the United Kingdom, and how to safeguard future health.

Mitochondrial donation techniques raise many questions, particularly the move to first human use, the role of mtDNA, and interactions between mtDNA and nDNA. First human use will always, to some extent, involve a "leap in the dark." Experimenting on more mature human embryos is often restricted (such as through the United Kingdom "14-day rule"—see, e.g., Cavaliere, 2017) and there have been recent calls to extend this time limit before embryo destruction or implantation in order to further scientific knowledge of the implications of genetic intervention. Mitochondrial donation has been used with mice and macaques (Sato et al., 2005; Tachibana et al., 2009) although a mismatch between animal models and human use is recognized (Craven et al., 2017a,b, also see Lewis et al., 2013). The UK government supported an extensive survey to examine the current state of the technology. In 2011, the DoH requested that the HFEA convene an independent expert panel to consider the safety and efficacy of the techniques. The expert panel reported their findings, based on invited contributions and deliberation (HFEA, 2011), meeting several times and producing further updates in 2013, 2014, and 2016.

One concern expressed throughout was the risk and implications of "carryover," where faulty mitochondria remain attached to the nucleus during the process of transfer. The panel concluded that although it is possible that carryover can occur, this would be such a small percentage that it is unlikely to be problematic. In 2016, the panel recognized that advances had been made, highlighting several studies, which demonstrated low levels of carryover (also see Craven et al., 2017a,b). Hyslop et al. (2016), for example, the first preclinical study "to address the safety and efficacy of PNT," focused on a "refined technique," ePNT, which resulted in less than 2% carryover. However, as the 2016 HFEA report noted, all of the studies also reported phenomena known as "reversion," "genetic drift," or "genetic instability," where the level of faulty mitochondria increased. Hyslop et al. (2016, p. 383) recognized this and meant that although PNT might reduce the risk of transmitting mitochondrial disease, "it may not guarantee prevention." Gómez-Tatay et al. (2017) noting the difficulties of prediction, suggest that carryover should be at least less than 3% to reduce the risk to the child. A mismatch between the mtDNA haplotype of the mitochondria donor and that of the intending mother has been identified as a potential cause (Burgstaller et al., 2014; Reinhardt et al., 2013). The HFEA suggested that licensed clinics could consider haplotype matching, which involves selecting a donor who is genetically compatible, as a precaution. But this issue might be addressed in the future by more innovative methods. As Poulton et al. (2017) suggest, "it might be possible to recruit the cell's own quality control mechanisms to target the pathogenic mutant mtDNAs."

The conclusion from the 2014 and the 2016 updates were that "the evidence it has seen does not suggest that these techniques are unsafe" (HFEA, 2014, p. 4). For many this pronouncement was unsatisfactory. It highlights the difficulties of legislating newly emerging reproductive techniques, and tensions over what counts as evidence and when a technology can be considered safe. The 2016 report differed from the others in that it was produced following the change in law and its task was to consider whether mitochondrial donation was suitable for clinical adoption. The report recommended "in specific circumstances, MST and PNT are cautiously adopted in clinical practice where inheritance of the disease is likely to cause death or serious disease and where there are no acceptable alternatives" (HFEA, 2016, p. 3). Overall, the reports on safety and efficacy concluded that the techniques were potentially useful for a specific group of people, that is, for women who want to have a genetically related child and who are at risk of having a child with severe mitochondrial disease.

Of course, the health of the child will only be known after a period of follow-up, and risks for future generations will not be known until the first cohort are old enough to have their own children (see, e.g., Poulton et al., 2017; Chinnery and Hudson, 2013; Morrow et al., 2015). Whereas mitochondrial donation was developed and legalized in the United Kingdom to prevent the transmission of serious mitochondrial disease, similar technologies have previously been used to address problems with infertility. Cytoplasm injection was developed in the United States and used as fertility treatment for older women, in the hope that "beneficial components" from the donor egg may "restore normal growth and viability" (Barritt et al., 2001, p. 513). Several children were born as a result, but by 2003 concerns about safety and the health of the children were emerging, and the technique was discouraged by the US FDA. One of the problems for current researchers in attempting to learn from this experience was that the children were not enrolled in any follow-up studies. Initial findings from studies initiated almost 15 years later confirmed that some health issues were reported (Chen et al., 2016). Interestingly, only a few were found to have traces of mtDNA from the donor (Cohen and Alikani, 2013), which suggests that cytoplasm injection would not prevent the transmission of mitochondrial disease (Craven et al., 2017a,b). The earliest cited example of PNT used for fertility was through a collaboration between the United States and China in 2003, resulting in a pregnancy with triplets but no live births. It was not until 2016 that scientific details about the case were published (Zhang et al., 2016a) with failure attributed not to the technique itself but to other factors, including local medical care and multipregnancy reduction (also see Gómez-Tatay et al., 2017).

The more recent birth of a baby born using MST through a US/Mexico collaboration so far provides the most relevant information about the efficacy of the technique (Hamzelou, 2016; Zhang et al., 2016b). The baby's birth was celebrated across many national newspapers as a "world first," and proof that the techniques can produce a live baby (González-Santos, 2017; Gonzalez-Santos et al., 2018), shown to have a low level of carryover ($<1\%$) and which appears healthy (Gómez-Tatay et al., 2017). However, the US doctor was widely criticized for operating outside of a legal remit (Lyon, 2017; Dimond and Stephens, 2018) and for only publishing details some months after the birth (Zhang et al., 2017). The concern of those commentating at the time advised follow-up care in order to track his health, particularly in tracking mtDNA segregation over time, although extensive testing has been rejected by the parents if without health benefits to the child (Zhang et al., 2017). Appleby (2015) suggests that follow-up is of such import that it should be made a condition of access to the technology, but follow-up raises questions about resources and the health-care system, as well as questions about consent and disclosure (see, e.g., Bredenoord and Braude, 2010; Dimond, 2013; Pennings, 2017). At the time of writing, further pregnancies had been announced in Ukraine (reported in Coghlan, 2017), although for fertility purposes and not to prevent the transmission of genetic disease. The UK DoH has recommended that any child born through the new techniques of mitochondrial donation be involved in clinical studies to monitor their current and future health. Whether it is ethically acceptable to test children, how long follow-up will last (e.g., whether follow-up will extend to future generations) and whether incentives will be needed to encourage long-term participation are, of course, yet to be decided.

6. ETHICAL ISSUES

Alongside the safety reviews was concern about the ethical implications of the techniques and their legalization. This section explores several key issues including "three-parent babies" and identity, genetic modification and the slippery slope, the risks for the mitochondrial donor, and the differences between PNT and MST.

7. "THREE-PARENT BABIES" AND IDENTITY

Richards et al. (2012) explored reproduction as a collaborative enterprise, distinguishing between the roles of a social mother and father, a biological father in the form of a sperm provider, a biological mother in the form of an egg provider, and a gestational mother in the form of a surrogate. The mitochondrial donor can be added as a sixth progenitor, who alongside the biological father and biological mother is also a genetic contributor. Central questions asked throughout the debates on mitochondrial donation are: "What role should the mitochondrial donor play?" and "What is the significance of her genetic contribution?"

Relationships produced through donation, and the meanings we give them, are dependent on the legal, social, and cultural context. Because of legal requirements and cultural expectations, we generally expect different relationships between donor and recipient depending on what kind of material has been donated. The United Kingdom operates a system of anonymity for donors of blood or tissue (apart from situations where donors are known to each other, e.g., in living kidney donation); since 2005, anonymity was lifted for donors of eggs, sperm, or gametes. This means that children born after 2005 as a result of egg, sperm, or gamete donation would be able to access information about the donor's identity at the age of 18. But we are increasingly being challenged by the capacity of new technology to create new relationships between our bodies and others (e.g., uterus donation challenges our definition of motherhood, and the production of laboratory sperm challenges our understanding of fatherhood, see, e.g., Mertes and Pennings, 2008; Palacios-González et al., 2014). Mitochondrial donation poses such a challenge. Mitochondrial donation involves the transfer of genetic but not nuclear material, and this has led to uncertainty as to whether it should be regulated as egg donation or as tissue donation, the nature of the relationship between child and donor, and how this should be managed. Baylis (2013) suggests that the language of "three parents" is accurate because the child would have a genetic contribution from three people irrespective of how much genetic material is transferred or where that genetic contribution is located. However, in the United Kingdom, the DoH took a different view. Based on the extent of the genetic contribution and the function of the genes involved, it did not accept that the child born through mitochondrial donation would have three parents:

> *Genetically, the child will, indeed, have DNA from three individuals but all available scientific evidence indicates that the genes contributing to personal characteristics and traits come solely from the nuclear DNA, which will only come from the proposed child's mother and father. The donated mitochondrial DNA will not affect those characteristics. (DoH, 2014, p. 15)*

Following the change in law (HFEA, 2015) the debate has, to some extent, been settled—at least within the United Kingdom. The relationship between child and donor was defined as one where there is no legal obligation toward each other. However, the UK government recognized that the child might want to know something about the donor, recommending the child has access to non-identifying information such as screening tests, family health, and personal information provided by the donor.

Identity was also a key issue although an even more "slippery" concept. In the context of discussions about the value of legalizing mitochondrial donation, many highlighted that our nuclear genes solely determine our character and physical appearance, with mtDNA playing a limited role. Within this framing, it would not be expected that altering mitochondrial genes would have a significant impact on the child. This was the perspective taken by many of the UK institutional reports, including from the DoH, the NCoB, and the HFEA. But many have rejected this kind of genetic essentialism, recognizing that while identity is difficult to define, it cannot or should not be reduced to our character and physical traits (Baylis, 2013). In addition, it is clear that mitochondria play a really significant role in the creation of mitochondrial disease; in which case, being born with or without mitochondrial disease would of course have a significant impact on the child and family. This was the case for Bredenoord et al. (2011) who highlight, "a person without a mtDNA disease will have a different life experience, a different biography, and perhaps also a different character." Some also questioned the appropriateness of allowing a child to have "three parents" and the implications this would have on how the child perceived themselves. But as many have pointed out, there are many ways in which people experience family. Being conceived through IVF or having three parents (e.g., through step-families or adoption) is not unusual, and as Scully (cited in NCoB, 2012, p. 72) highlights, what will be important is how society and families respond to mitochondrial donation as a "new kind of normal."

8.   GENETIC MODIFICATION AND THE "SLIPPERY SLOPE"

One of the issues that was strongly debated within the United Kingdom prior to the change in law was the impact of the techniques on the human genome, specifically whether mitochondrial donation could be classified as a form of genetic modification. The DoH identified that the techniques involved germline modification but importantly, they did not involve genetic modification:

> *There is no universally agreed definition of 'genetic modification' in humans— people who have organ transplants, blood donations, or even gene therapy are not generally regarded as being 'genetically modified'. While there is no universally agreed definition, the Government has decided to adopt a working definition for the purpose of taking forward these regulations. The working definition that we have adopted is that genetic modification involves the germ-line modification of nuclear DNA (in the chromosomes) that can be passed on to future generations. This will be kept under review. (DoH, 2014, p. 15)*

The UK government has strongly defended its decision to use the "working definition" of genetic modification. In this context, mitochondrial donation involves an exchange of mitochondrial genes, while leaving both nDNA and mtDNA intact. The move to use a working definition was controversial, and many leading scientists accused the government of dishonesty, misleading the public, and acting by stealth (see, e.g., newspaper coverage at the time, Connor, 2014). The result of ruling out mitochondrial donation as genetic modification was that it enabled mitochondrial donation to be seen as a distinct technology, which did not threaten strongly held views about genetic modification of the human genome; These boundaries were also strengthened through specifying eligibility to use the techniques—only women at risk of transmitting serious mitochondrial disease to their offspring. Marking mitochondrial donation as distinct from nuclear modification was a politically prudent move as it would have been unlikely that the public and government would have accepted attempts to approve the modification of the nuclear genome at that stage.

Imposing restrictions on who can use health technologies and under what circumstances can be ethically problematic. Palacios-González et al. (2014) suggest that if a technique is "morally unproblematic" for some uses, then access should not be prevented for others. As with other countries, the United Kingdom has historically prioritized some groups and uses over others; for example, new-born screening is encouraged for some diseases where early diagnosis can make a difference to the health of the child, late abortions are allowed in some cases (e.g., if an ongoing pregnancy threatens a mother's health), and access to free IVF treatment can depend on postcode and local policies. On this latter note, and in terms of access and resources, mitochondrial donation will be funded by the NHS for at least a few years, which means UK-based families should not be disadvantaged in accessing the technology according to where they live.

But inequality of access to reproductive technologies across the globe is a considerable issue and mitochondrial donation represents one of many challenges facing national and international regulatory systems. One of the unintended consequences of different regulatory regimes could be fertility tourism, where people cross geographical boundaries to access treatment that might not be legal or accessible in their home country. There are many reasons why medical or fertility tourism is a cause for concern, including questions about who provides aftercare, or as in the case of mitochondrial donation, how or whether the child would be engaged in any kind of follow-up monitoring.

The restrictions on use within the United Kingdom have to some extent addressed concerns about the "slippery slope" toward further genetic modification and "designer babies." Johnson (2013) highlights that concern about the slippery slope from medicine to enhancement is based on the value distinction between medical and social, similar arguments to those being made for increasing access to infertility treatment and for social egg freezing (Baldwin et al., 2014). The slippery slope argument is evoked for several reasons. There is a concern that legalization of mitochondrial donation could lead to greater acceptance of related techniques, such as modification of nDNA, or cloning. There is also a concern that legalization of one technique could lead to its "misuse," although what constitutes "misuse" will be dependent on cultural, societal, and legal contexts. But for some the slippery slope also represents a misplaced desire for technological development irrespective of need.

9. DONATING EGGS (RISKS AND BENEFITS)

One concern expressed about many kinds of assisted reproductive technologies is the requirement for egg donation and the potential exploitation of egg donors. Widespread support for mitochondrial donation was evident in the large number of women who came forward to donate their eggs for research following a public appeal by Newcastle University. Egg donors for mitochondrial donation will be compensated in line with UK policy for egg donation, but it remains the case that donating is a complex process with potential risks for the healthy donor, including intrusive tests, discomfort, the possibility of overstimulation, and risks to their own fertility (Baylis, 2013; Haimes, 2013; Haimes and Taylor, 2015, 2017), although these are also noted as issues that are relevant to all egg donors and not specific to mitochondrial donation. Several commentators have highlighted a disparity between the crucial role the donor plays within mitochondrial donation, yet the egg donors themselves and their health and safety were not a key consideration within debates (Dickenson, 2013; Rulli, 2016; Haimes and Taylor, 2015, 2017). Haimes and Taylor (2015), in their analysis of the UK debates, highlighted the political nature of positioning the mitochondrial donor, who they described as an egg provider rather than a mitochondrial donor (the terminology of "mitochondrial donor" implies that it was only mitochondria she provided, rather than other cellular structures in the egg). They document how the donor become an "absent presence" where her role was rendered invisible as part of a political strategy.

10. DIFFERENCES BETWEEN PNT AND MST

Both techniques of MST and PNT are legalized within the United Kingdom. In terms of safety and efficacy, the HFEA (2014, 2016) noted that there was no evidence to suggest that one technique was preferable to the other, and that both "were sufficiently safe to proceed cautiously and in restricted circumstances" (HFEA, 2016, p. 43). It highlighted that the decision as to which technique would be used would depend on the parents' circumstances and would be discussed between the family and the clinic. However as noted within the report, PNT is at a more developed stage in the United Kingdom, which means that in practice, PNT might be a more preferable option than MST. The report also noted that future technologies (such as polar body transfer techniques) "showed great promise" (p7) but would not be permitted under current UK legislation.

MST and PNT both involve the use of a donor egg, but the latter specifically requires the creation and destruction of an embryo in the process. Preference for MST over PNT was highlighted in the Mexico case, where it was reported that the parents demonstrated a moral preference for MST because they were Muslim (Zhang et al., 2017). Many people engaged in the UK debates also stressed a strong preference for one of the techniques over the other. This was the case for one parliamentarian, Fiona Bruce, a conservative member of parliament (MP):

Let me be straightforward: I do oppose these proposals in principle. However, that should not prevent my concerns regarding their safety from being given a fair hearing. One of the two procedures that we are being asked to sanction today—pro-nuclear transfer—involves the deliberate creation and destruction of at least two human embryos, and in practice probably more, to create a third embryo, which it is hoped will be free of human mitochondrial disease. Are we happy to sacrifice two early human lives to make a third life? (Fiona Bruce, Conservative MP for Congleton, House of Commons debate).

Many bioethical commentaries have also made ethical distinctions between the two techniques (see Wrigley et al., 2015 for a comprehensive review). Perspectives differ according to broad views about when life begins and the value attributed to an embryo. Palacios-González (2017), for example, explored differences in liberal, conservative, and gradualist perspectives. He suggested that if embryos are given high moral value, then the destruction necessary for PNT would make the two technologies distinct, with MST preferable over PNT. However, he also notes that both of the techniques involve the creation and destruction of embryos. Distinctions are also made on the basis of the implication for MST and PNT for what is brought into existence. MST can be understood as a form of "selective reproduction" (i.e., an embryo is created through the process) whereas PNT described as a "therapeutic intervention" (the embryo is first formed, and then "treated") (see, e.g., Palacios-González, 2017; Wrigley et al., 2015 but also the concerns of Rulli, 2016 about making "life saving" claims).

11. MITOCHONDRIAL DISEASE AS COMPLEX: DIAGNOSIS, PREDICTING RISK, AND GENETIC COUNSELING

Although evidence is slowly emerging (see, e.g., Herbrand, 2017; Herbrand and Dimond, 2017; Dimond, 2013), we know very little about how families at risk of mitochondrial disease make reproductive decisions. Yet despite this lack of evidence, throughout the UK debates, several assumptions were made about how families might respond to mitochondrial donation as a reproductive option. Mitochondrial donation was widely reported as a technological solution, which once legalized, could benefit thousands of women (see, e.g., Gorman et al., 2015 and subsequent newspaper coverage). Importantly, the suggestion that all those at risk would take up mitochondrial donation, and that this will "halt" or "eradicate" the disease in families are at contrast to the practical difficulties associated with mitochondrial disease and the ways in which families negotiate genetic disease.

Dimond (2013) highlights how genetic disease can become "normalized" within the family, particularly when other family members are affected and when symptoms are variable or late onset, and means that reproductive technology is not always sought when family planning. The extract below is a starting point to highlight the realities of family life in the context of a complex and variable genetic disease. The interview took place in the United Kingdom 3 years before mitochondrial donation was legalized. The woman was 33 years old, and had been diagnosed with mitochondrial disease many years before, along with other family members, and before she had her first son. Her decision to use prenatal diagnosis for a second pregnancy was triggered by the sudden illness of a close family member, and in the extract she describes her experiences of using prenatal diagnosis and the kinds of information, which informed her decisions:

It was a baby boy, but he had 68% mutation and my mother had a similar percentage so I was able to compare what could happen to my baby. We then made the decision [to abort] because I could see what my mother was like and also my wider family [...] We would be bringing a baby into the world knowing that at some point it wasn't going to have its health so we walked away and we thought, as much as we don't want to do this, it is the right thing to do. It was not an easy decision to make, and we discussed this with our whole family, everyone who is affected by this mutation [...] Our final decision came down to me and my husband saying if our eldest son ever says, well why did you not do anything with me, why did you bring me into the world—simple because we didn't know. We didn't understand and we didn't realise what could happen. But with that baby, we couldn't have done that. It would have been purely selfish reasons as to have just said, well no actually we're still going to let him come into the world. (Dimond, 2013, p. 8)

This is a highly personal account of making decisions on the basis of prenatal diagnostic information, specifically the identification of a mutation level of 68%. Diagnosis of mitochondrial disease itself is a complex disease classification, where the presence of mutated mitochondria does not necessarily lead to a diagnosis of mitochondrial disease and requires clinical judgment. A high ratio of faulty mitochondrial is generally, but not always, associated with increasing severity of disease (Craven et al., 2017a,b). For many types of mitochondrial disease, a "gray zone" exists where disease status, the risk of developing symptoms, and severity of symptoms are difficult to measure leading to difficulties in assessing reproductive risk and predicting the outcome (Bredenoord et al., 2010). The woman in the extract above makes sense of this ratio by comparing with other family members, and then made the decision not to continue with the pregnancy on the basis of predicting the future health of that child. But this was only possible because the woman knew about the health of other family members and their mutation levels (and raises questions about whether familial diagnostic information should be made available, see, e.g., Hallowell et al., 2003; Lucassen, 2007).

How individuals make sense of uncertainty is an important question when considering emerging technology, and this extract highlights many issues, including how genetic information is translated as genetic risk and the potential for parents to experience blame and guilt—whatever decisions they make. As the extract suggested, mitochondrial disease presents particular challenges for making sense of diagnosis, predicting genetic risk, and offering advice through genetic counseling (see Poulton et al., 2017; Craven et al., 2017a,b). To conclude this chapter, this section will focus on issues for genetic counseling and alternative reproductive options.

There are many reasons as to why reproductive decision-making in the context of mitochondrial disease, and providing patients with genetic counseling is complicated. This includes a complex relationship between genotype and phenotype, different mutation ratios in different organs, a wide range of symptoms and levels of severity and the potential for changes in mutation ratio over time (Bredenoord et al., 2010; Poulton et al., 2010; Nesbitt et al., 2014). The clinical and genetic complexity of mitochondrial disease poses challenges for patients and health professionals, and current UK advice is that reproductive choice needs to be attended to by "experienced healthcare professionals as part of a multidisciplinary team" (Poulton et al., 2017). Within the United Kingdom, The Rare Mitochondrial Disease Service for Adults and Children provides genetic counseling for women at risk of carrying a mitochondrial mutation and to assist with reproductive choices.

Importantly, the severity of disease in offspring cannot be predicted on the basis of the severity of the disease of the mother. One reason put forward for this is the "bottleneck theory," which affects the degree of heteroplasmy in a female's oocytes (Poulton et al., 2017). Put simply, replication of mtDNA between cells and redistribution during oocyte maturation can lead to extreme differences in levels of mutation. This means that an asymptomatic mother can have a very severely or fatally affected child, many not knowing that they were at risk of transmitting mitochondrial disease until the birth and diagnosis of their child.

12. ALTERNATIVE REPRODUCTIVE OPTIONS

One of the discussion points throughout the UK mitochondrial debates was the question of whether other suitable alternatives exist. There are several options available to women at risk of having a child with mitochondrial disease, including egg donation, prenatal genetic testing, and preimplantation genetic testing all of which have become often used, almost routine, technologies. Egg donation is widely used within assisted reproduction and will prevent the transmission of maternally related mitochondrial disease (Poulton and Oakeshott, 2012). However, egg donation does not offer the woman the opportunity to have a genetically related child (but also see Nordqvist and Smart, 2014 as an example of how families negotiate nongenetic relationships). It also relies on a supply of donor eggs and as described above, there are risks involved for the donors. Prenatal diagnosis and preimplantation genetic diagnosis are currently available for women with mitochondrial disease and offer the opportunity to reduce their risk of having a child with the disease (Poulton et al., 2010). But both are complicated because of the difficulties of diagnosis, including ensuring that the tissue tested represents the embryo and the ratio between healthy and faulty mitochondria could alter over time. Prenatal diagnosis is a test offered during pregnancy after 10 weeks gestation, which will detect and assess the risk for mitochondrial disease (Craven et al., 2017a,b). It is widely deemed reliable, and demand for prenatal diagnosis has been increasing (Nesbitt et al., 2014). The outcome if the ratio is high is that the family could opt for a termination (as was the case for the family described earlier), or could potentially be supported in order to prepare for a child with special needs. Preimplantation genetic diagnosis (PGD) involves the use of IVF to create embryos, and then the selection of the embryo with the lowest mutation rate. There have been several reports where this has been used successfully for mitochondrial disease (Craven et al., 2017a,b). Prenatal diagnosis and PGD would have limited success for those with homoplasmic mitochondrial disease. For PGD to be successful, embryos must be produced with low levels of faulty mitochondria, leading to the question of how to make decisions if all embryos carry a degree of mutation and whether it is ethical and clinically appropriate to implant an embryo with even a low level of mutation?

13.   POLAR BODY TRANSFER AND GENE EDITING

Newly emerging and more controversial technologies include polar body transfer and gene editing techniques using Crispr/Cas9. Following the 2014 report by the expert panel on safety and efficacy (HFEA, 2014), Wang et al. (2014) published details on polar body transfer, which was a new technique to prevent the transmission of mitochondrial disease. The expert panel met and concluded that this technique had potential but it required further work to be considered for clinical adoption, and it has not been made legal under UK law. A polar body is produced during egg maturation or just following fertilization, and as it mainly only contains a nucleus; it means that the whole polar body can be transferred into an enucleated egg. Gómez-Tatay et al. (2017) highlight the advantages of polar body transfer, particularly in terms of reducing the risk of carryover, although a number of limitations remain (see Craven et al., 2017a,b). Crispr/Cas9 is a gene-editing tool that can target and "edit" hundreds of genes at a time. It is quick, cheap, and has the potential to drastically alter the genetic landscape across many different areas including human health, food production and warfare. The world's first license for Crispr/Cas9 work on viable normal human embryos was granted to the Francis Crick Institute in early 2016, and several key institutions such as The NCoB (2016) and the National Academies of Science, Engineering, and Medicine (2017) have explored its ongoing potential. Gene editing such as Crispr/Cas9 could be used to prevent the transmission of mitochondrial disease, and there are advantages over mitochondrial donation: it does not bring the risk of "carryover" and its use would not be restricted only to those with maternally inherited mitochondrial disease (Gómez-Tatay et al., 2017). It also does not require the genetic contribution of a third person. However, further research is needed, and technical challenges remain such as reducing the risks of off-target effects (Craven et al., 2017a,b).

14. CONCLUSION

Mitochondrial donation was legalized in the United Kingdom following a process of scientific and ethical review, and it currently stands alone as the only country to explicitly legalize the techniques (Dimond and Stephens, 2018). For some, legalization was appropriate following extensive reviews, which found the techniques to be "not unsafe" and where the benefits outweigh the risks. For others, the process of review was rushed and ill considered, where the powerful emphasis on the "life saving benefits" for children was exaggerated and politically motivated. In this chapter, the UK situation was explored, and questions of safety, ethics, and reproductive alternatives addressed. What is clear throughout is that further research is needed alongside scientific development—on the national contexts in which newly emerging technology is being debated, embraced, or rejected, and the ways in which patients and families with mitochondrial disease approach reproductive decisions. Where the risks are known, how individuals make sense of uncertainty and complexity of mitochondrial disease, how they assess the risk of having a child with mitochondrial disease, and how they negotiate IVF technologies are important questions.

The United Kingdom's position on legalizing mitochondrial donation, the Mexican "first" and the Ukraine cases for fertility, all highlight international differences in terms of legal frameworks and access to assisted reproductive technologies. It is important to consider the United Kingdom's position as both permissive and restrictive. Although legalized in the United Kingdom, it would be illegal to use mitochondrial donation for purposes other than to avoid the transmission of mitochondrial disease, but there is no reason why other jurisdictions should follow suit, particularly if safety of the technique itself is confirmed. For example, as we have seen in the Ukraine, it could be possible to use the technique to enhance fertility. And if we accept that mitochondrial donation involves genetic inheritance from three individuals, lesbian couples could use the techniques so that the child has a genetic contribution from both women, adding to a range of technologies which currently facilitate lesbian comotherhood, including egg donation and surrogacy (Pennings, 2015). It might also be the case that other countries place a different emphasis on the role of the mitochondrial donor, embracing the possibility and benefits of having "three parents" for example, or invest their resources instead in the possibilities of the new gene-editing techniques of Crispr/Cas9 rather than mitochondrial donation. Whatever happens across the global landscape of assisted reproduction, the United Kingdom's decision to legalize mitochondrial donation will continue to be debated for some time.

REFERENCES

Adashi, E.Y., Cohen, I.G., 2017. Mitochondrial replacement therapy: unmade in the USA. JAMA 317 (6), 574–575.

Appleby, J.B., 2015. The ethical challenges of the clinical introduction of mitochondrial replacement techniques. Med. Health Care Philos. 18 (4), 501–514.

Araki, M., Ishii, T., 2014. International regulatory landscape and integration of corrective genome editing into in vitro fertilization. Reprod. Biol. Endocrinol. 12 (1), 108.

Baldwin, K., Culley, L., Hudson, N., Mitchell, H., 2014. Reproductive technology and the life course: current debates and research in social egg freezing. Hum. Fertil. 17 (3), 170–179.

Barritt, J.A., Brenner, C.A., Malter, H.E., Cohen, J., 2001. Mitochondria in human offspring derived from ooplasmic transplantation. Hum. Reprod. 16, 513–516.

Baylis, F., 2013. The ethics of creating children with three genetic parents. Reprod. BioMed. Online 26, 531–534.

Baylis, F., 2017. Human nuclear genome transfer (so-called mitochondrial replacement): clearing the underbrush. Bioethics 31 (1), 7–19.

Bredenoord, A.L., Braude, P., 2010. Ethics of mitochondrial gene replacement: from bench to bedside. Br. Med. J. 341, c6021.

Bredenoord, A.L., Dondorp, W., Pennings, G., De Wert, G., 2011. Ethics of modifying the mitochondrial genome. J. Med. Ethics 37 (2), 97–100.

Bredenoord, A.L., Krumeich, A., De Vries, M.C., Dondorp, W., et al., 2010. Reproductive decision-making in the context of mitochondrial DNA disorders: views and experiences of professionals. Clin. Genet. 77 (1), 10–17.

Burgstaller, J.P., Johnston, I.G., Jones, N.S., et al., 2014. mtDNA segregation in heteroplasmic tissues is common in vivo and modulated by haplotype differences and developmental stage. Cell Rep. (6), 2031–2041.

Carroll, J., Christodoulou, J., Egli, D., et al., 2015. Parliament should approve regulations for mitochondrial donation. Guardian. 30 January 2015, http://www.theguardian.com/science/2015/jan/30/parliament-should-approve-regulations-for-mitochondrial-donation. Accessed 22 March 2015.

Castro, R.J., 2016. Mitochondrial replacement therapy: the UK and US regulatory landscapes. pp. 726–735.

Cavaliere, G., 2017. A 14-day limit for bioethics: the debate over human embryo research. BMC Med. Ethics 18 (1), 38.

Chen, S.H., Pascale, C., Jackson, M., Szvetecz, M.A., et al., 2016. A limited survey-based uncontrolled follow-up study of children born after ooplasmic transplantation in a single center. Reprod. BioMed. Online 33, 737–744.

Chinnery, P.F., Hudson, G., 2013. Mitochondrial genetics. Br. Med. Bull. 106, 135–159.

Coghlan, A., 2017. First baby born using 3-parent technique to treat infertility. New Scientist 18th January 2017, https://www.newscientist.com/article/2118334-first-baby-born-using-3-parent-technique-to-treat-infertility. Accessed 27 November 2017.

Cohen, J., Alikani, M., 2013. The biological basis for defining bi-parental or tri-parental origin of offspring from cytoplasmic and spindle transfer. Reprod. BioMed. Online 26 (6), 535–537.

Connor, S 2014, Exclusive: scientists accuse government of dishonesty over GM babies in its regulation of new IVF technique, The Independent 28 July.

Craven, L., Alston, C.L., Taylor, R.W., Turnbull, D.M., 2017a. Recent advances in mitochondrial disease. Annu. Rev. Genom. Hum. Genet. 18, 257–275.

Craven, L., Tang, M.X., Gorman, G.S., De Sutter, P., et al., 2017b. Novel reproductive technologies to prevent mitochondrial disease. Hum. Reprod. Update 23 (5), 501–519.

Craven, L., Tuppen, H.A., Greggains, G.D., Harbottle, S.J., et al., 2010. Pronuclear transfer in human embryos to prevent transmission of mitochondrial DNA disease. Nature 465 (7294), 82–85.

Dickenson, D.L., 2013. The commercialization of human eggs in mitochondrial replacement research. N. Bioeth. 19, 18–29.

DiMauro, S., 2011. A history of mitochondrial diseases. J. Inherit. Metab. Dis. 34 (2), 276.

Dimond, R., 2013. Patient and family trajectories of mitochondrial disease: diversity, uncertainty and genetic risk. Life Sci. Soc. Policy 9, 2.

Dimond, R., Stephens, N., 2018. Legalising Mitochondrial Donation: Enacting Ethical Futures in UK Biomedical Politics. Palgrave Pivot.

DoH, 2014. Mitochondrial donation: government response to the consultation on draft regulations to permit the use of new treatment techniques to prevent the transmission of a serious mitochondrial disease from mother to child. https://www.gov.uk/government/consultations/serious-mitochondrial-disease-new-techniques-to-prevent-transmission. Accessed 27 November 2017.

Gómez-Tatay, L., Hernández-Andreu, J.M., Aznar, J., 2017. Mitochondrial modification techniques and ethical issues. J. Clin. Med. 6 (3), 25.

González-Santos, S.P., 2017. Shifting the focus in the legal analysis of the first MST case. J. Law Biosci. 4 (3), 611–616.

Gonzalez-Santos, S., Stephens, N., Dimond, R., 2018. Narrating the first "Three-Parent Baby": the initial press reactions from the United Kingdom, the United States, and Mexico. Sci. Commun. https://doi.org/10.1177/1075547018772312.

Gorman, G.S., Grady, J.P., Ng, Y., et al., 2015. Mitochondrial donation: how many women could benefit? N. Engl. J. Med. 372, 885–887.

Haimes, E., 2013. Juggling on a rollercoaster? Gains, loss and uncertainties in IVF patients' accounts of volunteering for a UK 'egg sharing for research' scheme. Soc. Sci. Med. 86, 45–51.

Haimes, E., Taylor, K., 2015. Rendered invisible? The absent presence of egg providers in UK debates on the acceptability of research and therapy for mitochondrial disease. Monash Bioeth. Rev. 33, 360–378.

Haimes, E., Taylor, K., 2017. Sharpening the cutting edge: additional considerations for the UK debates on embryonic interventions for mitochondrial diseases. Life Sci. Soc. Policy 13 (1), 1.

Hallowell, N., Foster, C., Eeles, R., Ardern-Jones, A., Murday, V., Watson, M., 2003. Balancing autonomy and responsibility: the ethics of generating and disclosing genetic information. J. Med. Eth. 29 (2), 74–79. discussion 80–73.

Hamzelou, J., 2016. Exclusive: World's first baby born with new '3 parent' technique. New Scientist. 27th September 2016, https://www.newscientist.com/article/2107219-exclusive-worlds-first-baby-born-with-new-3-parent-technique. Accessed 13 November 2017.

Hargreaves, L.P., 2014. Coenzyme Q10 as a therapy for mitochondrial disease. Int. J. Biochem. Cell Biol. 49, 105–111.

Herbrand, C., 2017. Mitochondrial replacement techniques: who are the potential users and will they benefit? Bioethics 31 (1), 46–54.

Herbrand, C., Dimond, R., 2017. Mitochondrial donation, patient engagement and narratives of hope. Sociol. Health Illn.

HFEA, 2011. Scientific review of the safety and efficacy of methods to avoid mitochondrial disease through assisted conception. April 2011, http://hfeaarchive.uksouth.cloudapp.azure.com/www.hfea.gov.uk/docs/2011-04-18_Mitochondria_review_-_final_report.pdf. Accessed 29 November 2017.

HFEA 2013, Mitochondria replacement consultation: advice to government, March 2013.

HFEA, 2014. Third scientific review of the safety and efficacy of methods to avoid mitochondrial disease through assisted conception. Update 2014, http://hfeaarchive.uksouth.cloudapp.azure.com/www.hfea.gov.uk/8807.html. Accessed 10 November 2017.

HFEA, 2015. The Human Fertilisation and Embryology (Mitochondrial Donation) Regulations 2015. SI 2015/572, HFEA, London.

HFEA, 2016. Scientific review of the safety and efficacy of methods to avoid mitochondrial disease through assisted conception: 2016 update. http://hfeaarchive.uksouth.cloudapp.azure.com/www.hfea.gov.uk/10557.html. Accessed 29 November 2017.

Hyslop, L.A., Blakeley, P., Craven, L., Richardson, J., et al., 2016. Towards clinical application of pronuclear transfer to prevent mitochondrial DNA disease. Nature 534 (7607), 383–386.

Ishii, T., 2015. Germline genome-editing research and its socioethical implications. Trends Mol. Med. 21 (8), 473–481.

Johnson, M.H., 2013. Tri-parenthood—a simply misleading term or an ethically misguided approach? Reprod. BioMed. Online 26 (6), 516–519.

Jones, C., Holme, I., 2013. Relatively (im)material: mtDNA and genetic relatedness in law and policy. Life Sci. Soc. Policy 9 (4), 1–14.

Kisler, J.E., Whittaker, R.G., McFarland, R., 2010. Mitochondrial diseases in childhood: a clinical approach to investigation and management. Dev. Med. Child Neurol. 52, 422–433.

Lewis, J., Atkinson, P., Harrington, J., et al., 2013. Representation and practical accomplishment in the laboratory: when is an animal model good-enough? Sociology 47, 776–792.

Lucassen, A., 2007. Head to head: should families own genetic information? Yes. Br. Med. J. 335 (7609), 22.

Lyon, J., 2017. Sanctioned UK trial of mitochondrial transfer nears. JAMA 317 (5), 462–464.

Manwaring, N., Jones, M.M., Wang, J.J., et al., 2007. Population prevalence of the MELAS A3243G mutation. Mitochondrion 7, 230–233.

Meldau, S., Riordan, G., Van der Westhuizen, F., Elson, J.L., Smuts, I., Pepper, M.S., Soodyall, H., 2016. Could we offer mitochondrial donation or similar assisted reproductive technology to South African patients with mitochondrial DNA disease? S. Afr. Med. J. 106 (3), 234–236.

Mertes, H., Pennings, G., 2008. Embryonic stem cell-derived gametes and genetic parenthood: a problematic relationship. Camb. Q. Healthc. Ethics 17, 7–14.

Morrow, E.H., Reinhardt, K., Wolff, J.N., Dowling, D.K., 2015. Risks inherent to mitochondrial replacement. EMBO Rep. 16 (5), 541–544.

National Academies of Science, Engineering, and Medicine, 2017. Human Genome Editing: Science, Ethics, and Governance. The National Academies Press.

National Academies of Sciences, Engineering, and Medicine, 2016. Mitochondrial Replacement Techniques: Ethical, Social, and Policy Considerations. The National Academies Press, Washington, DC.

NCoB, 2012. Novel Techniques for the Prevention of Mitochondrial DNA Disorders: An Ethical Review. Nuffield Council on Bioethics, London.

NCoB, 2016. Genome Editing: An Ethical Review. Nuffield Council on Bioethics, London.

Nesbitt, V., Alston, C.L., Blakely, E.L., Fratter, C., et al., 2014. A national perspective on prenatal testing for mitochondrial disease. Eur. J. Hum. Genet. 22 (11), 1255–1259.

Nordqvist, P., Smart, C., 2014. Relative Strangers: Family Life, Genes and Donor Conception. Palgrave Macmillan.

Palacios-González, C., 2017. Are there moral differences between maternal spindle transfer and pronuclear transfer? Med. Health Care Philos. 20 (4), 503–511.

Palacios-González, C., Harris, J., Testa, G., 2014. Multiplex parenting: IVG and the generations to come. J. Med. Ethics 40 (11), 752–758.

Parikh, S., Goldstein, A., Karaa, A., Koenig, M.K., Anselm, I., Brunel-Guitton, C., Christodoulou, J., Cohen, B.H., Dimmock, D., Enns, G.M., Falk, M.J., 2017. Patient care standards for primary mitochondrial disease: a consensus statement from the mitochondrial medicine society. Genet. Med. 19(12). https://doi.org/10.1038/gim.2017.107.

Pennings, G., 2015. Having a child together in lesbian families: combining gestation and genetics. Indian J. Med. Ethics. pp. medethics-2015.

Pennings, G., 2017. Disclosure of donor conception, age of disclosure and the well-being of donor offspring. Hum. Reprod. 32 (5), 969–973.

Poulton, J., Finsterer, J., Yu-Wai-Man, P., 2017. Genetic counseling for maternally inherited mitochondrial disorders. Mol. Diagn. Ther. 21 (4), 419–429.

Poulton, J., Oakeshott, P., 2012. Nuclear transfer to prevent maternal transmission of mitochondrial DNA disease. BMJ 345, e6651.

Poulton, J., et al., 2010. Transmission of mitochondrial DNA diseases and ways to prevent them. PLoS Genet. 6 (8), e1001066.

Reinhardt, K., Dowling, D.K., Morrow, E.H., 2013. Mitochondrial replacement, evolution, and the clinic. Science 341, 1345–1346.

Richards, M., Pennings, G., Appleby, J., 2012. Introduction. In: Richards, M., Pennings, G., Appleby, J. (Eds.), Reproductive Donation: Practice, Policy and Bioethics. Cambridge University Press, Cambridge, pp. 1–12.

Roccella, E., Buttiglione, R., Picchi, G., et al., 2015. Three-person DNA, The Times letter. February 20th 2015, http://www.thetimes.co.uk/tto/opinion/letters/article4360729.ece.

Rulli, T., 2016. What is the value of three-parent IVF? Hast. Cent. Rep. 46 (4), 38–47.

Sato, A., Kono, T., Nakada, K., et al., 2005. Gene therapy for progeny of mito-mice carrying pathogenic mtDNA by nuclear transplantation. Proc. Natl. Acad. Sci. U. S. A. 102, 16765–16770.

Schaefer, A.M., McFarland, R., Blakely, E.L., et al., 2008. Prevalence of mitochondrial DNA disease in adults. Ann. Neurol. 63, 35–39.

Schwartz, M., Vissing, J., 2002. Paternal inheritance of mitochondrial DNA. N. Engl. J. Med. 347, 576–580.

Tachibana, M., Sparman, M., Sritanaudomchai, H., et al., 2009. Mitochondrial gene replacement in primate offspring and embryonic stem cells. Nature 461, 367–372.

Varvaštian, S., 2015. UK's legalisation of mitochondrial donation in IVF treatment: a challenge to the international community or a promotion of life-saving medical innovation to be followed by others? Eur. J. Health Law 22 (5), 405–425.

Wang, T., Sha, H., Ji, D., Zhang, H.L., Chen, D., Cao, Y., Zhu, J., 2014. Polar body genome transfer for preventing the transmission of inherited mitochondrial diseases. Cell 157 (7), 1591–1604.

Wrigley, A., Wilkinson, S., Appleby, J.B., 2015. Mitochondrial replacement: ethics and identity. Bioethics 29 (9), 631–638.

Zhang, J., Liu, H., Luo, S., Chavez-Badiola, A., Liu, Z., Yang, M., Munné, S., Konstantinidis, M., Wells, D., Huang, T., 2016b. First live birth using human oocytes reconstituted by spindle nuclear transfer for mitochondrial DNA mutation causing Leigh syndrome. Fertil. Steril. 106 (3 Suppl), e375–e376.

Zhang, J., Liu, H., Luo, S., Lu, Z., Chávez-Badiola, A., Liu, Z., Yang, M., Merhi, Z., Silber, S.J., Munné, S., Konstandinidis, M., Wells, D., Huang, T., 2017. Live birth derived from oocyte spindle transfer to prevent mitochondrial disease. Reprod. Biomed. Online 34, 25–32.

Zhang, J., Zhuang, G., Zeng, Y., Grifo, J., et al., 2016a. Pregnancy derived from human zygote pronuclear transfer in a patient who had arrested embryos after IVF. Reprod. BioMed. Online 33 (4), 529–533.

FURTHER READING

Craven, L., Herbert, M., Murdoch, A., Murphy, J., et al., 2016. Research into policy: a brief history of mitochondrial donation. Stem Cells 34 (2), 265–267.

Dimond, R., Stephens, N., 2017. Three persons, three genetic contributors, three parents: mitochondrial donation, genetic parenting and the immutable grammar of the 'three x x'. Health.

Haran, J., Kitzinger, J., McNeil, M., O'Riordan, K., 2007. Human Cloning in the Media. Routledge, London.

House of Commons Science and Technology Committee, 2014. Oral evidence: mitochondrial donation. HC 730 22 October 2014, http://www.parliament.uk/business/committees/committees-a-z/commons-select/science-andtechnology-committee/inquiries/parliament-2010/mitochondrial-donation/. Accessed 25 March 2015.

Ishii, T., 2017. Germ line genome editing in clinics: the approaches, objectives and global society. Brief. Funct. Genomics 16 (1), 46–56.

Newson, A.J., Wrigley, A., 2017. Is mitochondrial donation germ-line gene therapy? Classifications and ethical implications. Bioethics 31 (1), 55–67.

Scully, J.L., 2017. A mitochondrial story: mitochondrial replacement, identity and narrative. Bioethics 31 (1), 37–45.

Chapter 3

Reproductive Technologies Used by Same Gender Couples

Diana Badiu* and Valentina Nastasel[†]
*Ovidius University, Constanta, Romania, [†]Kantonsspital Graubünden, Chur, Switzerland

1. INTRODUCTION

Same-sex marriage and the right of atypical family cores to have children have been highly debated issues in recent times. Many advances in assisted reproductive technologies (ARTs) allowed same-sex couples to have children, even though this has been often received with circumspection by the public, physicians, certain religious groups, or some specific population groups. Even if these issues are controversial, they currently belong, at least in some parts of the world, to "everyday life." A study conducted in the United States showed that more than one in three lesbians and one in six gay men had children (either genetically related or adopted), and that 65,000 or around 4% of all adopted children lived with, and were raised by gay or lesbian parents (Gates et al., 2007). Most of these children were born within previous heterosexual marriages; however, since the 1980s the number of children born to lesbians has surged, followed in the 1990s by children born from gay parents (DeLair, 2000), a phenomenon coined by the media as *lesbian baby boom, gay baby boom, gay-by boom* (Pressley and Andrews, 1992), and the "choosing children movement" (Benkov, 1994). In Norway, in 1999, a study was published, reported that approximately 10% of lesbians and gays have children (Hegna et al., 1999).

Isolated cases of same-sex marriages have been recorded by historians since ancient times. Nero (37–68 AD) was married to two men, on different occasions—a former slave named Pythagoras (Williams, 2010, pp. 1–512), with whom Nero played the bride, and Sporus, with whom the emperor played the groom (Woods, 2009, pp. 73–82). The child emperor Elagabalus (203–222 AD) also married a slave named Hierocles. It was only in 342 AD that the emperors Constantin the Second and Constans explicitly forbade same-sex marriages in the Roman Empire in a law known as the Theodosian Code (Kuefler, 2007, pp. 343–370).

Clinical Ethics at the Crossroads of Genetic and Reproductive Technologies.
https://doi.org/10.1016/B978-0-12-813764-2.00003-9

55

Even if the Christian Church explicitly forbade same-sex relationships, there were many ways to circumvent it in practice. In the early 18th century, numerous "molly houses" were founded in London; pubs for gays, occasionally with a "marrying room," were not exceptionally rare, even if they were initially without the Church's approval. That changed in 1810, when the Calvinist priest John Church started to officiate gay marriages in mock ceremonies at "The Swan" from Vere Street (Clark, 1988, pp. 56–69).

Denmark was the first country to institutionalize same-sex marriage in modern times, by introducing the "registered partnership" in 1989 (Barnes et al., 2004). It allowed some rights associated with heterosexual marriage, but forbade others, namely the right to adopt or obtain joint custody of a child (Brown, 2010). Today, various forms of same-sex marriage are present in 21 countries: Argentina, Belgium, Brazil, Canada, Colombia, Denmark, Finland, France, Iceland, Ireland, Luxembourg, Mexico (partially), the Netherlands, New Zealand (partially), Norway, Portugal, South Africa, Spain, Sweden, the United Kingdom (England, Scotland, Wales), the United States, and Uruguay, and will soon be legal in Germany, Malta, and Taiwan.

After obtaining the right to marry, a natural extension was the right to form families, and subsequently, the right to have children or, for those leaving a heterosexual marriage—to retain custodial rights (Rivers, 2010). The right of same-sex couples to have children was reachable through the following means: by retaining custodial rights to children from previous marriages/relationships, by adoption, or by fathering/mothering children with or without the genetic material of one or both the members of the couple.

An ever-increasing number of countries accept lesbian, gay, bisexual, transgender, and queer (LGBTQ) adoption through various mechanisms—antidiscrimination laws, public demand, legal rulings. Nowadays, joint adoption by same-sex parents is legal in Andorra, Argentina, Australia (partially), Austria, Belgium, Brazil, Canada, Colombia, Denmark, Finland, France, Germany, Iceland, Ireland, Israel, Luxembourg, Malta, Mexico (partially) the Netherlands, New Zeeland, Norway, Portugal, South Africa, Spain, Sweden, the United Kingdom, and the United States. Other countries allow step-child adoption, in which one partner can adopt the natural or adopted child of the other partner: Croatia, Estonia, Italy, Germany, Slovenia, and Switzerland.

Having their own children, either genetically related or not, is the third enlisted option for LGBTQ couples to form an extended family, one made possible almost entirely through the development of ARTs. ARTs—such as artificial insemination (AI), in vitro fertilization (IVF), egg donation, surrogacy, or genetic manipulation of reproductive material—are mostly used by heterosexual couples when one or both partners are infertile, but their usefulness was more recently expanded to other categories of patients as well, including same-gender couples and single parents (Greenfeld and Seli, 2011; Robertson, 2004; Greenfeld, 2007; De Wert et al., 2014). Once nontraditional couples started referring to ARTs, moral debates arose surrounding their access and conditions of access to these techniques. Requests for ARTs from same-gender couples increased often, especially from lesbian couples (Marina et al., 2010). ARTs are the only method for LGBTQ persons to potentially have genetically related offspring. For lesbian couples, if both partners are fertile, the decisions revolve around who will donate the ovum, who will bear the child, where to obtain the male gamete from, and whether the donor should be anonymous or not. Usually, the embryo is formed through AI, but alternative procedures can also be used (Falcone and Hurd, 2017). For gay couples, if both partners are fertile, the decisions revolve around who will provide the sperm, who will donate the ovum, whether the egg donor should be known or anonymous, and who will be the surrogate mother.

An underrepresented type of same-sex couple is that in which one or both partners are either transgender or transsexual. According to the fifth edition of the Diagnostic and Statistical Manual of Mental Disorders, "transgender refers to the broad spectrum of individuals who transiently or persistently identify with a gender different from their natal gender" (DSM-5, 2013). The desired gender identity is usually that of the opposite sex, but other instances can be identified as well, including agender, bigender, trigender, pangender, genderfluid, or nongender. Transgender should be confused neither with sexual orientation, as they can be heterosexual, homosexual, bisexual, asexual, nor with intersex (persons with physical characteristics not appertaining exclusively to one gender). A transsexual is an "individual who seeks, or has undergone, a social transition from male to female or female to male, which in many, but not all, cases involves a somatic transition by cross-sex hormone treatment and genital surgery (sex reassignment surgery)" (DSM-5, 2013). If for gay and lesbian couples more data is available regarding parenting and adoption, this is not the case for transgender couples (Biblarz and Savci, 2010). Some studies suggest that transgender persons face completely different social and psychological issues compared to gays and lesbians, mainly generated by the fact that their social circle must adapt to a change in sexual identity and not sexual orientation (Israel, 2005). This issue also generates increased difficulties regarding adoption—investigations conducted in the Unitd States showed that transgender persons who wished to adopt, either from private or public adoption agencies, encounter increased difficulties, generated not by legal barriers but rather by discrimination and pervasive bias (Levi and Monnin-Browder, 2012). Transphobia was a demonstrated negative predictor for perceived emotional stability and decreased willingness to grant custody of a child to a transsexual couple compared to a gay couple (Weiner and Zinner, 2015). In addition to external factors, fear that pregnancy or birth would disrupt the transitioning process or contradict the gender identity was also mentioned as a barrier to parenthood (Tornello and Bos, 2017). Requests for ARTs by couples in which at least one person is transsexual are encountered less frequently compared to those by gay or lesbian couples. In couples comprised of a female-to-male (FtM) transsexual and a cissexual man, the ART of choice is AI with partner or donor sperm. In couples comprised of a cissexual man and a male-to-female (MtF) transsexual, ART usually involves oocyte donation and surrogate motherhood.

2. PROCREATIVE AUTONOMY

The public debate surrounding same-gender couples' right to procreate uses four main ethical theories either for the defense or critique of this right. Under Kantian ethics, same-sex couple reproduction through ARTs could be viewed as immoral, as it instrumentalizes a biological process whose main purpose is to generate human life; however, this would apply to the use of ARTs in infertile heterosexual couples in general, whose decisions, although yet not universally accepted by society, are nevertheless better tolerated. Also, if we extend this doctrine, we could infer that all children born through any kind of instrumentalization would contradict the finality principle. Children born from arranged marriages could be seen as intrumentalized; or even children born out of love, which can be seen as a means through which nature dictates the actions of rational agents. But can the act of giving birth be considered immoral? No one should think so (or almost no one)—a moral agent is an end in itself as long as his/her telos is not instrumentalized, and not if its biological genesis was instrumentalized; therefore, we believe that the use of ARTs should be allowed for anyone who needs them, and within the requirements of the law; by not allowing subgroups of moral agents to access a medical procedure, we would discriminate, and therefore act immorally. Under the consequentialist ethics, we should look at the overall benefit generated by the procedure. Some authors argued in the past that children born out of these atypical unions are cognitively and sociopsychologically impaired, assumptions that have been invalidated by numerous social science studies, as detailed below. Additionally, bringing a child into the world can only bring happiness to the couples involved, their extended families and friends, not to mention the child, who gets a life. Pragmatic ethics contends that moral judgments are stated by society, not single individuals, and can change as society progresses. Under pragmatism, nontraditional couples' desire to procreate is ethical, because social norms have changed and society now more easily accepts LGBTQ persons and children born to single mothers, interracial couples, or out of wedlock. Under the principialist theory of biomedical ethics, the autonomy principle asserts that each individual has a right to health decisions in general and reproductive and sexual decisions in particular, derived from the fundamental and inalienable right to self-determination (Corvino, 2013). Most progresses in ensuring the right to a family have been made by successfully defending this last principle.

The right to family life is one of the most important human rights, clearly established in various international regulations. The Universal Declaration of

Human Rights (UDoHR) states, in Article 16, that *"(1) Men and women of full age, without any limitation due to race, nationality or religion, have the right to marry and to found a family. They are entitled to equal rights as to marriage, during marriage, and at its dissolution. (2) Marriage shall be entered into only with the free and full consent of the intending spouses. (3) The family is the natural and fundamental group unit of society and is entitled to protection by society and the State."* This article discusses the right of each individual to start a family, without explicitly limiting the family core to a heterosexual binomial structure. As stated in the UDoHR, the right to family life is negative—the signatory states are obliged to respect the founded family cores, and to *not* actively intervene in and limit the process of a couple to start a family. The European Convention on Human Rights (ECHR) also states, in Article 8, that: *"(1) Everyone has the right to respect for his private and family life, his home and his correspondence; (2) There shall be no interference by a public authority with the exercise of this right except such as is in accordance with the law and is necessary in a democratic society in the interests of national security, public safety or the economic well-being of the country, for the prevention of disorder or crime, for the protection of health or morals, or for the protection of the rights and freedoms of others."* Again, the right to form a family is seen as a negative one—the state should *not* intervene in forming the family core, and should protect it and its privacy. In article 12 the ECHR states that: *"Men and women of marriageable age have the right to marry and to found a family, according to the national laws governing the exercise of this right"* and in article 14: *"The enjoyment of the rights and freedoms set forth in this Convention shall be secured without discrimination on any ground such as sex, race, colour, language, religion, political or other opinion, national or social origin, association with a national minority, property, birth or other status."*

If the right to a family is clearly protected by the ECHR, the specifics of the right to marry are left to state legislatures. In the absence of these specific regulations, to deny same-sex couples the fundamental right to a family can be considered discriminatory (a gender-based and/or sexual orientation-based discrimination—see Art 14 from the ECHR).

Marriage confers legal and social rights and can contribute to the stabilization of same-sex unions. Same-sex marriages can also confer benefits to children in the same manner that heterosexual ones do, by ensuring financial well-being and continuity, by strengthening the parents' relationship and by increasing social acceptance and support for same-sex families (Meezan and Rauch, 2005; Artis, 2007; Brown, 2004; Carlson and Corcoran, 2001; Manning and Lamb, 2003; Videon, 2002). In addition to bestowing rights, same-gender marriage can also eliminate cumbersome legal procedures. It can facilitate the surrogacy process for the commissioning parents who currently must adopt their children born through such a technique, and it can similarly simplify or eliminate the need of the nonbiological parent to coadopt his/her child. And because marriage instability clearly affects children (Brown, 2004, 2010), one can reason that they also have the right to be and benefit from being raised in same-sex marriages.

Several gay rights activists have nevertheless criticized LGBTQ groups' use of the promarriage argument to fashion a strategy on lifting the ban on same-sex procreation, to the detriment of the former promoting of family pluralism and broader social justice.

Queer academic and activist Nancy Polikoff contends: "*While advocates for lesbian and gay parents once saw themselves as part of a larger movement to promote respect, nondiscrimination, and recognition of diverse family forms, some now appear to embrace a privileged position for marriage. They thus abandon a longstanding commitment to defining and evaluating families based on function rather than form, distancing themselves from single-parent and divorced families, extended families, and other stigmatized childrearing units*" (Polikoff, 2005, p. 573). The current debate is not centered as much on the hetero-/homosexual family dichotomy; instead, it remains confined within the bounds of heteronormalcy by insisting on marriage as the only type of union advantageous for children.

The right to have children includes parenthood through adoption as well as the right to reproductive autonomy. However, except cases in which children come from one of their parents' previous heterosexual marriage(s) or through adoption, same-sex couples cannot have offspring without the aid of a "facilitating other" (Mitchell and Green, 2007) and of special reproductive techniques and as such, without the active intervention of the State into the family life, through specific legal norms. Therefore the right of same-sex couples to a family should be considered a *positive right*. This brings to the forefront the heavily debated questions of homosexuals' fitness to parent and the implications on children's well-being and development.

The gay liberation movement of the 1970s saw judges forming professional opinions and settling custody disputes on the basis of "traditional common sense" or folk psychology, which dictates that "like produces like" (Cameron, 1999). The lack of empirical research data on homosexual parenting at that time sparked a theme-related upsurge in investigations and media coverage that continues to the present day.

Many studies have been conducted on same-sex parenting, but they have been criticized nevertheless for their small representative samples, corresponding weak statistical power and effect sizes, choice of comparison group, subject pool heterogeneity, and lack of methodological rigor (Meezan and Rauch, 2005; Amato, 2012). Cameron (2009) has also criticized the extrapolation of lesbian parenting investigations results to gay parenting.

Opponents of same-sex parenting come forth with arguments that can generally be divided into two groups: they either revolve around the immorality of the homosexual union and familial atmosphere in which the child will be raised, as seen through a religious or biological prism ("the Bible and the

Church condemn gay and lesbian parenting as sinful," "same-sex parenting is unnatural," "children are *designed* to grow up in a heterosexual, nuclear family," "homosexual unions and parenting are against the natural order,") or around the notion of the child's future psychological and sociological well-being. In both sets of arguments, the individual's or the group's responsibility for prejudicing against same-sex families is projected to a superior force or unit of organization (God, Nature, society) (Clarke, 2001).

Studies supporting an attitude against same-sex parenting have inferred that children of lesbian and gay parents are more vulnerable to sexual and physical abuse by their parents (Cameron and Cameron, 1996a; Cameron et al., 2017) and more inclined themselves to become homosexuals (Cameron, 2006). They also implied increased psychosocial and behavioral dysfunctions (Blankenhorn, 1995; Cameron and Cameron, 1996a,b; Wardle, 1997; Knight, 1997), and a higher probability to question their own sexual preference (O'Connell, 1993).

Recent meta-analyses (Allen and Burrell, 1997; Crowl et al., 2008; Fedewa et al., 2014) of numerous studies on homosexual vs heterosexual parent families have consistently indicated that with regards to gender development, psychological adjustment, school performance, and parenting quality, children from same-sex families fare just as well as those from traditional families. Although fewer studies have been conducted on gay parent families, and especially so on those formed through surrogacy, the prevailing literature agrees that children's psychological, cognitive, and gender development is not impaired in these family forms (Erich et al., 2005; Farr et al., 2010, Farr, 2017; Golombok et al., 2013) and that they can thrive in a supportive emotional environment (Bos et al., 2007; Golombok, 2015). Additionally, adoptive gay father families demonstrated more positive psychological well-being and parenting compared to heterosexual ones (Golombok et al., 2006; Farr et al., 2010).

Columbia Law School's "What We Know" portal, a public policy research project, identified and reviewed as of yet 79 investigations on the well-being of children raised in same-sex families, out of which 75 concluded that they fared no worse than others. The four remaining studies all shared a common flaw: all children came from initially heterosexual families and endured post-divorce stress, changes in family structure, and the coming out of their mother or father, so it would be erroneous to consider that they grew up from the beginning in a same-gender family (What We Know, 2017).

Contrary to the popular and entrenched belief that "families headed by married, biological parents are best for children" (Popenoe, 1993, pp. 527–541), studies indicate that the gender and sexual orientation of parents is of minor significance for children's psychosocial adjustment and success (Biblarz and Savci, 2010) and that the quality of parenthood is in reality more important and more predictive of children's social adjustment (MacCallum and Golombok, 2004).

In their turn, lesbian feminists have argued that lesbians' decisions to become mothers are either progressive or retrograde. Herman writes: *"Does the baby boom of the past decade indicate the success of gay liberation in increasing lesbian self-esteem and inaugurating an era of sexual and reproductive freedom? Or does it indicate lesbian surrender to the rules and regulations of femininity, maternity being first among them?"* Poet and radical feminist Adrienne Rich stated that the only sexual identity that is truly actively imposed on children is (compulsory) heterosexuality (Rich, 1981). Alice, Gordon, Debbie, and Mary, writing at the beginning of the positive lesbian and gay movement, viewed *"childcare as an enormous drain of energy on our still embryonic movement (...). We see no need to add more children at this time."* Similarly, for Firestone, women's emotional energies must "flow toward our sisters not backwards towards our oppressors" (Firestone, 1970). From the radical feminists' point of view, being a "real" lesbian automatically means shunning men and rejecting the idea of having (their) children, especially sons, because procreation is seen as an function exclusively associated with the domineering and patriarchal institution of heterosexual marriage and not tied to women's procreative consciousness per se. Motherhood is seen as *"too loaded with this patriarchal history and function to be an entirely different phenomenon just because lesbians are doing it"* and as a *"rejection of lesbian political activism"* (Polikoff, 1987). Jo argues that lesbian mothers *"are fulfilling a male-defined role of femininity and motherhood,"* that by receiving sperm, being pregnant, and giving birth they are committing "a heterosexual act," unacceptable to the lesbian society, and that *"Motherhood can be lethal (...) and ultimately a reactionary choice"* (Jo, 1984/1988, pp. 315–317).

Yet other feminists were more optimistic and viewed lesbian parenting as revolutionary. Copper viewed lesbian mothers as *"the only category of women sufficiently alienated from patriarchal tradition to sustain radical modification in the socialization of daughters"* (Copper, 1987/1988, pp. 306–321). For Rowen, "the influence of strong lesbian, feminist parents" is the only guarantee that sons won't grow up like their "oppressor cousins," who "will surely steal their souls" (Rowen, 1981, pp. 97–106).

Feminist literature on gay parenting on the other hand is scant. Gay parents are viewed as a paradox and a puzzling social category: they contest the mainstream precepts of (heterosexual) men avoiding parenthood responsibilities and that of the effervescent, freedom-loving gay subculture. This is compounded by their even stronger determination to become fathers, because they have to struggle harder and longer than heterosexual couples to become parents, and as such are more frequently criticized and stigmatized.

But is the issue of LGBTQ parenthood to be reduced to a dichotomous, "either/or" approach by feminists? One can approach the issue from a more centered, individualistic, reproductive perspective and not only from a sociopolitical one. Marsiglio and Hutchinson (2002, p. 288) originally coined the term "procreative consciousness" to refer to how men see themselves as procreative beings. The "procreative consciousness" or "identity" of gay and lesbians parents-to-be, as separate from their sexual orientation, has only more recently been investigated (Berkowitz and Marsiglio, 2007; Berkowitz, 2008), but references to its relevance date back to the 1970s: *"A "real" lesbian will be especially loath to keep male children. With that idea of victory, who needs defeat? Some of us may not want children. But that choice and avoiding motherhood because it is" unlesbian" are two entirely different things!"* (Wyland, 1977, pp. 1–36). For the majority of LGBTQ persons, especially gays, their sexual identity was for a long period of time synonymous with childlessness and lost custody (Murphy, 2013). The advent of ARTs and its regulated use by same-sex couples has meant that procreation was no longer a consequence of fate or nature (Strathern, 1999), but a choice facilitated by the governments and physicians.

So, as seen earlier, there are no significant arguments, either medical or related to the well-being of the child, that would restrict the access to ARTs for atypical family cores, and physicians should not, within the limits of national regulations, artificially limit it.

3. GAMETE AND EMBRYO DONATION

Performing ARTs requires a series of steps, depending on the type of procedure. However, most of them start with the requirement of obtaining a gamete from outside the assisted couple. Riggs (2008) found important differences regarding the reasons behind sperm donation by homosexual vs heterosexual persons. Homosexual persons were less likely to become parents, and more likely to donate in private arrangements, their main motivations being: the presence of a relationship with the recipient, the building of a genetic legacy, and a belief that children should determine their own best interests. Heterosexual persons were more likely to become parents, to donate anonymously, saw more often the donation as an altruistic gift, and believed that the adults should primarily be the ones to seek the best interests of the children (Riggs, 2008). Gay men donating sperm were more likely to wish being involved in the parenting of children born from their genetic material, and were more explicit regarding the characteristics they require in the recipient (mostly traits associated with "good mothering") (Ripper, 2008).

When it comes to the receiving end of the donor insemination process, lesbian mothers-to-be from a study conducted by Lingiardi et al. (Lingiardi et al., 2016) viewed their donors differently, with one overriding theme: none considered them as the fathers of their children. The donor was constructed into the family narrative in three ways: he was either completely absent (donor as an "entity" or "a ghost"), objectified as only a body part (donor as "medical process," "as a place"), or humanized and viewed in his entirety (donor "as a person," "as a kind man"). The first two viewpoints are facilitated by the couple's interaction with the physicians and clinics involved, who medically objectified the donor. They correlate with the mothers' emphasis on intentionality as the main driving force in creating a child, to the detriment of a genetic connection.

Lesbian couples can refer to a technique, known as "comaternity," "egg sharing," "reciprocal IVF," or Reception of Oocytes from Partner (ROPA) (Marina et al., 2010), wherein one partner's egg is harvested, fertilized, and then implanted into the other partner's uterus. This procedure is essentially a gamete donation and surrogacy agreement at the same time. It attempts to level the contributions and roles of each parent and to limit a third or, considering the use of donor sperm, a fourth party's intrusion into the family structure and life.

Lesbian mothers-to-be both have the biological potential to procreate and both highly value relationship equality (Pelka, 2009; Peplau et al., 1982; Pai, 2017; Dunne, 1997; Weeks et al., 2001). These traits make them more susceptible to jealousy and conflicts arising in case the child prefers one mother (usually the biological) over the other. This instability is further compounded by the nongenetical mother's social and legal invisibility. They may thus prefer the egg-sharing procedure over other types of ARTs. Indeed, IVF and adoption have been shown to reduce power imbalance and insecurities in lesbian families (Pelka, 2009).

The question arising is whether supporting and financing ROPA or IVF for medically fertile patients is warranted or not. ROPA and IVF implicitly are appraised for their ability to make same-sex couples' dreams of a family come true. Egg-sharing enables both lesbians to become more involved in the process of having a child than any type of direct insemination. At the same time, it is a more invasive and expensive procedure than intrauterine or intracervical insemination and less cost-effective when it comes to high-order multiple pregnancies (Goldfarb et al., 1996). Marina et al. (Marina et al., 2010) argue that egg sharing as applied to lesbian couples is ethical, because it encompasses three important principles: beneficence, nonmaleficence, and autonomy of the patient. The last principle is of interest because in many cases it enters in conflict with the physicians' paternalistic attitude and refusal to carry on with reciprocal IVF, in countries where it is legalized (Sandel, 2009).

Most lesbian and gay couples are medically fertile, but *socially infertile*. They belong to a group of patients who cannot procreate without medical intervention. This creates a debate regarding whether physicians have a duty to perform ARTs to these couples. As a general rule, physicians have a duty to perform medical interventions based on the medical needs of the patients (the principle of beneficence). In this case, there are no medical benefits for the couples that are medically fertile. Moreover, by allowing the procedure, we can argue that, based on the type of procedure, it may even generate breaches of the principle of justice, as they choose to use medical procedures aimed primarily toward infertile couples (which have a medical need); as the resources in most health-care systems are limited, by allowing access to these types of procedures to fertile couples, we limit the access to infertile couples to the treatments. Additionally, we may even argue that these procedures breach the nonmaleficence principle, as ARTs have higher risks compared to natural procreation, risks that may be considered unjustifiable in persons who, again, do not have a medical condition precluding them to reproduce naturally. The main counterarguments that favor these procedures are: (1) the principle of justice once again, as, by not allowing patients to have access to a medical procedure based on their sexual orientation, we are actually discriminating them, and (2) sometimes medical procedures are not done solely for the medical benefit of the patients, but rather for other types of benefits, such as social, economic, psychological, a principle named beneficence as satisfaction. This principle is often used to justify aesthetic procedures, but can easily be implemented in ARTs for atypical family cores. The procedure should be allowed, based on this last principle, as it generates significant benefits for the couple from a social and psychological perspective.

4. SURROGACY AGREEMENT

The concept of "surrogacy" involves the agreement between the surrogate mother—and her husband or partner, if present—on one side and a couple or a single person on the other for the surrogate to carry, deliver, and then hand over the child to the latter, most commonly referred to as the commissioning or intended parent/s.

Ethically and legally, surrogacy still galvanizes public and scientific opinion. Its historical roots were documented as early as around 2000 BC in the Old Testament (Holy Bible) and on an Assyrian clay tablet of marriage (Turp et al., 2017), but only in the last decades did it switch from an exclusively genetic form of surrogacy, where the surrogate acts both as an egg donor and as a carrier, to a predominantly gestational one. The advent of gestational surrogacy and its contemporary use by same-sex couples has stretched even further society's traditional beliefs about the nuclear, sexual family model, parenthood, and the well-being and new social reality of children born under these structures. The expansion of fertility clinics, legal advice services, and online surrogate arrangement programs tailored to hetero- as well as homosexual couples and singles alike has imposed surrogacy as a new, albeit very expensive, standard in ARTs (Hostiuc et al., 2016).

In the context of gestational surrogacy and LGBTQ couples, the majority of commissioning parents are affluent gay men, for whom an egg donor and a surrogate mother are mandatory. Lesbian couples can also refer to this technique (see earlier). Their biological advantage as women makes them however better "equipped" to procreate with the aid of AI and less likely to request the help and services of a surrogate. Gay couples will as such be the main focus of our discussion in the following paragraphs.

With the increased globalization and consequent reproductive tourism, many couples residing in countries where surrogacy is not available, illegal, or prohibitively expensive choose to cross borders to surrogacy-friendly countries like the United Kingdom, the United States, Ukraine, Georgia, and India in order to achieve parenthood (Inhorn and Patrizio, 2015). Legislations and societal beliefs have not, however, kept up with the flurry of surrogacy centers and offers of recent years. Currently most European Union countries, including Switzerland, Germany, Austria, France, Spain, Italy, Sweden, Finland, Iceland, and Bulgaria, ban surrogacy arrangements, either commercial or altruistic. Others countries, like the Czech Republic, Poland, and Romania, informally allow it by failing to properly regulate it, while the few remaining—Portugal, Greece, Belgium, Denmark, the Netherlands, the United Kingdom, and Hungary—only legalize altruistic gestational surrogacy agreements. The lack of legal uniformity concerning parentage, surrogacy, and same-sex marriage in the European Union countries has further hindered gay couples' hopes and efforts to form a family. The legal loopholes created by these differences in regulations have the potential to create legal and diplomatic conflicts for both parties of the surrogate contract, as well as for the physicians and the clinics involved in the process. Physicians involved in cross-border reproductive care (CBRC) (Inhorn and Patrizio, 2015) must exercise caution and thorough knowledge of the surrogacy and citizenship laws of the countries in which they practice and from which their prospective patients come from before accepting their case. Although not part of their formal attributions, they could also counsel their patients to hire an attorney with experience in reproductive law before embarking on the process. The consequences of not knowing the enacted ART and adoption laws can lead to broken families, missed bonding opportunities, and tens of thousands of euros in legal costs. Numerous cases of CBRC have been documented where couples were denied parentage of their offspring born through surrogacy, leaving the children stateless, or the surrogate and her partner as the unwilling legal parents (Deomampo, 2014). Although more surrogacy cases have been ascribed to heterosexual couples, homophobia and the lack of ART regulations still present in many countries leave same-sex couples as the more vulnerable category (Woodward and Norton, 2006).

The ethical challenge posed by surrogacy is twofold; on one hand, surrogacy per se, used by hetero- and homosexual couples and singles is viewed as exploitative, misogynistic, and degrading of women and their natural maternal function (Greenfeld and Seli, 2016b; Watson, 2016). A practice often compared with "a commercial trade" and "a market for wombs and children," it affords wealthy customers, after a carefully executed cost-benefit analysis, the opportunity to choose an unknown woman as the most affordable, appealing, and fit carrier mother-to-be. On the other hand, the rise of this technique amid same-sex couples reinforces the notion of the nuclear family.

After lobbying for marriage equality, it comes as a corollary that family equality is the next proposal on the table of discussion. This would also include nondiscriminatory access to ARTs. In the United States, Maryland and California currently have enacted laws that ensure unmarried and gay couples coverage for infertility treatments (California Legislative Information, 2017; General

Assembly of Maryland, 2017). As author of the Californian law, Assemblyman Tom Ammiano, stated: "Reproductive medicine is for everybody's benefit (…) and to restrict fertility coverage solely to heterosexual married couples violates California's non-discrimination laws."

But do these changes truly contribute to the reshaping of modern society or do they reinforce the golden standard form of family, biologically bonded? The actors might have changed, but the play remains the same.

If public and scientific discourses argue for the restructuring of families from those based on biology to those based on function and choice, then one could argue that the rising trend in IVF parenthood and *enforcement* of health-care service plans to provide financial coverage for medically fertile patients could be considered contradictory.

ARTs assert the opposite concept, that of the bionormative or Platonic form of family (Witt, 2014, pp. 49–63), and contribute to maintaining the superiority of families with children genetically related to their parents, to the detriment of abandoned children currently living in residential care and of families with adopted children. This idea is furthered even more by some gay couples' choice to have twins genetically related to each of them and to each other; by fertilizing two eggs from the same donor with sperm from each partner and then implanting the resulting embryos into the surrogate's uterus, they level each other's genetic contribution and attempt to remove the status of nonbiologic parent. At the same time, they desire no contact or shared responsibility with the genetic and/or surrogate mother (Ekman, 2014; Stonewall, 2015). This procedure is similar in reasoning to lesbian couples' egg-sharing procedure, if we consider that the carrying mother has an epi*genetic* contribution to the fetus's development (Bjorklund, 2006).

As Julie Bindel, a lesbian feminist, notes: *"If gay couples want children, why on earth do they have to go down this exploitative route rather than adopting a child? The answer raises a profoundly troubling question about the attitudes of too many gay and lesbian couples. Fixated by vanity, imbued with overweening self-regard, they want to create a child in their own image, meeting a checklist of ideal characteristics.(…) I campaigned for years for gays and lesbians to be allowed to adopt children, not only because of our own fundamental human rights to have a family but also because of the need to give secure, loving homes to vulnerable children"* (Bindel, 2015). Not surprisingly, gay men, as well as lesbians and heterosexual couples, often choose AR over adoption because one or both partners want to have a genetic connection with the child (Greenfeld and Seli, 2016a), a further argument that procreation is still strongly tied to genetic identity and that not "only love makes a family" (Kaeser and Gillespie, 1999, pp. 1–280).

Additionally, adopting would ensure that a third or fourth party does not threaten the family by contesting the legal parentage, as is the case with many surrogacy arrangements that end up in court over child custody disputes between the intended and surrogate parents.

Having children is known to have one of the most important stabilizing effects on relationships. Therefore one may argue that having genetically related children can be a stabilizing factor for atypical family cores. For example, Johnson et al. (1999, pp. 160–177) suggested that marital stability is the result of three main elements: (1) personal commitment—wanting to continue the relationship, (2) structural commitment—barriers to leaving the relationships, such as social pressure, irretrievable investments such as housing, and (3) moral commitment, such as honor vows to remain married, or safeguarding the well-being of the children at any cost (Johnson et al., 1999). Adopting can stabilize same-sex couples in a similar manner to procreation (Golombok et al., 2006; Perrin et al., 2002; Crouch et al., 2014; Patterson, 1995) making the need to stabilize a couple less relevant for the debate regarding the ethical acceptability of ARTs in atypical family cores.

Another invoked argument is that surrogacy children have no choice but to relinquish their rights to know their gestational/genetic mothers, who might wish to remain involved in their lives, a violation of Articles 7 and 8 of the UN Convention on the Rights of the Child (OHCHR, 2017; Iona Institute, 2017). By having an "either/or" approach in establishing legal parenthood, judges and legislators, in particular, and society, in general, reduce the complexity of modern parenthood structures and can rob children of knowing their full, nuanced biological identity. A proposed solution is to include all participants from the act of procreation into a larger web of closely knit providers, by pursuing the kinship care model. An integral part of the child welfare system, this model has its roots in the African community and its collective role in child rearing. It implies the concerted efforts of family members, neighbors, and friends to raise a child when his/her parents are not present. An African proverb contends, "It takes a whole village to raise a child" and confirms the adaptable nature of extended family care. In the context of a model of parenthood based on function and choice rather than biology alone, this proverb seems to fit right in.

5. TRANSGENDER PARENTHOOD

Transgender couples have become in recent years more outspoken about their wish for parenthood (De Sutter et al., 2001; Wierckx et al., 2011; Tornello and Bos, 2017). They thus join the ranks of a growing number of LGBTQ and infertile heterosexual couples, unable to procreate without the aid of ARTs.

A transsexual couple can be socially infertile even before transitioning if it consists of an MtF transsexual and a cissexual man, an FtM transsexual, and a cissexual woman, or two persons of the same-gender who have both transitioned to the other gender. It is known that sex reassignment surgery or "transitioning" is usually associated with the loss of reproductive capacities. Irrespective of their social fertility status, almost all transsexual couples are then *medically* infertile. This combination of statuses could justify their requests for assisted reproduction and respect both the principles of beneficence and justice discussed above. One could further argue that their infertility is chosen, or a "price to pay" (De Sutter et al., 2001), but similar to cancer patients, it is a direct consequence of therapy and/or surgery undergone for an illness. Cancer patients are routinely counseled before entering chemotherapy about fertility preserving techniques, so why should transsexuals be excluded from such a consultation? Although the World Professional Association for Transgender Health (World Professional Association for Transgender Health, 2011) and Clinical practice guidelines of the Endocrine Society (Hembree et al., 2009) currently recommend that transsexuals be counseled about the effects of transitioning on fertility and possibilities of fertility preservation, research suggests that not all are informed (De Sutter et al., 2001).

Which raises the next question: in what measure is transsexualism an illness incompatible with parenthood? According to contemporary psychiatry, transsexualism is a severe form of gender dysphoria, wherein a person seeks a social transition to the opposite gender. In itself, gender dysphoria is not associated with cognitive impairment and does not preclude one from evaluating the risks of a pregnancy or the responsibilities of raising children, and from making an informed and conscious decision in this regard. In the measure in which sex reassignment surgery is a symptomatic cure that leads to increased quality of life in a majority of patients (Rakic et al., 1996; De Sutter, 2001; Cuypere et al., 2006; Lobato et al., 2006; Papadopulos et al., 2017), it can be argued that transsexuals are in an even better position to parent than before transitioning. Yet other researchers were more circumspect in their findings, contending that transsexuals had increased rates of psychiatric morbidity and death by suicide posttransitioning (Dhejne et al., 2011) and significantly lower quality of life in several domains (Kuhn et al., 2009; Lingiardi et al., 2016). Optimal results of the operation, family, and social support were listed as critical factors for good postoperative functioning (Berg and Gustafsson, 1997).

Research on transgender parenting is still scant. A meta-analysis by Stotzer et al. (2014) concluded that children's psychological development, gender identity, and sexual orientation are not negatively impacted from being raised in transgender families and that they do not experience significant levels of social stigmatizing. On the contrary, some studies showed that, following their parent's shift in gender identity, children learn to become more tolerant and accepting of others' diversity (Goff et al., 2007; Pyne, 2012). Nevertheless, most children come from heterosexual families and not from planned transgender ones; they have to pass through the "coming out" and conversion of their former mother or father, their family's dissolution, and the subsequent legal battles for custody, important factors that can influence study results, a situation similar to that described in lesbian and gay family studies.

Another point worth noting is that most adolescent and adult males and females with early-onset gender dysphoria self-identify as androphilic (sexually attracted to men) and gynephilic (sexually attracted to women), respectively

(DSM-5, 2013; Cerwenka et al., 2012), although recent research points to a more fluid sexual orientation (Kuper et al., 2011; Katz-Wise et al., 2017). After transitioning, they would form typical, heterosexual couples, in which case their desire for a family would realign with the socially prevalent bigendered family model.

Murphy (2012) also points to another dilemma emerging when transsexuals choose to preserve their fertility pretransitioning, which leads to a collision between their genetic identity as mothers or fathers and their newly established and opposite social identity. Indeed, a minority of transsexual men reported feeling conflicted about their opposing identities (De Sutter et al., 2001; Tornello and Bos, 2017). The issue of genetic vs social identity can also crop up in cases of anonymous gamete donation pretransitioning. Adults conceived through anonymous gamete donation who want to contact their genetic parents for social or medical reasons can find their searches slowed or blocked because the donor now has another name and gender posttransitioning, a change that most likely is not documented.

6. CONSCIENCE CLAUSE[*]

Sometimes, national regulations may give specific rights to patients who can enter in conflict with the personal belief system of the attending physicians. When this is the case, they recognize the right of the physician to refuse performing certain procedures, based on religious, moral, or ethical grounds, unless the refusal generates a harm to the patient that can only be avoided through a positive action from the physician, a concept coined "conscience clause" (Hostiuc et al., 2017).

Usually, the conscience clause has been analyzed not only in the context of reproductive technologies, such as sterilization, contraception, stem cell-based therapies, or abortion, but also in the context of other controversial medical issues, including aiding terrorists, transplantation from brain-dead patients (Veatch, 1999), or euthanasia (Harrington, 2007).

White-Domain argues that all conscience clauses must be understood through four axes: (1) the individuals/entities protected by the law (physicians, hospitals, insurance companies); (2) the health-care services that form the basis of the protected objection (abortion, contraception, in vitro fertilization); (3) the activities that a person/entity can refuse (e.g., performing the abortion); and (4) the reason given for the objection (religious, moral, ethical) (2009). Mason Pope approached the issue in a similar manner and argued that conscientious objection laws, irrespective of their type, have four main properties: (1) they affect certain types of health-care providers; (2) specific categories of health-care services; (3) have specific patient circumstances; and (4) certain conditions under which a right or obligation can be triggered (2010).

---

[*]An earlier version of this chapter has been partially published in Hostiuc et al. (2017).

ARTs should be made available to everyone in need, if the national regulations allow them. However, sometimes this right has been denied to patients due to their marital status or sexual orientation, based on the conscience clause. Such an example was presented in the North Coast Women's Care Medical Group v Ct.App. 4/1 D045438 San Diego County Superior Court. The plaintiff, Guadalupe Benitez, was a lesbian living with her partner, Joanne Clark, in San Diego, California. They wanted GB to become pregnant, and decided to use intravaginal self-insemination with sperm obtained from a sperm bank. After several unsuccessful attempts, GB was diagnosed with polycystic ovarian syndrome and was referred to North Coast Women's Care Medical Group for fertility treatment. Her obstetrician and gynecologist, Christine Brody, explained to the patient that she might need intrauterine insemination, but that she could not go forward with the procedure for religious reasons, as the patient previously informed her of her sexual orientation. However, she recommended two other physicians, from the same medical institution, who could perform it. For the procedure, GB wished to use fresh sperm from a friend, a procedure that was not routinely done at the clinic, due to legal issues. As a consequence, GB accepted to use sperm from a sperm bank for the procedure. CB went on a vacation and the case was taken over by Dr. Douglas Fenton, who was also against the procedure. DF was not informed of the decision to use sperm from a sperm bank, and believed that the parties agreed to use fresh sperm; he was the only physician from the Center who was licensed to perform the needed tasks for in vitro fertilization with fresh sperm. As a consequence, he referred the patient to Dr. Kettle, from another center, who finally performed the procedure. Soon after, GB sued the North Coast Center and its physicians based on several issues, including sexual orientation discrimination (Leonard and Cain, 2009; Storrow, 2009). The court ruled against the physicians and excluded from the decision-making process any arguments appertaining to medical ethics. More explicitly, the conscience clause was considered as not being applicable in this instance because (1) the Center accepted GB as the patient, and therefore entered in a contractual relationship and (2) the Center did not have a specific issue with the procedure (intrauterine insemination), but with the sexual orientation of the plaintiff (Storrow, 2009). Therefore, from an ethical point of view, we should regard the conscience clause as a combination of three principles— personal autonomy of the physician vs reproductive autonomy of the patient vs justice—which should be closely balanced in clinical practice to minimize the risks of potential malpractice suits. The personal autonomy of the physician can be manifested in two main areas in his/her relationship with the patient: (1) in accepting to enter a physician-patient relationship and (2) in accepting or refusing to perform various medical procedures. The acceptance of the initiation of the physician-patient relationship is voluntary, unless there is an emergency, in which this relationship enters in force automatically (the physician has an absolute duty to aid those who, without a prompt medical intervention, will have a high risk or mortality or other significant adverse health-related consequences).

Once the physician-patient relationship has been established, the physician has a moral duty to aid his/her patient, as much as she/he possibly can, from a medical point of view. If something appears during the physician-patient relationship that renders the physician unable to fulfill this moral duty (such as medical noncompetence, diverging opinions regarding the therapeutic management, or issues appertaining to the conscience clause), she/he can terminate, in certain conditions, the relationship. In this case, however, the first physician delayed significantly this procedure, which potentially delayed the pregnancy, which can be interpreted as maleficence, and therefore against the established norms of medical ethics.

According to the principle of justice, we should treat equals equally, and unequals unequally (Aristotel, 2012, pp. 1–293); the inequality in treatment should, however, be based on elements that are relevant to the generation of an unequal treatment, and not on subjective criteria such as race, sexual orientation, gender, political or religious affiliation. In the above-mentioned case, the physicians did not accept to perform the procedure based on the sexual orientation of the patient, and accepted to perform similar procedures in heterosexual couples. In this case, the unequal treatment was not generated by a morally relevant inequality (i.e., which could have been, e.g., the presence of additional pathologies that would have been able to decrease the success odds). Therefore we believe that the principle of justice was not respected in this instance. Regarding procreative autonomy, as seen above, there are no relevant limits to ARTs appertaining to clinical ethics (Hostiuc et al., 2017). Within the keyword "procreative autonomy," autonomy refers to the autonomy of the couple, and it should be respected in reproductive medicine, in a similar way in which autonomy of the patient is respected in all other medical disciplines.

7. CONCLUSIONS

In the last 40 years, access of LGBTQ couples to ARTs and planned parenthood increased significantly, especially in the Western world. In this chapter, we showed that for same-sex couples, reproduction should not be denied based on sexual orientation, but that there are some particularities which should be taken into account by physicians. Today, the practicing clinician faces a myriad of clinical scenarios, complicated by globalization and increased medical tourism. In the context of regulated access of LGBTQ couples to ARTs, a good clinical approach involves a balance between the physician's personal and the patient's reproductive autonomy.

More research needs to be conducted, especially on transgender and polyamorous families, to prove, without a reasonable doubt, the absence (or presence) of objective reasons to liberalize, as much as possible, their procreative rights.

REFERENCES

Allen, M., Burrell, N., 1997. Comparing the impact of homosexual and heterosexual parents on children: meta-analysis of existing research. J. Homosex. 32, 19–35.

Amato, P., 2012. The well-being of children with gay and lesbian parents. Soc. Sci. Res. 41, 771–774.

Aristotel, 2012. Politica, 1–293.

Artis, J., 2007. Maternal cohabitation and child well-being among kindergarten children. J. Marriage Fam. 69 (1), 222–236.

Barnes, J., Sutcliffe, A.G., Kristogeersen, I., Loft, A., Wennerholm, U., Tarlatzis, B.C., Kantaris, X., Nekkebroeck, J., Hagberg, B.S., Madsen, S.V., Bonduelle, M., 2004. The influence of assisted reproduction on family functioning and childrens' socio-emotional development: results from a European study. Hum. Reprod. 19, 1480–1487.

Benkov, L., 1994. Reinventing the Family: The Emerging Story of Lesbian and Gay Parents. Crown Publishers, Inc., New York, pp. 1–289.

Berg, J.E.A., Gustafsson, M., 1997. Long term follow up after sex reassignment surgery. Scand. J. Plast. Reconstr. Surg. Hand Surg. 31 (1), 39–45.

Berkowitz, D., 2008. A sociohistorical analysis of gay men's procreative consciousness. J. GLBT Fam. Stud. 3 (2–3), 157–190.

Berkowitz, D., Marsiglio, W., 2007. Gay men: negotiating procreative, father, and family identities. J. Marriage Fam. 69 (2), 366–381.

Biblarz, T.J., Savci, E., 2010. Lesbian, gay, bisexual, and transgender families. J. Marriage Fam. 72, 480–497.

Bindel, J., 2015. Surrogacy and gay couples. New Feminism. [Online]. Available at: http://www.newfeminism.co/2015/06/surrogacy-and-gay-couples/. Accessed 2 October 2017.

Bjorklund, D., 2006. Mother knows best: epigenetic inheritance, maternal effects, and the evolution of human intelligence. Dev. Rev. 26 (2), 213–242.

Blankenhorn, D., 1995. Fatherless America: Cofronting Our Most Urgent Social Problem. Basic, A Division of Harper Collins Publisher, New York, pp. 1–222.

Bos, H., Van Balen, F., Van Den Boom, D., 2007. Child adjustment and parenting in planned lesbian-parent families. Am. J. Orthopsychiatry 77 (1), 38–48.

Brown, S., 2004. Family structure and child well-being: the significance of parental cohabitation. J. Marriage Fam. 66 (2), 351–367.

Brown, S., 2010. Marriage and child well-being: research and policy perspectives. J. Marriage Fam. 72 (5), 1059–1077.

California Legislative Information, 2017. Bill text—AB-460 health care coverage: infertility. [Online]. Available from: http://leginfo.legislature.ca.gov/faces/billNavClient.xhtml?bill_id=201320140AB460. Accessed 2 October 2017.

Cameron, P., 1999. Homosexual parents: testing "common sense"—a literature review emphasizing the Golombok and Tasker longitudinal study of lesbians' children. Psychol. Rep. 85 (1), 282–322.

Cameron, P., 2006. Children of homosexuals and transsexuals more apt to be homosexual. J. Biosoc. Sci. 38 (3), 413–438.

Cameron, P., 2009. Gay fathers' effects on children: a review. Psychol. Rep. 104 (2), 649–659.

Cameron, P., Cameron, K., 1996a. Homosexual parents. Adolescence 31 (124), 757–776.

Cameron, P., Cameron, K., 1996b. Do homosexual teachers pose a risk to pupils? J. Psychol. 130 (6), 603–613.

Cameron, P., Cameron, K., Proctor, K., 2017. Children of homosexuals more apt to become homosexual and experience parental molestation: surveys over three decades. Marriage Fam. Rev. 53 (5), 429–433.

Carlson, M., Corcoran, M., 2001. Family structure and children's behavioral and cognitive outcomes. J. Marriage Fam. 63 (3), 779–792.

Cerwenka, S., Nieder, T., Richter-Appelt, H., 2012. Sexuelle orientierung und Partnerwahl transsexueller Frauen und Männer vor körpermedizinischen geschlechtsanpassenden Maßnahmen. Psychother. Psychosom. Med. Psychol. 62, 214–222.

Clark, A., 1988. The sexual crisis and popular religion in London, 1770–1820. Int. Labor Working-Class Hist. 34, 56–69.

Clarke, V., 2001. What about the children? Arguments against lesbian and gay parenting. Women's Stud. Int. Forum 24, 555–570.

Copper, B., 1987/1988. Mothers and daughters of invention. In: Alpert, H. (Ed.), We Are Everywhere: Writings by and About Lesbian Parents. The Crossing Press, Freedom, CA, pp. 306–321. Originally published in 1987.

Corvino, J., 2013. Same-sex marriage. In: LaFollette, H. (Ed.), The International Encyclopedia of Ethics. Wiley-Blackwell, Hoboken, NJ.

Crouch, S.R., Waters, E., McNair, R., Power, J., Davis, E., 2014. Parent-reported measures of child health and wellbeing in same-sex parent families: a cross-sectional survey. BMC Public Health 14, 635.

Crowl, A., Ahn, S., Baker, J., 2008. A meta-analysis of developmental outcomes for children of same-sex and heterosexual parents. J. GLBT Fam. Stud. 4 (3), 385–407.

Cuypere, G.D., Elaut, E., Heylens, G., Maele, G.V., Selvaggi, G., Tsjoen, G., Rubens, R., Hoebeke, P., Monstrey, S., 2006. Long-term follow-up: psychosocial outcome of Belgian transsexuals after sex reassignment surgery. Theol. Sex. 15, 126–133.

De Sutter, P., 2001. Gender reassignment and assisted reproduction: present and future reproductive options for transsexual people. Hum. Reprod. 16, 612–614.

De Sutter, P., et al., 2001. The desire to have children and the preservation of fertility in transsexual women: a survey. Int. J. Transgenderism 6 (3), 215–221.

De Wert, G., Dondorp, W., Shenfield, F., Barri, P., Devroey, P., Diedrich, K., Tarlatzis, B., Provoost, V., Pennings, G., 2014. ESHRE task force on ethics and law 23: medically assisted reproduction in singles, lesbian and gay couples, and transsexual people. Hum. Reprod. 29, 1859–1865.

DeLair, C., 2000. Ethical, moral, economic and legal barriers to assisted reproductive technologies employed by gay men and lesbian women. DePaul J Health Care L 4, 147.

Deomampo, D., 2014. Defining parents, making citizens: nationality and citizenship in transnational surrogacy. Med. Anthropol. 34, 210–225.

Dhejne, C., et al., 2011. Long-term follow-up of transsexual persons undergoing sex reassignment surgery: cohort study in Sweden. PLoS One 6 (2), e16885.

Diagnostic and statistical manual of mental disorders, 2013. Arlington, VA: American Psychiatric Association.

Dunne, G.A., 1997. Lesbian Lifestyles: Women's Work and the Politics of Sexuality. Macmillan Press, Basingstoke, UK, pp. 1–258.

Ekman, K.E., 2014. Being and Being Bought: Prostitution, Surrogacy and the Split Self. Spinifex Press, Victoria, Australia.

Erich, S., Leung, P., Kindle, P., 2005. A comparative analysis of adoptive family functioning with gay, lesbian, and heterosexual parents and their children. J. GLBT Fam. Stud. 1 (4), 43–60.

Falcone, T., Hurd, W., 2017. Clinical Reproductive Medicine and Surgery: A Practical Guide, third ed. Springer International Publishing AG, Basel, Switzerland, pp. 1–489.

Farr, R., 2017. Does parental sexual orientation matter? A longitudinal follow-up of adoptive families with school-age children. Dev. Psychol. 53 (2), 252–264.

Farr, R., Forsell, S., Patterson, C., 2010. Parenting and child development in adoptive families: does parental sexual orientation matter? Appl. Dev. Sci. 14 (3), 164–178.

Fedewa, A., Black, W., Ahn, S., 2014. Children and adolescents with same-gender parents: a meta-analytic approach in assessing outcomes. J. GLBT Fam. Stud. 11 (1), 1–34.

Firestone, S., 1970. The Dialectic of Sex: The Case for Feminist Revolution. Farrar, Strauss & Giroux, New York.

Gates, G.J., Badgett, M., Macomber, J.E., Chambers, K., 2007. Adoption and foster care by gay and lesbian parents in the United States. [Online]. The Williams Institute. Available from: https://williamsinstitute.law.ucla.edu/wp-content/uploads/Gates-Badgett-Macomber-Chambers-Final-Adoption-Report-Mar-2007.pdf. Accessed 12 December 2017.

General Assembly of Maryland, 2017. Health insurance—coverage for infertility services. [Online]. Available at: http://mgaleg.maryland.gov/2015RS/Chapters_noln/CH_483_hb0838t.pdf. Accessed 2 October 2017.

Goff, B.S.N., Crow, J.R., Reisbig, A.M.J., Hamilton, S., 2007. The impact of individual trauma symptoms of deployed soldiers on relationship satisfaction. J. Fam. Psychol. 21 (3), 344–353.

Goldfarb, J.M., Austin, C., Lisbona, H., Peskin, B., Clapp, M., 1996. Cost-effectiveness of in vitro fertilization. Obstet. Gynecol. 87 (1), 18–21.

Golombok, S., 2015. Modern Families: Parents and Children in New Family Forms. Cambridge Uiversity Press, Cambridge, UK.

Golombok, S., Mellish, L., Jennings, S., Casey, P., Tasker, F., Lamb, M., 2013. Adoptive gay father families: parent-child relationships and children's psychological adjustment. Child Dev. 85 (2), 456–468.

Golombok, S., Murray, C., Jadva, V., Lycett, E., MacCallum, F., Rust, J., 2006. Non-genetic and non-gestational parenthood: consequences for parent–child relationships and the psychological well-being of mothers, fathers and children at age 3. Hum. Reprod. 21, 1918–1924.

Greenfeld, D.A., 2007. Gay male couples and assisted reproduction: should we assist? Fertil. Steril. 88 (1), 18–20.

Greenfeld, D.A., Seli, E., 2011. Gay men choosing parenthood through assisted reproduction: medical and psychosocial considerations. Fertil. Steril. 95, 225–229.

Greenfeld, D.A., Seli, E., 2016a. Same-sex reproduction. Curr. Opin. Obstet. Gynecol. 28 (3), 202–205.

Greenfeld, D.A., Seli, E., 2016b. Assisted reproduction in same sex couples. In: Kunse, H., Schublenk, U., Singer, P. (Eds.), Bioethics: An Anthology. John Wiley & Sons, Inc, pp. 74–84.

Harrington, M.M., 2007. The ever-expanding health care conscience clause: the quest for immunity in the struggle between professional duties and moral beliefs. Fla. St. UL Rev. 34.

Hegna, K., Kristiansen, HW., Moseng, BU., 1999. Levekår og livskvalitet blant lesbiske kvinner og homofile menn. Norsk institutt for forskning om oppvekst, velferd og aldring Oslo, 1–333.

Hembree, W., Cohen-Kettenis, P., Delemarre-van de Waal, H., Gooren, L., Meyer, W., Spack, N., Tangpricha, V., Montori, V., 2009. Endocrine treatment of transsexual persons: an endocrine society clinical practice guideline. J. Clin. Endocrinol. Metab. 94 (9), 3132–3154.

Hostiuc, S., Badiu, D., Nastasel, V., Hangan, T., Marinescu, M., 2017. Conscience clause and its potential applicability in reproductive medicine. Gineco.eu 13 (4), 162–164.

Hostiuc, S., Iancu, C., Rentea, I., Drima, E., Aluas, M., Hangan, L.T., Badiu, D., Navolan, D., Vladareanu, S., Nastasel, V., 2016. Ethical controversies in maternal surrogacy. Gineco.eu 12 (2), 99–102.

Inhorn, M.C., Patrizio, P., 2015. Infertility around the globe: new thinking on gender, reproductive technologies and global movements in the 21st century. Hum. Reprod. Update 21 (4), 411–426.

Iona Institute, 2017. The ethical case against surrogate motherhood: what we can learn from the law of other European countries. [Online]. Available at: http://ionainstitute.org/assets/files/Surrogacy%20final%20PDF.pdf. Accessed 1 December 2017.

Israel, G.E., 2005. Translove: transgender persons and their families. J. GLBT Fam. Stud. 1, 53–67.

Jo, B., 1984/1988. For women who call themselves lesbians—are you thinking of getting pregnant? In: Hoagland, S.L., Penelope, J. (Eds.), For Lesbians Only—A Separatist Anthology. Onlywomen Press, London, UK, pp. 315–317. Originally published in 1984.

Johnson, M.P., Caughlin, J.P., Huston, T.L., 1999. The tripartite nature of marital commitment: personal, moral, and structural reasons to stay married. J. Marriage Fam. 160–177.

Kaeser, G., Gillespie, P., 1999. Love Makes a Family: Portraits of Lesbian, Gay, Bisexual, and Transgender Parents and Their Families. University of Massachusetts Press, Amherst, MA, pp. 1–280.

Katz-Wise, S.L., Rosario, M., Calzo, J.P., Scherer, E.A., Sarda, V., Austin, S.B., 2017. Endorsement and timing of sexual orientation developmental milestones among sexual minority young adults in the growing up today study. J. Sex Res. 54 (2), 172–185.

Knight, R.H., 1997. Homosexual Parents Are Not in a child's Best Interests. In: Roleff, T.L. (Ed.), Gay Rights. Greenhaven Press, San Diego, CA, pp. 84–89.

Kuefler, M., 2007. The marriage revolution in late antiquity: the Theodosian code and later Roman marriage law. J. Fam. Hist. 32, 343–370.

Kuhn, A., Bodmer, C., Stadlmayr, W., Kuhn, P., Mueller, M.D., Birkhäuser, M., 2009. Quality of life 15 years after sex reassignment surgery for transsexualism. Fertil. Steril. 92 (5), 1685–1689.

Kuper, L.E., Nussbaum, R., Mustanski, B., 2011. Exploring the diversity of gender and sexual orientation identities in an online sample of transgender individuals. J. Sex Res. 49 (2–3), 244–254.

Leonard, A.S., Cain, P.A., 2009. Sexuality Law, second ed. Carolina Academic Press, Durham, NC, pp. 1–938.

Levi, J., Monnin-Browder, E.E., 2012. Transgender Family Law: A Guide to Effective Advocacy. AuthorHouse, Bloomington, IN, pp. 1–232.

Lingiardi, V., Carone, N., Morelli, M., Baiocco, R., 2016. 'Its a bit too much fathering this seed': the meaning-making of the sperm donor in Italian lesbian mother families. Reprod. BioMed. Online 33 (3), 412–424.

Lobato, M.I.I., Koff, W.J., Manenti, C., da Fonseca Seger, D., Salvador, J., da Graça Borges Fortes, M., Petry, A.R., Silveira, E., Henriques, A.A., 2006. Follow-up of sex reassignment surgery in transsexuals: a Brazilian cohort. Arch. Sex. Behav. 35 (6), 711–715.

MacCallum, F., Golombok, S., 2004. Children raised in fatherless families from infancy: a follow-up of children of lesbian and single heterosexual mothers at early adolescence. J. Child Psychol. Psychiatry 45 (8), 1407–1419.

Manning, W., Lamb, K., 2003. Adolescent well-being in cohabiting, married, and single-parent families. J. Marriage Fam. 65 (4), 876–893.

Marina, S., Marina, D., Marina, F., Fosas, N., Galiana, N., Jove, I., 2010. Sharing motherhood: biological lesbian co-mothers, a new IVF indication. Hum. Reprod. 25, 938–941.

Marsiglio, W., Hutchinson, S., 2002. Sex, Men, and Babies: Stories of Awareness and Responsibility. New York, New York University Press.

Meezan, W., Rauch, J., 2005. Gay marriage, same-sex parenting, and America's children. Futur. Child. 15 (2), 97–113.

Mitchell, V., Green, R.J., 2007. Different storks for different folks. J. GLBT Fam. Stud. 3, 81–104.

Murphy, D.A., 2013. The desire for parenthood. J. Fam. 34, 1104–1124.

Murphy, T.F., 2012. The ethics of fertility preservation in transgender body modifications. J Bioeth Inq. 9 (3), 311–316.

O'Connell, A., 1993. Voices from the Heart: The Developmental Impact of a Mother's Lesbianism on Her Adolescent Children. Smith College Studies in Social Work 63, pp. 281–299.

Office of the United Nations High Commissioner for Human Rights (OHCHR), 2017. Convention on the rights of the child. [Online] Available at: http://www.ohchr.org/Documents/ProfessionalInterest/crc.pdf. Accessed 1 December 2017.

Pai, I.E.Y., 2017. Egalitarian Lesbian Relationships? In: Sexual Identity and Lesbian Family Life. Palgrave Macmillan, Basingstoke, UK, pp. 211–263.

Papadopulos, N.A., Zavlin, D., Lellé, J.D., Herschbach, P., Henrich, G., Kovacs, L., Ehrenberger, B., Machens, H.G., Schaff, J., 2017. Male-to-female sex reassignment surgery using the combined technique leads to increased quality of life in a prospective study. Plast. Reconstr. Surg. 140 (2), 286–294.

Patterson, C.J., 1995. Lesbian mothers, gay fathers, and their children. In: Lesbian, Gay, and Bisexual Identities Over the Lifespan: Psychological Perspectives. Oxford University Press, New York, pp. 262–290.

Pelka, S., 2009. Sharing motherhood: maternal jealousy among lesbian co-mothers. J. Homosex. 56 (2), 195–217.

Peplau, L.A., Padesky, C., Hamilton, M., 1982. Satisfaction in lesbian relationships. J. Homosex. 8 (2), 23–35.

Perrin, E.C., Committee on Psychosocial Aspects of Child and Family Health Pediatrics, 2002. Technical report: Coparent or second-parent adoption by same-sex parents. Pediatrics 109 (2), 341–344.

Polikoff, N.D., 1987. Lesbians Choosing Children: The Personal Is Political. In: Pollack, S., Vaughn, J. (Eds.), Politics of the Heart: A Lesbian Parenting Anthology. Firebrand Books, Ithaca, NY, pp. 48–52.

Polikoff, N.D., 2005. For the sake of all children: opponents and supporters of same-sex marriage both miss the mark. N. Y. City L Rev. 8, 573.

Pope, A.L., Murray, C.E., Mobley, A.K., 2010. Personal, relational, and contextual resources and relationship satisfaction in same-sex couples. Fam. J. 18 (2), 163–168.

Popenoe, D., 1993. American family decline, 1960-1990: A review and appraisal. J. Marriage Fam. 55, 527–541.

Pressley, S.A., Andrews, N., 1992. For gay couples, the nursery becomes the new frontier. Wash Post. 20 (A1), A22–A23.

Pyne, J., 2012. Transforming Family: Trans Parents and Their Struggles, Strategies, and Strengths. LGBTQ Parenting Network, Sherbourne Health Clinic, Toronto, pp. 10–39.

Rakic, Z., Starcevic, V., Maric, J., Kelin, K., 1996. The outcome of sex reassignment surgery in Belgrade: 32 patients of both sexes. Arch. Sex. Behav. 25 (5), 515–525.

Rich, A., 1981. Compulsory heterosexuality and lesbian existence. Women Sex Sexuality 5 (4), 631–660.

Riggs, D., 2008. Using multinomial logistic regression analysis to develop a model of Australian gay and heterosexual sperm donors' motivations and beliefs. Int. J. Emerg. Technol. Soc. 6 (2), 106–123.

Ripper, M., 2008. Australian sperm donors: public image and private motives of gay, bisexual and heterosexual donors. Health Sociol. Rev. 17, 313–325.

Rivers, D., 2010. "in the best interests of the child": lesbian and gay parenting custody cases, 1967–1985. J. Soc. Hist. 43 (4), 917–943.

Robertson, J.A., 2004. Gay and lesbian access to assisted reproductive technology. Case W Res L Rev. 55, 323.

Rowen, A.J., 1981. Notes from the battlefield: lesbian custody. Common Lives Lesb. Lives 2, 97–106.

Sandel, J.M., 2009. Justice. What's the Right Thing to Do? Farrar, Straus and Giroux, New York.

Stonewall, 2015. Surrogacy. [Online]. Available at: http://www.stonewall.org.uk/help-advice/parenting-rights/surrogacy-1. Accessed 1 December 2017.

Storrow, R.F., 2009. Medical conscience and the policing of parenthood. Wm. Mary J. Women L. 16, 369.

Stotzer, R.L., Herman, J.L., Hasenbush, A., 2014. Transgender Parenting: A Review of Existing Research, The Williams Institute. pp. 1–25.

Strathern, M., 1999. After Nature: English Kinship in the Late Twentieth Century. Cambridge University Press, Cambridge, pp. 10–153.

Tornello, S.L., Bos, H., 2017. Parenting intentions among transgender individuals. LGBT Health 4, 115–120.

Turp, A.B., Guler, I., Bozkurt, N., Uysal, A., Yilmaz, B., Demir, M., Karabacak, O., 2017. Infertility and surrogacy first mentioned on a 4000-year-old Assyrian clay tablet of marriage contract in Turkey. Gynecol. Endocrinol. 34 (1), 25–27.

Veatch, R.M., 1999. The Conscience Clause: How Much Individual Choice in Defining Death Can Our Society Tolerate? In: Youngner, S.J., Arnold, R.M., Schapiro, R. (Eds.), The Definition of Death: Contemporary Controversies. Johns Hopkins University Press, Baltimore, MD, pp. 137–160.

Videon, T., 2002. The effects of parent-adolescent relationships and parental separation on adolescent well-being. J. Marriage Fam. 64 (2), 489–503.

Wardle, L., 1997. The Potential Impact of Homosexual Parenting on Children. University of Illinois Law Review. pp. 833–919.

Watson, C., 2016. Womb rentals and baby-selling: does surrogacy undermine the human dignity and rights of the surrogate mother and child? N. Bioethics 22, 212–228.

Weeks, J., Heaphy, B., Donovan, C., 2001. Same Sex Intimacies. Routledge, London.

Weiner, B.A., Zinner, L., 2015. Attitudes toward straight, gay male, and transsexual parenting. J. Homosex. 62, 327–339.

What We Know, 2017. What does the scholarly research say about the wellbeing of children with gay or lesbian parents? [Online]. Available at: http://whatweknow.law.columbia.edu/topics/lgbt-equality/what-does-the-scholarly-research-say-about-the-wellbeing-of-children-with-gay-or-lesbian-parents/. Accessed 2 October 2017.

White-Domain, R., 2009. Making rules and unmaking choice: federal conscience clauses, the provider conscience regulation, and the war on reproductive freedom. DePaul L. Rev. 59, 1249.

Wierckx, K., Caenegem, E.V., Pennings, G., Elaut, E., Dedecker, D., Peer, F.V.D., Weyers, S., Sutter, P.D., Tsjoen, G., 2011. Reproductive wish in transsexual men. Hum. Reprod. 27, 483–487.

Williams, C.A., 2010. Roman Homosexuality. Oxford University Press, Oxford, UK.

Witt, C., 2014. A critique of the bionormative concept of the family. In: Baylis, F., McLeod, C. (Eds.), Family-Making. Contemporary Ethical Challenges. Oxford University Press, Oxford, UK, pp. 49–63.

Woods, D., 2009. Nero and sporus. Latomus Rev. d'etudes latines 68 (1), 73–82.

Woodward, B.J., Norton, W.J., 2006. Lesbian intra-partner oocyte donation: a possible shake-upin the Garden of Eden? Hum. Fertil. 9 (4), 217–222.

World Professional Association for Transgender Health. Standards of care for the health of transsexual, transgender and gender variant people. Version 7. [Online]. Available at: http://www.wpath.org/site_page.cfm?pk_association_webpage_menu¼1351&pk_association_web-page¼3926; 2011. (Accessed October 29, 2017).

Wyland, F., 1977. Motherhood, lesbianism and child custody. Wages due lesbians and failing. Wall Press, pp. 1–36.

FURTHER READING

Brinsden, P.R., 2016. Surrogacy's Past, Present, and Future. In: Handbook of Gestational Surrogacy. Cambridge University Press.

Holy Bible, Genesis. 16: 1–15; 17: 15–19; 21: 1–4.

Lindqvist, E.K., Sigurjonsson, H., Möllermark, C., Rinder, J., Farnebo, F., Lundgren, T.K., 2017. Quality of life improves early after gender reassignment surgery in transgender women. Eur. J. Plast. Surg. 40 (3), 223–226.

T'Sjoen, G., Van Caenegem, E., Wierckx, K., 2013. Transgenderism and reproduction. Curr. Opin. Endocrinol. Diab. Obes. 20 (6), 575–579.

Weiss, M.D., 2008. Gay shame and BDSM pride: neoliberalism, privacy, and sexual politics. Radic. Hist. Rev. 100, 87–101.

White, C., 2015. Surrogacy and same-sex marriage: a tale of two countries. Public Discourse. [Online]. Available from: http://www.thepublicdiscourse.com/2015/07/15362/. Accessed 12 January 2017.

Chapter 4

Ethical Issues Raised by Multiparents[☆]

Maria Aluas[*,†]
[*]*Iuliu Haţieganu University of Medicine and Pharmacy, Cluj-Napoca, Romania,* [†]*Babes-Bolyai University, Cluj-Napoca, Romania*

1. INTRODUCTION

In the last decade, assisted reproductive techniques (ARTs) have become an integral part of medicine as it allowed infertile, older couples the possibility to have children, as well as nontraditional couples, who could not procreate naturally. Those who cannot procreate due to various reasons, and who consider that having a child is a main aim of their life, are desperately trying to find solutions. Physicians, or more broadly, medical personnel, are the first who are called upon to help, putting them at the core of various ethical debates in reproductive medicine. This affects doctors from different specialties, such as obstetrics, gynecology, neonatology, pediatrics, and clinical genetics. After the deployment of ARTs, the major question becomes, especially in cases in which there are more than two parties involved in the birth of the baby who actually are the real parents of the newborn.

1.1 Key Terms in the Debate: Filiation, Kinship, and Parenthood

In order to clarify some grey areas of the debates on this topic, it is necessary to define and clarify the concepts of filiation, kinship, and parenthood. The use of ARTs has forced a reconceptualization of filiation in the new societal and cultural context.

In a broad sense, **filiation** is the existing legal relationship of a person to his or her ascendants; in a narrow sense, the same notion designates the relationship of a person's progeny to his or her parents, the direct and immediate link between children and parents (Florian, 2011).

[☆] *"Truth exists, only lies have to be invented"*. L.M. Montgomery, The Selected Journals of L.M. Montgomery, Vol. 3, 1921–1929.

Clinical Ethics at the Crossroads of Genetic and Reproductive Technologies.
https://doi.org/10.1016/B978-0-12-813764-2.00004-0

Kinship is the "social construction of natural facts" (Strathern, 1992), and it refers to the biological axis of the filiation (Mathieu, 2014), namely the ties between parents and children born directly through the biological act of procreation, which is increasingly challenged nowadays by ARTs. The legal aspect of the kinship will not arise, automatically or through a legal act, unless one of the legal hypotheses regarding the establishment of maternal or paternal filiation is achieved. Therefore, in the legal sense, biology alone does not equates filiation if it is not recognized from a legal point of view.

Parenthood refers to the educational and socio-emotional axis and to the de facto exercise of the parental authority, through education and care (Mathieu, 2014). It refers to a particular type of human relationship, encompassing nurturance, shared experience, interdependency, and responsibility rather than merely biology (Mykitiuk, 2001).

New medical ARTs have fragmented the biological process of human reproduction and have eroded the traditional concepts of family, "characterized by transmission by the two family branches, paternal and maternal, and by blood ideology" (Courduriès and Herbrand, 2014). By using ARTs, it became possible to separate the genetic, gestational, and social motherhood, and the genetic, legal, and social fatherhood.

2. MULTIPARENTS OF CHILDREN BORN THROUGH ARTs

A child born through ARTs has a different biological and sometimes legal family structure compared to one born in a traditional manner. The child born in a traditional manner has a family, and the other has alternative families. At first glance, at the question *of whose child will be the one born by ARTs, with gametes from a donor or with a surrogate*, we are tempted to consider as parents those who have a parental project with this child. But, from a legal and ethical point of view, the situation is often unclear or unregulated and we should make some clarifications. In the traditional meaning of a family, a child has a mother and a father. A child normally knows who his/her parents are. But using ARTs, there is more than one mother or one father for a particular child. A child could have up to three different mothers at the same time, a genetic (biological), a social, and a legal (gestational) mother, with the mention that various sources coin them differently. The legitimate question is: who is actually the real mother of this child?

2.1 Maternity

Mater est quam gestation demonstrat (the mother is demonstrated by gestation). As a general rule and from a legal point of view, the mother is usually considered to be the one who gives birth, according the *mater in jure semper certa est* principle. Therefore maternal lineage results from the material fact of childbirth by a certain woman. The acknowledging birth certificate proves both the fact of birth and the child's identity. It shows that a certain woman had a baby at the indicated time and place, as well as the fact that the owner of the birth certificate, and not another, was born by that woman (Florian, 2011).

The same rule is applied when the child is born through ARTs. The woman who gave birth to the baby is, *ipso facto*, his/her mother, as proved by the birth certificate issued in the civil register. What is always unknown is the *biological* father. However it is presumed that mother's husband (although he is not mandatory to be the natural one) is the father of the child.

The genetic **mother** is the egg donor, or more exactly, the one who donates the nuclear genetic material (as there have been cases in which the nuclear material is transmitted to another ovum, or in which the mitochondrial DNA was replaced to minimize the risk of the child to have mitochondrial disorders). She can be the **social (gestational) mother**, namely with the one who will give birth, or may be an anonymous donor. In the case of an anonymous donor, although the biological child's patrimony entitles her to be his/her mother, he/she is usually bound to comply with the rule that anonymous donors should remain unknown, with the accepted exception (in some countries) of the discovery of a significant health-related risk for the child or its descendants (United Nations Convention on the Rights of the Child, 1989, article 7 and 8). When the gestational mother is a **surrogate**, she will appear as the legal mother of the child on the child's birth certificate, irrespective of the agreements between the "want to be" parents and the social mother (Knoppers, le Bris, 1993). Surrogacy generates a separation between pregnancy and maternity. Moreover, in a surrogacy agreement, the egg may originate from the social mother, a third-party donor, or the surrogate mother who will be, in this last case, both *genitrix* and *gestatix* (Mahoux, 2008), the **genetic mother** and the *surrogate mother* of the child. This separation or fragmentation occurs within the maternity institution, decomposing it into its genetic, uterine, and socio-affective components, an issue that also happens within the paternal institution when using a third party donor. This maternal and paternal fragmentation favored by using ARTs, generates many ethical and legal dilemmas, such as the following example.

The wife of a couple, who already had a daughter, underwent a complete hysterectomy. The couple wanted another baby, and turned to the wife's sister (married, with two children) to be their surrogate mother. They signed an agreement, and the in vitro fertilization was performed with genetic material from the husband of the initial couple. The surrogate mother—the sister—gave birth to twins and she and her husband became the legal parents of the children on their birth certificates. The first couple took the children under their care as their own children (Hostiuc et al., 2016a,b).

Having two mothers and two fathers at the same time has, fragmented these children's familial identity, the biological not being equated with the legal one. Moreover, the legal parents are relatives with the gestational mother, further confusing the family ties.

2.2 Paternity

Pater est quem nuptia demonstrant (The father is he who is married to the mother). Regarding paternity, in most cases in which the woman is sterile, the genetic material is obtained from the husband, who is therefore the biological father of the child. However, when he is also unable to procreate for various reasons, it is possible that the biological father is not the same as the social one, leading to a dissociation or fragmentation of the fatherhood. The situation is further complicated when the surrogate mother is married; in this case, from a legal point of view, her husband is the father of the child, if paternity will not be contested, even if the genetic material is from the social father and not the husband of the surrogate. If the social father is not the biological father, but an anonymous donor, things get even more complicated (Knoppers, le Bris, 1993). Finally, the main question is: who should be considered as the **real father of the child**: the **biological** one, the **legal** one, the **social** one, or the **one who wants to have the child**?

3. WHERE DO ALL THESE PARENTS COME FROM?

If the family is fragmented by the existence of several parents (mothers and fathers), the child's identity can be seen as fragmented. However, there are numerous recent studies showing that this is not the case, as children born from atypical family cores have at least the same chances and opportunities in life as those born in traditional couples (Mathieu, 2014). However, the child will never have a *full, complete* mother, encompassing all its dimensions. Below we will discuss three ARTs that are the most frequently used in the field: IVF with gametes donation, IVF with embryo donation, and IVF with a surrogate mother.

3.1 Sperm Donation

Sperm donation is used for the artificial insemination with a third party donor (AID). The donor becomes the *biological father* of the child, but he will not be considered the legal or social father. As a consequence, the child will have three parents, two fathers and a mother (Fig. 1).

The sperm is kept in a medical facility called a *sperm bank*. Sperm donation can be used for intraconjugal artificial insemination (AIC) or, occasionally, for research or other scientific protocols. The first mention of this technique was done in 1909, when *Medical World* detailed an artificial insemination that was conducted in Philadephia in 1884 with sperm harvested from a donor. William Pancoast used the sperm of a *best looking member of the class* (Hard, 1909; Snowden, 2006). The practice was severely condemned during that time, and is the reason for which there are only a few mentions of this technique in the literature.

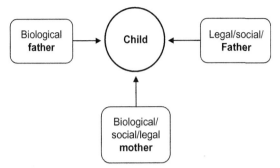

FIG. 1 IVF with sperm donor.

In 1934, in a study published by Professor Herman Rohleder (1934), the following criteria for donors' selection were proposed: (a) the sperm of the donor has to be examined; (b) the donor has to undergo a general examination; (c) the donor should preferably be unmarried; (d) the husband himself has to select the donor, paying particular attention to the state of his physical and mental health. Mentions regarding donors' remuneration are found in 1941 in an article published by Seymour and Koerner, where the conditions of the free donation act are talked about, illustrating the existence of a debate on this issue at that time (Seymour and Koerner, 1941).

3.2 Egg Donation

This procedure is the feminine alternative to sperm donation, in which a fertile woman agrees to have one or more of her eggs harvested (Leroy, 2001). As a consequence, the child will have three parents, two mothers and a father (Fig. 2).

Medically speaking, egg donation was mainly aimed at solving female infertility, due to a lack of functional ovaries (early menopause, surgery that led to the removal of ovaries). More recently, the method became used when the woman wanting to have a child had a high probability to transmit a genetic

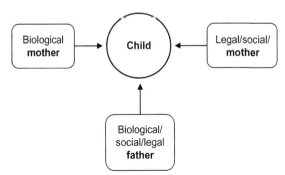

FIG. 2 IVF with egg donor.

disease to her offspring. Egg donation can also be made for purely scientific purposes.

Egg donation can be done using many methods (Leroy, 2001), and includes the sampling of oocytes followed by in vitro fertilization with the partner's sperm (1) by embryo transfer (ET); (2) through the artificial insemination of the donor with the sperm of the prospective social mother's partner, followed after 6 days by a uterine lavage to recover the embryo to be transferred to the mother, (3) through donor insemination that will lead the pregnancy to term, in which case she will also be the surrogate mother.

The first variant, embryonic transfer, is the most frequent; the first success-ful attempts being made by Professor A. Trounson in Australia, in 1983 (Trounson and Mohr, 1983). The second was proposed by a team from Califor-nia led by doctors Bustillo and Buster (1984), but it was not propagated because of both technical and ethical reasons. Beyond the fact that the success rate of this method was low, it carried a risk of unintended pregnancy to the donor, in the situation in which the couple or one of the partners could change their mind and did not wish the intervention. On the other hand, there was no guarantee that the embryo was fertilized with the sperm from the donor, as the donor's sexual rela-tions could interfere with this procedure (Leroy, 2001).

3.3 Embryo Donation

Supernumerary frozen embryos, which are no longer used by couples, after a certain period can be donated to another couple, can be destroyed, or used in scientific research. Embryo donation takes two forms: donation to another infer-tile couple, appearing as an antenatal adoption, and donation for research. As a consequence, in this first instance the child has four parents—two mothers and two fathers (Fig. 3).

The first form seems acceptable, at least ethically, because it is closer to the purpose for which these embryos are created, namely to help sterile couples.

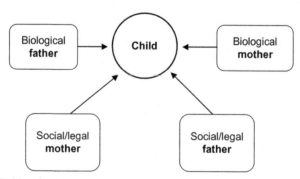

FIG. 3 IVF with embryo donors.

But ethical and legal conflicts seem irreconcilable. Details regarding them will be presented in the following subsections.

In both cases, embryo donation should respect certain conditions: only unused supernumerary embryos should be donated and with the *biological parents'* consent, or at least the one with whom the embryos have a genetic link, anonymous donors being excluded from this discussion (Deleury, 1993).

If donation is made for research purposes, the research is limited to a better understanding of embryonic development or to improvement of *IVF and ET* techniques. Experimentation should not be extended after the fourteenth day of embryonic development (Massager, 2001).

3.4 Surrogacy

Surrogacy is a procedure in which a woman accepts to carry out a pregnancy to term for a couple or for another woman, either altruistically or for financial gains. She pledges to abandon the child at birth, which is to be adopted by the woman who will become the legal mother (Knoppers, le Bris, 1993). Female carrier, surrogate mother, maternity substitute, pregnancy agreement, pregnancy for another, uterus location, delegation pregnancy, surrogate motherhood, and gestation for another are all expressions to denominate this practice.

The main indications for surrogacy include: the absence of the uterus, or uterine abnormalities, either congenital or acquired (e.g., a hysterectomy for an oncological disorder), repeated in vitro fertilization (IVF) failures, medical conditions for which pregnancy is contraindicated (e.g., severe thrombophilia, antiphospholipid syndrome, end stage renal diseases, or vasculitis), single males, and same sex male couples (Machtinger et al., 2017). There are also other instances in which surrogacy could be recommended, including the situation in which the woman cannot carry the pregnancy due to objective reasons, such as a high risk of impaired fetal development, the impossibility of the mother to take life-saving medications during pregnancy (such as chemotherapy), when there is a high maternal risk generated by the pregnancy itself (i.e., catastrophic antiphospholipid syndrome, active systemic lupus erythematosus, previous cerebral vascular hemorrhage, and pulmonary emboli), or when the intended mothers have contraindications for regular gonadotropins doses for ovarian stimulation in order to avoid any significant increase of estradiol levels as well as the risk for ovarian hyperstimulation syndrome (Machtinger et al., 2017).

In this situation, the child could have until six parents, three mothers and three fathers, if the surrogate is also married (Fig. 4).

Even if nowadays surrogacy is seen as a modern form of ART, it is actually an ancient practice. In Ancient Rome, abdomen rental—*ventrem locare*—was a common practice. Therefore, a man whose wife was fertile was able to *give* her temporary to another man, whose wife was sterile or was giving birth to dead children. There is also a biblical example—Sara, her servant Agar, and

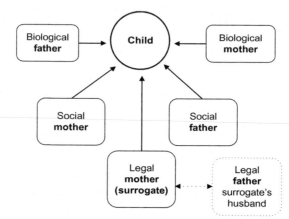

FIG. 4   IVF with embryo donors and surrogate mother.

Abraham (Bible, Genesis, 16). Sara, who is aware that she is sterile, asks Abraham to have a child with Agar, who gave birth to Ismail (Hostiuc et al., 2016a,b). The difference between these practices and the current concept of surrogacy is that in the above-mentioned cases fertilization was done naturally, its aim being the birth of a child for another party. In the current concept of surrogacy, a woman agrees to bear a child for another woman, a child that will be abandoned at birth, either altruistically or for a certain amount of money (it is worth mentioning that in the European Union altruistic surrogacy is usually the only accepted method), the conception being done through the doctor's intervention. After we have described the ARTs, we will present the ethical issues and difficult situations raised by consequences of using ARTs.

4.   ETHICAL ISSUES

The first ethical challenge generated by multiparent families that will be discussed here is the fragmentation of the child's identity: who are the real parents? And does the child have a right to know his or her biological identity? Children are curious and they seek their origins. Those who need information about their own identity transform this in a right: the right to know who they are, their origins, and the circumstances of their birth.

In recent years, the media presented several cases of adults (Kermalvezen, 2014) who were claiming the right to know the identity of their parents. Also, they are asking state authorities for the right to know their biological identity. The European Court of Human Rights (ECHR) recognized the right to know one's origins, the identity of biological parents, and the circumstances of the birth (Council of Europe, Tulkens, 2008), in several cases, such as, the case of Jäggi v. Switzerland (2006); the case of Pascaud v. France (2011); and the case of Godelli v. Italy (2012), all three decisions being "in favor of people asking to know their ancestry."

We are facing a very serious conflict between two important rights: the right of the child to know his or her parentage, and the right of parents—mother or father in the case of gametes donation, or both in the case of embryo donation—to preserve their anonymity.

Also, if the identity of the biological parents and the circumstances of the birth are revealed, as a consequence of a court decision, the donors should be informed about this possibility at the time when they decide to donate gametes or embryos or to carry out a pregnancy. The question is: who should inform them about this possible consequence of their choice? Have they been informed in an appropriate way in order to give an informed consent to the donation? In the following paragraphs we will develop these issues.

4.1 The Right to Know One's Origins

The etymology of the term *origin* (lat. *oriri*) means "to arise, to emerge," but also "to begin, to originate from, to be born of, to descend." It designates both the beginning of a thing and that which produced the thing, from which it came (Mathieu, 2014). As a human being, the term "origins" refers not only to the event of our coming into the world, our birth, but also to the causal antecedent process, the genesis of this moment. The search for origins naturally aims to obtain information about those who made it possible for us to come into the world. The quest for genetic origins has been greatly accentuated in recent years because of developments in science. Scientific progress has made it possible, on one hand, to multistake the interveners in the procreation process (gamete donor, embryo donor, recourse to a surrogate mother), on the other hand, to verify with virtual certainty the biological origin of a person through genetic expertise (or DNA test).

Several international legal frameworks regulate the right of a child to have access to his or her biological origins, but this right is not recognized and implemented in all countries and in all situations. Sweden is the first country where this right was regulated (1984), followed by Austria (1992), Switzerland (in force since 2001), Norway (2003), the Netherlands (in force since 2004), the United Kingdom (2004), and Finland (2006). Outside Europe, this is a right in force in the Australian state of Victoria (1995), Western Australia and New Zealand (2004), New South Wales (2007), Southern Australia (2010), Uruguay (2013), and Argentina (2014) (Amorós, 2015).

The right to know one's identity has been enforced with the adoption of the Convention on the Rights of the Child, by United Nations in the 1989 in New York, in articles 7 and 8.

Thus, article 7 recognizes, as possible, the right of a child to know his parents and be cared for by them: "The child shall be registered immediately after birth and shall have the right to a name, the right to acquire a nationality and, as far as possible, the right to know and be cared for by his or her parents." And article 8 states that: "State Parties undertake to respect the right of the child to preserve his or her identity, including nationality, name, and family relations as recognized by law without unlawful interference. Where a child is illegally deprived of some or all of the elements of his or her identity, States Parties shall provide appropriate assistance and protection, with a view to reestablishing speedily his or her identity." As this legal framework is aimed to protect the child, articles 7 and 8 cannot be used by a person to search his or her origins if he/she became major, the issue being left to national courts and, in the European Union, to the European Court of Human Rights (Mathieu, 2014).

The second European legal framework is the European Convention for the Protection of Human Rights and Fundamental Freedoms (ECPHRFF, 1950). Article 8 of this Convention defends the right to respect private and family life: "There shall be no interference by a public authority with the exercise of this right except such as is in accordance with the law and is necessary in a democratic society in the interests of national security, public safety or the economic wellbeing of the country, for the prevention of disorder or crime, for the protection of health or morals, or for the protection of the rights and freedoms of others." For the court, the right to identity is a core of the right to privacy, in that it is an essential condition of the right to autonomy and self-determination.

The Hague Convention on Protection of Children and Cooperation in Respect of Inter-country Adoption (1993) states in article 30 that "the competent authorities of the contracting State ensure the retention of information in their possession about a child's origins, especially those relating to the mother and father's identity (...). The states ensure a child's or his representative's access to this information, with appropriated advises to the extent permitted by law and State."

However, these international conventions do not stipulate a mandatory right regarding the access to the biological origins of a citizen. They only set the objectives through which signatory states' legislations should have the framework to recognize it. In addition, their content remains limited. *The UNCRC* registers the right of knowing the biological origins 'as possibility' and the Hague Convention only provides the conservation of a child's origin information, but it does not allow an unrestricted access of the holder. To illustrate the complexity of this right, we present the following case.

Audrey Kermalvezen, a young French lawyer, found out when she was 29 years old that she and her brother were born thanks to a sperm donation (Kermalvezen, 2014). She asked French authorities many times to have access to data about her origins, but they refused because the law did not allow access. In 2016, she had filed an appeal in the European Court of Human Rights, claiming the right of everyone to know their origins. She is still waiting for a decision. Challenging French authorities, Arthur Kermalvezen, 34, Audrey's husband, has done his own investigation that enabled him to find, on December 25, 2017, the man who potentially was her biological father. After unsuccessful political and judicial recourse to find the sperm donor, he decided in September 2017, to perform a DNA test via the Internet through the society *23andMe*. The results revealed a common origin with 'a young Franco-British man named Larry,' that he found on social networks. The website gave Arthur Kermalvezen access to the family tree of the French part of Larry's family, and he extracted "a single profile that could match a man old enough to could be his donor." He wrote a letter to this man telling him the story. Then, on December 25, he received a call from the man, saying: "Bravo for finding me" (Leclair, 2018). This case proves that even if the law does not allow access, in reality people can still manage to find this kind of information.

As mentioned earlier, the current trend in Europe is to lift the anonymity of the donor, in the name of the best interest of the child and especially the right to access his or her personal origins. But the real challenge is now the way in which to change mentalities and to accept a real fact: a link that reflects the truth of the origins and the circumstances of birth.

4.2 Donors' Anonymity

"Anonymity," defined as total absence of the identity between donors and beneficiaries, is the most applied rule in artificial insemination with third donor. The advantages and disadvantages of the anonymity rule were classically presented and analyzed in *Warnock Report* (Warnock, 1984), published in 1984.

The main advantages consist in facilitating the recruitment and donor's protection against establishing a lineage, which may have important legal, social, and personal consequences. For the receiver of the genetic material, anonymity protects information regarding a legal or social father's sterility and also diminishes the feeling of guilt for using such a method that others consider to be *adulterous*.

The disadvantages are birth certificate *falsification*, depriving a child's right to know his or her biological parent, and the possible medical problems inherited from the donor. The Warnock Report concludes, however, that the advantages exceed the disadvantages, but also suggests that the strict anonymity's situation could be modified, being allowed the access to genetic, physical, and social data of donors (Englert, 2001).

There is also the possibility for the child to break the anonymity at the age of adulthood. Such a regulation was first adopted in Sweden followed by the other countries as mentioned above, where a national register of donors' identity is available to adult children. In Germany, a child conceived with donor sperm is allowed to repudiate his legal father, and ask to be adopted by the sperm donor, at the age of 18 (MAP Report, 2008). That is why some parents prefer to keep silent.

When anonymity is not maintained anymore, the problem of fragmentation of the family institution is increased, which can have consequences for the child with a couple of *fathers*, namely confusions and difficulties within the family, which does not have a clear repartition of the parental responsibilities. These difficulties can sometimes become major or insurmountable; the Warnock Report classifies them as private arrangements, outside the general regulation framework and depending on the degree of maturity of those involved (Englert, 2001).

With regard to egg donors and maintaining anonymity, a few other problematic aspects are to be mentioned. One of the limits of these methods is gametes' scarcity, especially of oocytes. This disadvantage can be circumvented by appealing to a relative or friend of the sterile couple, or by using cross-gamete donations between two or even more couples or a couple's relatives/friends, in a fashion similar with organ transplants from living donors. Donor's oocytes known by the couple are exchanged with those of another donor of another couple, in order to maintain some anonymity, without it being fully enforced. To overcome this obstacle some solutions are proposed: with the consent of all concerned, the practitioner may proceed to anonymous permutation of donors between two couples of their choice. Psychologically, this approach proves, however, to be difficult to be applied. Donors often insist for their ova go to the couple they have chosen and not to strangers. Similarly, the welcoming mother prefers to have the egg of a sister or friend, rather than of an anonymous donor.

4.3 The Consent of Participants

The etymology of the verb *to consent* (Lat. cum and *sentire*) leads to the idea to agree on a feeling (de Broca, 2014). Consenting to something supposes having full information that allows participants or patients to make reasonable, informed decisions about their own care, based on likely outcomes balanced with risk of harm. "Informed consent" is defined as the decision to participate in research or to accept a medical intervention, taken by a competent individual who has received the necessary information; who has adequately understood the information; and who, after considering the information, has arrived at a decision without having been subjected to coercion, undue influence or inducement, or intimidation (CIOMS, 2002).

Who should be informed? Who are the participants to ARTs? Participants are the couple or the person who has a parental project, different donors, and the surrogate.

In order to sign the informed consent, patients/participants are informed about all the elements of the treatment. The "doctrine" of informed consent is that providers must give patients detailed information about the intervention, including: the type of the intervention, alternatives, including no treatment, the likelihood of success and risks, including the fact that some may currently be unknown. Also, patients should have: verbal and written explanation, in plain language, sufficient time to consider; voluntariness of decision; competence to decide (Thornton, 2000).

Regarding participants in ARTs, it is important for them to receive full information about alternative procedures available to manage their specific infertility problems, including procedures that are not performed by the treating center, and nonmedical options such as adoption and nontreatment (American Society for Reproductive Medicine, 2006).

When patients are actually gametes donors or surrogates, physicians should also discuss the implications of their decisions. Information should cover not only the short-term risks and side effects, but also long-term and nonmedical consequences. They should understand the uncertainty and the complexity of legal issues involved in having a child, or donating gametes or embryos. The question is: who should inform donors and surrogates in order for them to be able to make an informed decision and choice? The reality is that in some fertility centers donors do not receive all relevant information about, for example, the right of the child to know his or her kinship and the possibility for their identity to be revealed, as future donors could refuse to donate or to carry out a pregnancy. In France, for example, the anonymity rules are fully respected. Anyone who discloses the identity of a donor registered in France is, in principle, liable to criminal prosecution. Children have no right to access documents revealing the biological mother's name, and the possibility of establishing any bond between the mother and the child is generally prohibited (Clark, 2012).

How does one avoid conflicts and confusion between participants in these practices? The only way to do so is by informing all participants and advising them about the different consequences of their own choices.

Disputes in this area especially arise when couples and partners separate or divorce, and have jointly created embryos. As an example, we present a case judged by the ECHR in 2000, the Case of Evans v. The United Kingdom (2006).

In 2000, Natalie Evans was diagnosed with an ovarian tumor and was recommended a surgical intervention to remove the ovaries. Before the intervention was carried out, she decided, by mutual agreement with her partner, to conceive, in vitro, an embryo that would be preserved in view of its subsequent implantation. After that, the surgical intervention was carried out, followed by the medical recommendation to wait at least 2 years before transferring the embryo into the uterus. The couple separated in the meantime and the former partner did not renew his consent for the transfer of the embryo. The British Courts rejected women's claims to receive the embryo, because the English legislation in the matter does not authorize the implantation of the embryo unless both parties consent to the transfer (Aluas et al., 2017).

On April 10, 2007, Ms. Evans filed a complaint to ECHR, invoking the violation of article 2 (the Right to life) of the ECPHRFF. As the English legislation does not recognize the quality of autonomous subject of an embryo, article 2 was not applicable, considered the judges of the Court, so her embryo will never become a child.

This case involves one partner who wishes to reproduce, and the other who does not, but the ECHR thus far have not forced the unwilling partner to become a parent. The first observation is that we are faced with a conflict between two *autonomies*: Ms. Evans' desire to become a parent and her ex-partner's refusal to become a parent. The second is that this embryo represents Ms Evans' only chance to have a child that is biologically related to her, a fact that the judges did not take into consideration.

Cases of ineffectiveness of full right of consent are also related to the death of one of the partners, situations in which it is presumed that the parental project will cease to exist.

5. SECONDARY ETHICAL ISSUES

5.1 Medical Tourism

This phrase is used to describe peoples' displacement to another country, other than the one in which they are citizens or residents, in order to receive/acquire a health service that is unavailable or illegal in their country. Medical tourism represents, generally speaking, the act of traveling abroad in order to do certain medical treatments that cannot be performed in their own country (Tremblay, 2012).

There is no unitary definition for medical tourism (Lunt et al., 2011). Some authors insist on the touristic aspect, planned activities including certain treatments (Alsharif et al., 2011) and others adopt a new approach that emphasizes the economic aspect or the private care one (Crooks and Snyder, 2011).

There are also other terms used to describe this practice. In literature we find the use of terms such as: medical voyage, health voyage, health tourism, tourism for treatment, abroad treatments market, medically motivated travel, "out-of-pocket medical procedure." The last example refers to the costs supported exclusively by the patient or beneficiary.

In Canada, the United States, Belgium, Great Britain, Thailand, and Singapore the phrase "medical tourism" is used to differentiate from the medical treatments carried out abroad and that are unplanned by patient.

In case of ARTs, the couples or single persons can access procedures such as surrogacy, which are only allowed in some countries, subsequently becoming the destination in medical tourism. There are known cases of couples from European countries going to the United States (Mennesson, France, 2014), to return home with a child. For this reason traveling with the intention of using ARTs in other countries, is called *procreative* tourism or *reproductive business*.

These practices are contested because they can affect the internal laws of the countries of these citizens as it is an illegal way to have a child. They can also lead to pressure from public opinion to legalize practices considered incompatible with the culture, ethics, or internal law system. If doctors recommend them, they may become liable to civil or criminal charges.

5.2   Slippery Slope

This expression refers to the risk of engaging without intent, in a process that leads to consequences increasingly more difficult to control, that will prejudice the community or the society. In the case of gamete donors, the donors may only want to donate, not to have children of their own, and once taken, this decision opens a whole Pandora's Box with major consequences for all the involved parties, including becoming entrapped in kinship with children they did not want. We should judge if the actions or the adopted policies are acceptable after analyzing susceptible consequences that could result from these actions.

In Bioethics, the slippery slope principle is used to awaken our sensibility and to convince us, with rational arguments, that some practices, as the previous presented, could give rise to negative, unforeseen consequences, unknown and unstoppable. Therefore it is prudent to set clear guidelines for this procedures, to minimize the risk of breaches, and the generation of unforeseen consequences that can potentially be facilitated by gaps in regulations (Parizeau and Kash, 2006).

6.   FINAL CONSIDERATIONS

In this chapter, we tried to familiarize professionals who work in medical setting and all interested parties with the consequences of ARTs concerning multiparents: definitions of terms, concepts, ethical, and legal issues such as: the fragmentation of the family and of the parental authority through the existence of several parents for a single child. We also presented several serious human rights in conflict, such as the donors right to anonymity and the child's right to know one's origins. As a general conclusion, we can say that ARTs are successful but have also created many controversies, debates, and conflicts, both for scientists and society. These techniques and their potential consequences can be difficult to manage. They are confusing as they reach deep sensibilities and touch a dimension of human life that is charged with extreme emotions and symbols. Having a child or a healthy child remains elusive as a right in the context of multiparents. The state is not in a position to respond to all citizens' requests concerning their own choices in the field of reproductive health but it should try to optimize access, of course within the cultural, societal, and economic limits.

REFERENCES

Alsharif, M.J., Labonté, R., Zuxun, L., 2011. "Patients beyond borders: a study of medical tourists in four countries", in Global Social Policy, vol. 10, n° 3, (315–335).

Aluas, M., Gherman, C., Dumitrescu, C.I., 2017. Is the human embryo legally defined and protected? Causes and consequences. Romanian Journal of Morphology and Embryology 58 (2), 1–6.

American Society for Reproductive Medicine, 2006. Elements to be considered in obtaining informed consent for ART, in Fertility and Sterility, Volume 86, Issue 5, Supplement, (s272–s273).

Amorós, E.F., 2015. Donor anonymity, or the right to know one's origins? Butlletí del Comitè de Bioètica de Catalunya 15, 1–11.

Bustillo, M., Buster, J., 1984. Nonsurgical ovum transfer as a treatment in Infertil women. Journal of the American Medical Association 252(1171–1173).

Case of Evans v. The United Kingdom (2006, No. 6339/05).

Case of Godelli v. Italy (2012, n. 33783/09).

Case of Jäggi v. Switzerland (2006, 58757/00).

Case of Mennesson v. France (2014, 65192/11).

Case of Pascaud v. France (2011, no 19535/08).

Clark, B., 2012. A balancing act? The rights of donor-conceived children to know their biological origins. Georgia Journal Of International And Comparative Law 40 (3), 621–661.

Courduriès, J., Herbrand, C., 2014. Gender, Kinship and Assisted Reproductive Technologies: Future Directions After 30 Years of Research in Enfances Familles Générations. 21 | 2014, online since 15 November 2015, last accessed on the 25 January 2018, http://journals.openedition.org/efg/493.

Crooks, V.A., Snyder, J., 2011. Tourisme médical: Ce que les médecins de familles canadiens doivent savoir. Commentaire, Le Médecin de famille canadien, vol. 57, e151. Available at: http://www.cfp.ca/content/57/5/e151.full.pdf, last accessed on 13.09.2017.

De Broca, A., 2014. Le soin est un éthique. Les enjeux du consentement ou du refus de soins et de l'obstination déraisonnable. SeliArslan, Paris.

Deleury, E., 1993. Don d'embryon. In: Hottois, G., Parizeau, M.-H. (Eds.), Les mots de la Bioéthique. Un vocabulaire encyclopédique.De Boeck Universite, Bruxelles.

Englert, Y., 2001. Don de sperme. In: Hottois, G., Missa, J.-N. (Eds.), Nouvelle Encyclopédie de Bioéthique. DeBoeck Université, Bruxelles.

Florian, E., 2011. Family's Law (Dreptul Familiei), fourth ed. C.H. Beck, Bucharest.

Hard, A.D., 1909. Artificial impregnation. Med. World 27, 163.

Hostiuc, S., Iancu, C.B., Năştăşel, V., Aluaş, M., Rențea, I., 2016a. Maternal filiation in surrogacy. Legal consequences in Romanian context and the role of the genetic report for establishing kinship. Romanian Journal of Legal Medicine 24, 47–51.

Hostiuc, S., Iancu, C.B., Rentea, I., Drima, E., Aluas, M., Hangan, T.L., Badiu, D., Navolan, D., Vladareanu, S., Nastasel, V., 2016b. Ethical controversies in maternal surrogacy. Gineco.eu [12], 99–102.

Kermalvezen, A., 2014. Mes origines: Une affaire d'état. Max Milo, Paris. 364 p.

Knoppers, M.M., le Bris, S., 1993. Maternité de substitution. In: Hottois, G., Parizeau, M.-H. (Eds.), Les mots de la Bioéthique. Un vocabulaire encyclopédique. De Boeck Universite, Bruxelles.

Leclair, A., 2018. L'incroyable enquête d'Arthur, né d'un don anonyme, pour retrouver son géniteur, Le Figaro (15/01/2018).

Leroy, F., 2001. Don d'ovule. In: Hottois, G., Missa, J.-N. (Eds.), Nouvelle Encyclopédie de Bioéthique. DeBoeck Université, Bruxelles.

Lunt, N., Smith, R., Exworthy, M., Green, S., Horsfall, T., Mannion, R., 2011. Medical Tourism: Treatments, Markets and Health System Implications: A Scoping Review, Organisation for Economic Co-Operation and Development (OECD/OCDE), Available at: http://www.oecd.org/dataoecd/51/11/48723982.pdf, accessed on 12.09.2017.

Machtinger, R., Duvdevani, N.R., Lebovitz, O., 2017. Outcome of gestational surrogacy according to IVF protocol. J Assist Reprod Genet. 34 (4), 445–449. https://doi.org/10.1007/s10815-017-0877-7.

Mahoux, P., 2008. Proposition de loi relative à la maternité pour autrui, Sénat de Belgique, 13. MARS 2008.

MAP (Medically Assisted Procreation) Procréation Medicalement Assistée, 2008. Enjeux et défis éthiques. In: compte rendu de l'audition publique ouverte à la presse du mardi 10 juin.

Massager, N., 2001. Don'embryon. In: Hottois, G., Missa, J.-N. (Eds.), Nouvelle Encyclopédie de Bioéthique. DeBoeck Université, Bruxelles.

Mathieu, G., 2014. Le secret des origines en droit de la filiation. Kluwer, Waterloo.

Mykitiuk, R., 2001. Beyond conception: legal determinations of filiation in the context of assisted reproductive technologies, Osgoode Hall Law Journal 39.4: 771–815. http://digitalcommons. osgoode.yorku.ca/ohlj/vol39/iss4/2, last accessed on the January 12, 2018.

Parizeau, M.-H., Kash, S. (Eds.), 2006. Néoracisme et dérives génétiques, Sainte-Foy, Presses de l'Université Laval.

Rohleder, H., 1934. Test Tube Babies. A History of the Artificial Impregnation of Human Beings. The Panurge Press, New York.

Seymour, F.I., Koerner, A., 1941. Artificial insemination: present status in the United States as shown by a recent survey. Journal of the American Medical Association 116, 2747.

Snowden, B., 2006. New Reproductive Tehnologies. Children's Rights and Human Fertilization and Embryology Act 1990. In: John, M. (Ed.), Children in Our Charge. The Child's Rights to Resources. Jessica Kingsley Publisher, London. p. 155.

Strathern, M., 1992. Reproducing the Future: Essays on Anthropology, Kinship and the New Reproductive Technologies. Manchester University Press, Manchester.

Thornton, R.G., 2000. Informed consent. Proc (Bayl Univ Med Cent) 13 (2), 187–190.

Tremblay, M., 2012. Tourisme médical: quel rôle pour l'État?, Québec, Laboratoire d'étude sur les politiques publiques et la mondialisation, ENAP. (Rapport évolutif. Analyse des impacts de la mondialisation sur la santé au Québec; Rapport 9).

Trounson, A., Mohr, L., 1983. Human pregnancy following cryopreservation, thawing and transfer of an eight-cell embryo. Nature 305, 707–709.

Warnock, M., 1984. Report of the Committee of Inquiry into Human Fertilisation and Embryology, Available at: http://www.hfea.gov.uk/docs/Warnock_Report_of_the_Committee_of_Inquiry_ into_Human_Fertilisation_and_Embryology_1984.pdf, last accessed on 1.09.2017.

FURTHER READING

Deloitte, 2011. Medical tourism: update and implications, Deloitte Center for health solutions. Available at: http://www.deloitte.com/assets/Dcom-UnitedStates/Local%20Assets/Documents/ us_chs_MedicalTourism_102609.pdfDeloitte 2011, accessed on 12.09.2017.

Parizeau, M.-H., 1993. Consentement. In: Hottois, G., Parizeau, M.-H. (Eds.), Les mots de la Bioéthique. Un vocabulaire encyclopédique. De Boeck Universite, Bruxelles.

Chapter 5

Revisiting the Nondirective Principle of Genetic Counseling in Prenatal Screening

Sandrine de Montgolfier*,†

*Institut de Recherche Interdisciplinaire sur les enjeux Sociaux (IRIS), Bobigny, France,
†Université Paris Est Créteil (UPEC), Créteil, France

1. INTRODUCTION

The popular view of nondirective genetic counseling limits the counselor's role to providing information to patients and assists them and their families in making decisions in a morally neutral fashion. In prenatal screening, it is important to consider whether this nondirective principle is truly implemented, and whether we need to review how it is applied. These questions are prompted by controversies raised in different countries concerning the organization and goal of prenatal screening, and the real autonomy of the women taking these tests in choosing between pursuing or terminating pregnancy.

In this chapter, we will discuss these controversies and set out the different arguments put forward on this subject. Our purpose is to debate the ethical implications of these arguments. To this end we (1) look back at the history of the implementation of prenatal screening programs, (2) examine their associated ethical consequences, (3) review current practice in prenatal genetic counseling, (4) analyze the implications of prenatal diagnosis for the information provided to women, and (5) discuss the criteria underlying an effective shared decision, and how a decision-making support tool might help. We refer throughout to women only, but as stated in a recent joint recommendation of ESHG/ASHG, "decisions about prenatal screening or its outcomes will often be shared by pregnant women with their partners and that these partners (mostly the biological father-to-be) do have an interest in knowing a diagnosis in their future child. Moreover, depending on the conditions tested, some outcomes may also have implications for the father himself" (Dondorp et al., 2015). However, the ultimate decision whether to undergo screening is unequivocally the pregnant woman's alone.

Clinical Ethics at the Crossroads of Genetic and Reproductive Technologies.
https://doi.org/10.1016/B978-0-12-813764-2.00005-2

2. HISTORICAL BACKGROUND AND IMPLEMENTATION OF PUBLIC POLICY FOR PRENATAL SCREENING

To discuss what a prenatal consultation is and could be, we start by looking back at the advent of screening and prenatal diagnosis, and the public policy behind it. We recall that health screening was a part of medical preventive action concerned with strategies to detect the risks of contracting diseases and intervene to mitigate them by early care, or before the onset of symptoms by taking preventive actions. The approach to screening was to look for a concomitance of evidence and risk factors and decide in this light on how to best protect a person's health. In prenatal screening, the frame is more specific: the tests provide information about an unborn child whose parents, together with health-care professionals, will decide either to pursue the pregnancy with appropriate care provisions, prepare for disability support or palliative support for the newborn, or terminate the pregnancy. Some refer to these options as preventive actions—improperly, because abortion can hardly be considered as preventive health care, even if aborting embryos or fetuses with an incurable disability is recognized as acceptable (Mattei et al., 2008). A pregnant woman always faces a difficult choice if her test results are unfavorable.

The first screening provisions used ultrasound scanning, which spread widely in the 1970s in France and throughout the industrialized world, and led to the first prenatal screening policies (Vassy, 2011). This technique permitted the development of screening for a chromosomal abnormalities such as trisomy 21, responsible for Down syndrome (DS), a disability characterized by morphological abnormalities, cognitive problems, deficiencies in expression, and autistic tendencies, although studies also show that individuals with DS can have fulfilling lives and enjoy meaningful relationships (Hurford et al., 2013; Scott et al., 2013). In the 1970s, this test was based on a correlation found between certain fetal measurements and a risk of trisomy 21. In the 1980s, a more accurate measurement of fetal nuchal translucency between the 11th and 13th weeks, in relation to the woman's age, gave a chance ratio. Later in the 1980s, the assay of serum markers, i.e., analysis of biochemical markers in the maternal blood, was added to make a more accurate estimation of the risk. Today the risk is most often calculated by combining the mother's age, the results of nuchal translucency measurement, and the assay of serum markers. The estimated risk is usually confirmed or rebutted by an invasive diagnostic test involving amniocentesis followed by a karyotype with a 1% and 0.5% risk of miscarriage: this test verifies whether the fetus has the DS chromosomal abnormality trisomy 21, but cannot estimate how severe the mental disability is likely to be. Other abnormalities can be detected by this screening, but DS is the most frequently found, affecting one pregnancy in 500 (Greydanus and Pratt, 2005). It is acknowledged that these screening and diagnostic tests detect some 85%–90% of chromosomal abnormalities, with a false-positive rate of 5% (Larion et al., 2014).

To ensure fairness, generalized access to screening for all pregnant women has thus been set in place and is financed as a public health policy in many European countries. In some other countries, e.g. the United States, prenatal screening is made available to self-paying or insured patients in accordance with professional guidelines (Gekas et al., 2016).

Depending on the country, different screening provisions and different estimated risk thresholds have been set. Here we describe three provisions that illustrate the diversity of policy choices, and the range of criteria used according to the context, drawing from the work of Vassy et al. (2014).

In the United Kingdom, the serum marker assay was developed in the 1980s, but to ensure free nationwide accessibility, a screening program, open to all women, was launched in 2003. The health-care professionals involved were mostly midwives trained in testing and in providing women with relevant information nonpersuasively (NICE, 2003). The combined test was set up in 2010 with a chance threshold of 1/150 for proposing a diagnostic test. In 2010, 61% of pregnant women took the test, of whom 2% underwent the amniocentesis test (Ward and Soothill, 2011).

In France, the adoption of the serum markers developed in the United Kingdom came much sooner, spurred by professional organizations of gynecologists, whose advocacy won social security cover for all women in 1997, with an obligation to sign a consent form. In 2010, 84% of pregnant women took this test: despite it being formally optional, the Chairman of the National Ethics Advisory Committee declared it to be "virtually compulsory" (Sicard, 2007). The threshold of 1/250 was chosen, resulting in 9% of women tested undergoing a confirmatory diagnosis procedure (Agence de la Biomédecine (ABM), 2016). The health-care professionals involved in France were more varied than in the United Kingdom, and were comprised of general practitioners, midwives, gynecologists, and ultrasound scan specialists (Vassy, 2011). The combined test was set up in 2009, making it possible to halve the rate of diagnostic tests, which fell to 4% in 2015 (ABM, 2016).

In the Netherlands, the introduction of prenatal screening for DS sparked major public controversy (Popkema and Harbers, 2005). For many years there had been no screening program, but only local experimental initiatives. The arguments against development included concern about the test's efficacy, subjecting women to unnecessary anxiety, a possible drift toward liberal eugenics, and over-medical pregnancy. In 2007, the Population Screening Act placed screening for trisomy 21 under the aegis of eight university hospitals and instituted a professional training program. If the diagnosis was positive, abortion was permitted before the 24th week, in contrast to the United Kingdom and France, which allowed it up to full term. In 2009, 91% of pregnant women were informed that screening was available, but only 26% asked for it (Vassy, 2011). The threshold for subsequent diagnosis was set at 1/200.

3. ETHICAL REFERENCES IN THE DEVELOPMENT OF SCREENING TESTS

In this part we examine the ethical criteria underlying the prenatal screening provision. To forestall criticism of unequal access to the test, risks of discrimination against persons with disabilities, and the risk of insidious eugenics, women's reproductive autonomy must be emphasized, so that women are truly free to make fully informed decisions.

Prenatal screening practices have been criticized by various actors in the countries where they are promoted by public policy for being potentially eugenic in practice if a carefully balanced choice between abortion and giving birth to a child with a disability is not ensured (Mattei et al., 2008; Petrogiannis et al., 2001). Countries that have opted to favor screening thus have a duty to set in place logistic and financial provisions to help parents make this final choice whether or not to accept a child with a disability. Another criticism is that discrimination against persons with the disabilities tested for by the screening procedures might increase. In contemporary western societies that overvalue physical and intellectual performance, normed body aesthetics, and competitive success, it may be difficult for a person who does not fit these criteria to enjoy social acceptance; this concern may dissuade women from giving birth to a child with a disability (Asch, 1999). A further criticism is that women who are warned of a high risk after screening usually have to wait 1 or 2 weeks after the diagnostic test has been taken before they know the final result. One study thus estimated that of 10,000 women undergoing screening, 415 would be told they had a high chance of bearing a chromosomally abnormal fetus. If these 415 then went on to take a diagnostic test, 400 would be found to have normal fetuses, and so would have suffered several weeks of deep anxiety for nothing (Leiva Portocarrero et al., 2015).

To respond to this criticism, two ethical references were instituted to control screening practice, the first based on the concept of reproductive autonomy, and the second on women's informed consent. The various interactions between the health-care professionals involved and the women concerned are designed to ensure that the women can make a fully informed decision at each step in the process. This is made possible by providing information before of screening and diagnosis that enables women to decide whether to take the tests, and consultations after results are known, to discuss whether to pursue the pregnancy or have it medically terminated. Many public health options are based on this principle, despite being openly criticized in many human and social science studies: is decisional autonomy possible, and if so under what conditions? In this case, for example, is the information provided truly complete, balanced, and comprehensible? Might not conferring reproductive autonomy be, in effect, an incitement to abort a fetus with a disability. Given the high percentages of the choice to terminate the pregnancy after positive diagnosis (e.g., in the United Kingdom in 2012, 92% of women diagnosed positive made this choice), this could be the case.

These questions are particularly pressing because a new test is coming into general use that highlights certain issues: noninvasive prenatal testing (NIPT) requires only a simple blood sample from the mother; it uses sequencing of cell-free fetal DNA found in the maternal blood, and stands between serum screening and invasive diagnostic testing (Allyse et al., 2015). This test is currently offered as second-tier screening to women who are identified from initial screening tests as having a high risk of carrying a fetus with aneuploidy. Noninvasive prenatal screening is more accurate than traditional screening. For example, it detects 99% of trisomy 21 cases against only 85% with traditional screening and is a more specific screening, i.e., with fewer false-positive results <1% (Gekas et al., 2016).

This test should thus theoretically reduce the number of diagnostic tests, and so the number of fetuses lost by miscarriage as a result. In the United Kingdom, the test is available in the private sector as a self-paying service, although it is being evaluated for inclusion in National Health Service cover. In the Netherlands, it has been in place since 2014 for women found at high risk after taking a combined test (Health Council of the Netherlands, 2016). In France, a Law from May 2017 allows this test for prenatal diagnosis, but its cost is not yet covered by the social security. A report by the French Public Health Authority (HAS, Haute Autorité de Santé) published in April 2017 militates for offering this test to all women with a risk ratio after the combined test between 1/50 and 1/1000. This choice of threshold is based on the aim of finding more fetuses with trisomy 21: 84%, against 78% with a threshold at 1/250, i.e., 165 DS fetuses per year in France. This choice will raise the number of women subjected to anxiety to 58,000 although there were only 18,500 since the implementation of the combined test. Clearly, this lowering of the threshold from 1/250 to 1/1000 would mean also a high outlay for the social security system, estimated at 18 million euros to find 165 additional fetuses with trisomy 21 (HAS, 2017). No final decisions have yet been made in France, and at present women wishing to take this test pay about 350 euros. Public health strategies can thus impinge directly and strongly on the persons involved. The HAS report recognizes that "the modifications made to the screening procedure could influence how pregnant women and healthcare professionals understand the risk of finding fetal trisomy 21, and the resulting anxiety elicited among pregnant women" (HAS, 2017, p249). We should add that this test opens the way to a more systematic detection of other abnormalities in the future. It offers women a wider choice concerning a greater number of fetal disorders, and thus again raises the issue of what information is passed on during consultations upstream and downstream of prenatal screening and diagnosis. In the Netherlands, the latest recommendations restate, "good counseling is important to ensure that pregnant women can decide voluntarily and are well-informed about whether to participate in prenatal screening. The committee believes that the current practice of counseling needs to be improved. For example, sufficient time is not always taken, counseling is not always tailored to the needs of the pregnant woman and her partner and some pregnant women do not have sufficient support in the decision" (Health Council of the Netherlands, 2016).

4. GENETIC CONSULTATION IN PRENATAL SCREENING: NONDIRECTIVE AND NEUTRAL?

What should a prenatal consultation comprise? To answer this question, we first look again at the criteria governing a consultation, and then describe some results of human and social sciences research on prenatal screening and diagnosis consultations. We go on to discuss how the development of new noninvasive tests makes this process even more fragile.

Since the early 1970s, the ethical norm-governing counselors involved in testing and screening for genetic conditions related to reproduction has been strict neutrality. Counseling about reproductive genetics had to be patient-centered, but nondirective (Caplan, 2015). What do we mean by nondirective, neutral genetic counseling? How can we facilitate informed decision-making without influencing the decider? A health-care professional must not suggest that any one option is preferable to another, but must give all the information needed for the woman (along with the child's father) to reach an informed decision. Concretely, what information is given, and what tools are used upstream of the test to decide whether to take it, and downstream of it to interpret the results and so decide on what action is best? Many recommendations (WHO, 1998; ACOG, 2007) and the ethics literature on the subject (Bombard et al., 2010; Cartier et al., 2012) agree on the importance of communicating three sorts of information: (1) a description of the health disorder for which the screening is offered, (2) the main characteristics of the test, and (3) the possible outcomes according to the test results. Despite some criticism of its use, written consent is generalized as one of the means to ensure that the information has been delivered, and that the woman is expressing an explicit choice.

Many empirical studies have shown that the concept of consent is fragile, and so also is the patient's autonomy in this process. Here we examine some parts of a review by Champenois-Rousseau and Vassy (2012) on this subject. For example, few women know that the trisomy 21 screening test is not compulsory, and that they can choose whether to take it (Favre et al., 2007; Marteau et al., 1992; Press and Browner, 1997). Once the test has been taken, if the chance of disability proves to be high, few women will have thought beforehand about what decision they will take, and so suffer great anxiety (Agence de la Biomédecine (ABM), 2016; Heyman et al., 2006; Katz-Rothman, 2001). Some studies have also shown that owing to their poor understanding of the test, or because they think they are helping the patient, some professionals acknowledge that they fail to apply the principle of nondirectiveness, and give their opinion on what the patient ought to do if the tests are positive: slippage between choice and coercion may occur (Favre et al., 2007; Pilnick, 2008; Williams et al., 2002). The information about trisomy 21, if it is transmitted, is primarily medical, which may at first sight seems perfectly normal in a medical context. But is this sufficient to give the patient some idea of what it might be like for a person to live with this disorder? Little positive information on this syndrome (e.g., warmth and cheerfulness of DS children, possible ability to follow primary schooling) is given during the genetic consultation, in the belief that neutrality means focusing narrowly on medical information, although some would consider that medical information can only partly tell us what the life of a person with this disorder might be like. To summarize the difficulties encountered in counseling, we can first note that a proper consent process is resource-intensive. Yet, in reality, discussions preceding screening are short, and rarely conducted by trained genetic counselors. Second, clinicians' fear of possible litigation and liability following the birth of a child with a disorder leads them to prefer routine screening and so to present it to women as an obvious choice that requires little thought. Third, pregnant women themselves tend to prefer the narrative of "screening to ensure the health of my baby," to facing the fear of an unwelcome result, given that their only options are preparing for birth or terminating an otherwise desired pregnancy. These barriers to informed consent are a significant challenge to reproductive autonomy.

The current development of a noninvasive procedure (NIPT), because it addresses circulating fetal DNA, further undermines the free expression of a woman's informed consent to undergo screening (HAS, 2017; Van den Heuvel et al., 2010).

The possibility of it becoming widespread is high, and indeed some consider that it should be offered as a routine test, labeled "Prenatal reflex DNA screening for trisomy 21, 18, and 13," and that when samples for the blood marker assays are being taken, an extra tube for a possible fetal DNA test should be systematically filled in case the chance calculated with the first combined test is high (Wald et al., 2017). This routine risks making sampling a mere formality, and so might hinder the delivery of information to women that would support their informed decision, and so impair true reproductive autonomy.

5. NONDIRECTIVE CONSULTATION CHALLENGED OR AN OPPORTUNITY TO SUPPORT REPRODUCTIVE AUTONOMY BY COMPLETE BALANCED INFORMATION?

The worldwide controversy concerning the type of information that should be delivered to women shows how delicate and sensitive this subject is. We look at "Chloe's Law" in the United States and at the revised law on bioethics in France, and set out the relevant arguments. We conclude with the case of the United Kingdom, which long ago developed information comprising both medical and nonmedical aspects, and both positive and nonpositive aspects of the disorders being screened.

In view of the results obtained from empirical studies based on the sociology of screening practices, strengthened by much advocacy in favor of persons with

DS, some countries have undertaken to review the information given to women eligible for screening. The question most often asked is: should the positive aspects of living with DS be spoken about during a screening consultation, and should women be invited to contact associations of DS children's parents?

Thus, in the United States, first state legislation (Chloe's Law, 2014), and then a federal law requires women who receive a prenatal diagnosis of DS to be given some positive information about the condition. Some actors of prenatal screening considered that the United States law had thereby expunged the nominal goal of value-neutral counseling (Caplan, 2015). Other actors, such as ACMG, expressed their concern that these laws might have the effect of restricting access to voluntary termination of a pregnancy after a prenatal diagnosis (ACMG, 2013). Let us examine closely the content of the United States legislation: United States law requires the Department of Health to make available "up-to-date, evidence-based information about Down syndrome that has been reviewed by medical experts and national Down syndrome organizations," including information on treatment options, support services, hotlines specific to DS, relevant resource centers, clearing houses, and national and local DS organizations. Judging from this excerpt, the neutrality of the genetic council does not appear to be challenged in any way: information on the development of persons with DS and on the support structures that exist to help them seems an obvious requirement to let a woman decide on whether to have a child with this disability.

In France, this question has also been discussed, but no such strict directives on the information to be given have been issued. When the bioethics legislation was last revised in 2011, a sentence was added extending the information given to at-risk women concerning the medical aspects, and measures of care and prevention for the disorder, and stating that "a list of specialized organizations certified for the support of patients with the suspected disorder and their families" should be offered (French Bioethics Law, 2011). This addendum was included after strong advocacy by associations of families with DS members, and after discussions with physicians and entities concerned with professional and ethical standards who were demanding better information for women: "The national statistics show us that the great majority of pregnant women willingly undergo screening, but that their decisional autonomy is jeopardized by a lack of information. Let us add that to these factors, which contribute to hinder this decisional autonomy, is added a strong social pressure leading to an increasingly systematic rejection of the difference. This results in a very high rate of abortion in the case of antenatal diagnosis of trisomy 21." In all, 92% of fetuses with trisomy 21 are detected, of which 96% are aborted (Grangé et al., 2011). Parliamentary debate when this sentence was added also revealed widely divergent opinions on whether the information should include what it is like to live with the disorder diagnosed in the fetus. The vote was in favor of giving such information on the grounds that neutral complete information could not be provided solely by a medical expert, but that support groups expressing the points of view of families living with the disability could inform on aspects of a DS person's life that were other than purely medical.

In sum, controversy mainly arises from the content of the information delivered to the woman by the health-care professionals to enable her first to decide whether to undergo screening, and then, depending on the test results, to exercise true decisional autonomy in deciding whether to pursue pregnancy according to the estimated chance of having an abnormal fetus. This information cannot focus solely on the medical aspects (support, care, and prevention), but must also consider the patient's and her family's quality of life, which only families who have experienced such a situation first hand are qualified to appraise.

These new changes should not be viewed as calling into question the neutrality of genetic counseling, but rather as an opportunity to improve practice. Some of the studies cited above have shown that the information delivered to women, and the corresponding options proposed, were not always sufficient to support a fully informed decision. These new laws are thus a major step forward in balancing the information given and so in helping women reach a decision. We note that in some countries, the information already includes these items, e.g., in the United Kingdom on the site of the National Health Service (England), where additional information on living with DS and support given to families with DS children is given, together with links to websites of support groups and parents' forums discussing their choice of whether to pursue pregnancy (NHS, 2017; ARC, 2017).

6. TAKING REPRODUCTIVE AUTONOMY FURTHER: INNOVATIONS IN INFORMATION AND AIDED DECISION-MAKING TOOLS

Integrating all this information is possible only if multiple supports are deployed. But information alone is not enough to enable true decisional autonomy. Helping women to make decisions is one objective of prenatal consultations. One way to achieve this goal could be to use decision-making support tools that help women make decisions in line with their personal values.

To meet these demands for information, health-care professionals must be supported and trained to deliver it: websites, multilingual videos, and full information brochures must be designed and validated by all the actors involved (professionals, representatives of pregnant women, and patient associations).

However, information alone is not enough to empower women, along with their partners, to make a choice according to both the information given to them and their personal values and preferences. A Canadian study of all the documents supplied to women showed that they did not present a "comprehensive balanced portrayal of DS and thus are not adequate for supporting shared decision making" (Lawson et al., 2012). Yet many studies have highlighted the importance of maintaining a shared decision process throughout prenatal screening and diagnosis (Seror and Ville, 2010). According to the International Patient Decision Aid standards collaboration, "decision aids are tools designed to help people participate in decision-making about health care options by not only providing information on the options but also by helping them clarify and communicate the personal values they associate with the different options" (Leiva Portocarrero et al., 2015, p76). These tools have proved their efficacy in having the patients concerned participate more actively and in fostering a better understanding, in particular when results are expressed in terms of probabilities. They would also favor firmer decisions with stronger commitment, and reduce the anxiety generated in this difficult prenatal situation.

7. CONCLUSION

To conclude, we can consider that the new laws cited here, though they may seem awkwardly drafted, are indicators to favor better practice that are necessary to accompany ever-broadening screening provisions with the development of NIPT. ESHG and ASHG hope that "NIPT has the potential of helping the practice better achieve its aim of facilitating autonomous reproductive choices, provided that balanced pretest information and nondirective counseling are available as part of the screening offer" (Dondorp et al., 2015). This implies that women must be able to make a true choice of whether to undergo each successive step in the screening process offered, and that they are fully aware of what decisions are theirs to make at every stage. In France, the latest HAS report thus recommends that the information to be given to all pregnant women must include what trisomy 21 is, what support is provided for persons with trisomy 21 and their families (coordinated medical, social, educational, and psychological care), what screening is available, the advantages and disadvantages of the tests proposed, what at-risk means, and the difference between high chance and diagnostic certainty, what sampling options are available for prenatal diagnosis, and what actions during pregnancy are possible if fetal trisomy 21 is diagnosed. Neither prenatal screening nor medical termination of pregnancy must ever be presented as an obligation. This data should allow informed choices made by pregnant women at three decision stages (screening, diagnosis, and whether to pursue pregnancy if fetal trisomy 21 has been diagnosed). It would be useful to offer pregnant women an educational information tool explaining the different screening steps (HAS, 2017). At the same time there is need to develop information support and aided decision-making tools enabling all the actors to find a balanced role throughout this complex process. It would be desirable to favor research in the human and social sciences to help build such supports and tools and evaluate them to promote improvements in practice.

Lastly, improvement in the consultation process at the individual level can truly advance women's greater reproductive autonomy only if policy choices at the collective level favor the integration and support of persons with the disorders and disabilities that these tests screen for. This is an essential prerequisite for every woman and every family to feel free and able to make an informed decision in accordance with their own values (Mattei et al., 2008; Ravitsky, 2017).

REFERENCES

ACMG, 2013. ACMG Releases Statement on Access to Reproductive Options After Prenatal Diagnosis. Retrieved from: https://www.acmg.net/docs/Reproductive_Rights_News_Release.PDF.

ACOG Committee, 2007. Patient testing: ethical issues in selection and counseling. Opinion No. 363. American College of Obstetricians and Gynecologists. Obstet. Gynecol. 109, 1021–1023.

Agence de la biomédecine (ABM), 2016. Le rapport médical et scientifique de l'assistance médicale à la procréation et de la génétique humaines en France. Saint Denis, France. Retrieved from: https://www.agence-biomedecine.fr/annexes/bilan2016/donnees/sommaire-proc.htm.

Allyse, M., Minear, M.A., Berson, E., Sridhar, S., Rote, M., Hung, A., Chandrasekharan, S., 2015. Non-invasive prenatal testing: a review of international implementation and challenges. Int. J. Women's Health 7, 113–126. https://doi.org/10.2147/IJWH.S67124.

ARC, 2017. Helping parents and professionals through antenatal screening. http://www.arc-uk.org/home.

Asch, A., 1999. Prenatal diagnosis and selective abortion: a challenge to practice and policy. Am. J. Public Health 89 (11), 1649–1657.

Bombard, Y., Miller, F.A., Hayeems, R.Z., Avard, D., Knoppers, B.M., 2010. Reconsidering reproductive benefit through newborn screening: a systematic review of guidelines on preconception, prenatal and newborn screening. Eur. J. Hum. Genet. 18 (7), 751–760. https://doi.org/10.1038/ejhg.2010.13.

Caplan, A.L., 2015. Chloe's law: a powerful legislative movement challenging a core ethical norm of genetic testing. PLoS Biol. 13 (8). https://doi.org/10.1371/journal.pbio.1002219.

Cartier, L., Murphy-Kaulbeck, L., Genetics Committee, 2012. Counseling considerations for prenatal genetic screening. J. Obst. Gynaecol. Canada 34 (5), 489–493. https://doi.org/10.1016/S1701-2163(16)35248-3.

Champenois-Rousseau, B., Vassy, C., 2012. Ultrasonographers confronted with prenatal screening of Down's syndrome. A difficult arbitration between professional excellence and ethics of consent. Sciences sociales et santé 30 (4), 39–63.

H. A. de Santé (HAS). (2017). Recommandation en santé publique "Place des tests ADN libre circulant dans le sang maternel dans le dépistage de la trisomie 21 fœtale" (p. 339). France: HAS Retrieved from: https://www.has-sante.fr/portail/jcms/c_2768510/fr/place-des-tests-adn-libre-circulant-dans-le-sang-maternel-dans-le-depistage-de-la-trisomie-21-foetale.

Dondorp, W., de Wert, G., Bombard, Y., Bianchi, D.W., Bergmann, C., Borry, P., Chitty, L.S., Fellmann,F., Forzano, F., Hall, A., Henneman, L., Howard, H.C., Lucassen, A., Ormond, K., Peterlin, B., Radojkovic, D., Rogowski, W., Soller, M., Tibben, A., Tranebjærq, L., van EI, C.G., Cornel, M.C., 2015. Non-invasive prenatal testing for aneuploidy and beyond: challenges of responsible innovation in prenatal screening. Eur. J. Hum. Genet. 23 (11), 1438–1450. https://doi.org/10.1038/ejhg.2015.57.

Down Syndrome Prenatal Education Act ("Chloe's Law"). Pub. L. 2450, No. 130, Cl. 35. (Jul 18, 2014). http://www.legis.state.pa.us/cfdocs/billinfo/billinfo.cfm?syear=2013&sind=0&body=H&type=B&bn=2111.

Favre, R., Duchange, N., Vayssière, C., Kohler, M., Bouffard, N., Hunsinger, M.-C., Kohler, A., Mager, C., Neumann, M., Vayssière, C., Viville, B., Hervé, C., Moutel, G., 2007. How important is consent in maternal serum screening for Down syndrome in France? Prenat. Diagn. 27 (3), 197–205. https://doi.org/10.1002/pd.1656.

French Bioethics Law No. 2011-814 dated 7 July 2011 - part. 2. (2011). Code de la santé publique. L1131-1. JORF. https://www.legifrance.gouv.fr/.

Gekas, J., Langlois, S., Ravitsky, V., Audibert, F., van den Berg, D.G., Haidar, H., Rousseau, F., 2016. Non-invasive prenatal testing for fetal chromosome abnormalities: review of clinical and ethical issues. Appl. Clin. Genet. 9, 15–26. https://doi.org/10.2147/TACG.S85361.

Grangé, G., Azria E., Berger-Aubry M-C., Cabrol D., Favre R. et al. (2011). Le projet de loi de bioéthique concerne toutes les femmes enceintes. Le Monde. Retrieved from: http://www.lemonde.fr/idees/article/2011/02/08/le-projet-de-loi-de-bioethique-concerne-toutes-les-femmes-enceintes_1476744_3232.html

Greydanus, D.E., Pratt, H.D., 2005. Syndromes and disorders associated with mental retardation. Indian J. Pediatr. 72 (10), 859–864.

Health Council of the Netherlands, 2016. Prenatale screening. Gezondheidsraad, Den Haag.

Heyman, B., Hundt, G., Sandall, J., Spencer, K., Williams, C., Grellier, R., Pitson, L., 2006. On being at higher risk: a qualitative study of prenatal screening for chromosomal anomalies. Soc. Sci. Med. 62 (10), 2360–2372. https://doi.org/10.1016/j.socscimed.2005.10.018.

Hurford, E., Hawkins, A., Hudgins, L., Taylor, J., 2013. The decision to continue a pregnancy affected by Down syndrome: timing of decision and satisfaction with receiving a prenatal diagnosis. J. Genet. Couns. 22 (5), 587–593. https://doi.org/10.1007/s10897-013-9590-6.

Katz-Rothman, B., 2001. Spoiling the pregnancy: prenatal diagnosis in the Netherlands. In: De Vries, R., Benoit, C., van Teijlingen, E., Wrede, S. (Eds.), Birth by Design: Pregnancy, Maternity Care, and Midwifery in North America and Europe. Routledge, New York, pp. 180–198.

Larion, S., Warsof, S.L., Romary, L., Mlynarczyk, M., Peleg, D., Abuhamad, A.Z., 2014. Association of combined first-trimester screen and noninvasive prenatal testing on diagnostic procedures. Obstet. Gynecol. 123 (6), 1303–1310. https://doi.org/10.1097/AOG.0000000000000275.

Lawson, K.L., Carlson, K., Shynkaruk, J.M., 2012. The portrayal of down syndrome in prenatal screening information pamphlets. J. Obstet. Gynaecol. Can. 34 (8), 760–768. https://doi.org/10.1016/S1701-2163(16)35340-3.

Leiva Portocarrero, M.E., Garvelink, M.M., Becerra Perez, M.M., Giguère, A., Robitaille, H., Wilson, B.J., Rousseau, F., Légaré, F., 2015. Decision aids that support decisions about prenatal testing for Down syndrome: an environmental scan. BMC Med. Inform. Decis. Mak. 15, 76. https://doi.org/10.1186/s12911-015-0199-6.

Marteau, T.M., Slack, J., Kidd, J., Shaw, R.W., 1992. Presenting a routine screening test in antenatal care: practice observed. Public Health 106 (2), 131–141. https://doi.org/10.1016/S0033-3506(05)80390-7.

Mattei, J.-F., Harlé, J.-R., Coz, P.L., Malzac, P., 2008. Questions d'éthique biomédicale. Médecine Sciences Publications, Paris.

National Institute for Clinical Excellence (NICE). (2003). Antenatal Care: Routine Care for the Healthy Pregnant Woman (No. 41). London. Retrieved from: http://www.crd.york.ac.uk/crdweb/ShowRecord.asp?ID=32003001247

NHS, 2017. Screening for Down's, Edwards' and Patau's syndromes. https://www.nhs.uk/Conditions/pregnancy-and-baby/screening-amniocentesis-downs-syndrome/.

Petrogiannis, K., Tymstra, T., Jallinoja, P., Ettorre, E., 2001. Review of policy, law and ethics. In: Before Birth: Understanding Prenatal Screening. Ashgate, Aldershot; Burlington, USA.

Pilnick, A., 2008. 'It's something for you both to think about': choice and decision making in nuchal translucency screening for Down's syndrome: choice and decision making in antenatal screening for Down's syndrome. Sociol. Health Illn. 30 (4), 511–530. https://doi.org/10.1111/j.1467-9566.2007.01071.x.

Popkema, M., Harbers, H., 2005. The cultural politics of prenatal screening. In: Inside the Politics of Technology. Amsterdam University Press, Amsterdam, pp. 229–256.

Press, N., Browner, C.H., 1997. Why women say yes to prenatal diagnosis. Soc. Sci. Med. 45 (7), 979–989. https://doi.org/10.1016/S0277-9536(97)00011-7.

Ravitsky, V. (2017). Choice, not "Reflex": Routine Prenatal Screening. Retrieved from: https://impactethics.ca/2017/12/01/choice-not-reflex-routine-prenatal-screening/.

Scott, C.J., Futter, M., Wonkam, A., 2013. Prenatal diagnosis and termination of pregnancy: perspectives of South African parents of children with Down syndrome. J Commun. Genet. 4 (1), 87–97. https://doi.org/10.1007/s12687-012-0122-0.

Seror, V., Ville, Y., 2010. Women's attitudes to the successive decisions possibly involved in prenatal screening for Down syndrome: how consistent with their actual decisions? Prenat. Diagn. 30 (11), 1086–1093. https://doi.org/10.1002/pd.2616.

Sicard, D. (2007). La France au risque de l'eugénisme. Le Monde.fr. Retrieved from: http://www.lemonde.fr/planete/article/2007/02/03/la-france-au-risque-de-l-eugenisme_863262_3244.html.

Van den Heuvel, A., Chitty, L., Dormandy, E., Newson, A., Deans, Z., Attwood, S., Haynes, S., Marteau, T.M., 2010. Will the introduction of non-invasive prenatal diagnostic testing erode informed choices? An experimental study of health care professionals. Patient Educ. Couns. 78 (1), 24–28. https://doi.org/10.1016/j.pec.2009.05.014.

Vassy, C., 2011. De l'innovation biomédicale à la pratique de masse: le dépistage prénatal de la trisomie 21 en Angleterre et en France. Sciences sociales et santé 29 (3), 5. https://doi.org/10.3917/sss.293.0005.

Vassy, C., Rosman, S., Rousseau, B., 2014. From policy making to service use. Down's syndrome antenatal screening in England, France and the Netherlands. Soc. Sci. Med. 106, 67–74. https://doi.org/10.1016/j.socscimed.2014.01.046.

Wald, N.J., Huttly, W.J., Bestwick, J.P., Old, R., Morris, J.K., Cheng, R., Aquilina, J., Peregrine, E., Roberts, D., Alfirevic, Z., 2017. Prenatal reflex DNA screening for trisomies 21, 18, and 13. Genet. Med. https://doi.org/10.1038/gim.2017.188.

Ward, P., Soothill, P., 2011. Fetal anomaly ultrasound scanning: the development of a national programme for England. Obstetr. Gynaecol. 13 (4), 211–217. https://doi.org/10.1576/toag.13.4.211.27685.

WHO, 1998. Proposed International Guidelines on Ethical Issues in Medical Genetics and Genetic Services (document WHO/HGN/GL/ETH/98.1).

Williams, C., Alderson, P., Farsides, B., 2002. Is nondirectiveness possible within the context of antenatal screening and testing? Soc. Sci. Med. 54 (3), 339–347. https://doi.org/10.1016/S0277-9536(01)00032-6.

Chapter 6

Sex Selection, Gender Selection, and Sexism

Iva Rinčić*, Amir Muzur* and Stephen O. Sodeke†

*Faculty of Medicine and Faculty of Health Studies, University of Rijeka, Rijeka, Croatia,
†Tuskegee University National Center for Bioethics in Research and Health Care, Tuskegee,
AL, United States

1. INTRODUCTION

Reproduction is considered to be one of the most intriguing processes in the human life span. Despite the scientific and technological advancements and discoveries resulting in new knowledge about conception, pregnancy, and birth, there is no doubt that the label "miracle of life" will remain in popular conversations about the topic for a long time. This is primarily due to the traditional understanding of reproduction as a merely "*chancy* enterprise," an act beyond human control and will (Malmquist, 2008). What that means is that once a person decides to try to have a child (if we presume the act was decided freely and decisional capacity was present), it is beyond his/her power if, when, and how it is going to happen. It remains debatable whether, and to what extent, various religious traditions and truths (such as Christianity and immaculate conception) have been used by believers of such truths to make sense of this standpoint.

There is not a persuasive argument that can be made to deter and jeopardize the human desire and intention to be involved in the creation of an offspring. To be sure, the decision to have a child has always been considered as a desirable and natural goal of individuals and families. This goal is promoted both by dominant and suppressed cultures, nations, and ethnicities during their history, and across several cultures.[1]

---

1. Under the slogan "Kosovo, the young Europeans," the former Yougoslav region Kosovo celebrates its independence (gained from Serbia on February 17, 2008). It is not only the symbol of a short period of independence in Europe, but also a reflection of several decades of a silent revolution of reproduction within the local population, mainly Albanians. After the World War II the percentage of Albanians in Kosovo was 68% (total 498,244) and Serbs around 24% (176,718). Around 90% of Albanians of that period were illiterate. In the year 2006, there were 92% Albanians (1,932,000) and 5.3% Serbs (1,113,000). It was and still is one of the highest population increases in Europe, and mean citizen age is less than 30.

Clinical Ethics at the Crossroads of Genetic and Reproductive Technologies.
https://doi.org/10.1016/B978-0-12-813764-2.00006-4
113

The "human embryo transferring idea" from the womb of a mother to another's can be found in some historical accounts. One such example is found in the Jain religion. The religious leader Mahavira, born c.599 BCE, was initially conceived by the less desirable Brahman couple Rshabhadatta and Devananda, but due to Sahra, king of the gods, the embryo was transferred into more appropriate parents from the Kshatriya caste (holders of kshatra, or authority) (Biggers, 2012; Matthews, 2013).

By the middle of the 19th century, scientists were familiar with the fact that "blocked ovarium oviducts resulted in sterility" (Churchill, 1846; Biggers, 2012), which lead to the development of numerous research and treatments trying with variable success to remedy this issue. Embryo transfer was first attempted by the Austrian researcher Samuel-Leopold Schenk in 1887. He used the procedure of in vitro fertilization (IVF) on rabbits and guinea pigs, without success (Schenk, 1887). The first successful ova transfer was reported in 1891 by Walter Heape, who transferred two ova from an Angora doe rabbit to a Belgian hare recipient, obtaining four young of Belgian and two of Angora phenotype. Other experiments by Edouard Van Beneden, Oscar Hertwig, and many others, performed around the same time, used sea urchins, starfish, or rabbits (Biggers, 2012).

The first part of the 20th century was marked by the development of tissue culture methods, leaving unwittingly ectogenesis[2] as a marginal scientific issue, in great part due its potential to spark intensive religious debates. In 1932, ectogenesis became a fashionable term due to Aldous Huxley's book *Brave New World*. Realistic descriptions provided in the book are probably the result of Huxley's discussions with Gregory Pincus, a Cambridge-based researcher who later became known as the father of the oral contraceptive pill, and who also played a crucial role in the development of IVF, by later performing experiments using rabbits (Biggers, 2012; Ingle, 1971; Speroff, 2009).[3]

Although the first report of a fertilized human ova by John Rock and Miriam Merkin occurred in 1944, during the World War II, it became publicized later, when it gained the headline of the newspaper *Boston Globe,* claiming the possibility of treating infertility. Microsurgery procedures as a technique of infertility treatment were widely used until 1970s, when IVF and embryo transfer technology took a huge step forward, resulting in Louise Brown's birth in July 25, 1978 in Oldhelm General Hospital Manchester.[4] Soon after this success in

2. Development outside the body; especially development of a mammalian embryo in an artificial environment (https://www.merriam-webster.com/dictionary/ectogenesis).

3. Goodwin Pincus (1903–1967) got in 1931, a position of Assistant Professor at Harvard University. Negative publicity of in vitro fertilization experiments with rabbits resulted in losing his tenure (https://embryo.asu.edu/pages/gregory-goodwin-pincus-1903–1967).

4. The birth of first tube test baby was carefully planned and even guarded. After the birth, the Brown family underwent difficult moments, receiving "menacing notes, blood-spattered letters, a broken glass test tube, and a plastic fetus" (http://www.dailymail.co.uk/health/article-3173446/World-s-test-tube-baby-reveals-mother-received-blood-splattered-HATE-MAIL-born-including-letter-containing-plastic-foetus.html#ixzz4ysCIGWoA).

the United Kingdom, other countries followed the same example (Australia, 1980; USA, 1981; Sweden, 1982[5]).

The first preimplantation genetic diagnosis (PGD) was performed in 1989. The purpose behind PGD was primarily therapeutic, namely "to identify and deselect genetically abnormal embryos prior to implantation, thus providing parents who are at high-risk of producing children with genetic abnormalities with another option besides the often troublesome experience of in-utero diagnosis and abortion or, alternatively, the birth of a genetically abnormal child." (Remaley, 2000, p. 252). Today, PGD is used to diagnose numerous diseases, with the list increasing rapidly. At the same time, PGD was recognized as a shortcut for selecting desirable characteristics, especially regarding sex selection as, according to Remaley, Nisker, and Jones, "science does not operate in a vacuum; context influences how a technology is used" (Remaley, 2000, p. 253, according to Nisker and Jones, 1997).

Another technique for sex selection that attracted public interest was sperm sorting, in which sperm is sorted before fertilization, depending on the presence of a viable X or Y chromosome, and then used for artificial insemination and IVF for conceiving a child of desirable sex (Mcmillan, 2004). The success rate in sperm sorting is lower than in IVF with PGD, but it is seen as less ethically problematic. It is also less expensive, less harmful, and potentially less of a problem because the choice to sort is made before the moment of conception.[6] Techniques such as Microsort have been confirmed to be successful, especially when the couple wished to conceive a girl (Fugger et al., 1998).

Due to the scientific advances in reproductive technologies in the last decades, a shift of their use is becoming apparent, from being merely a method of conception to a method of selecting certain desired traits in the offspring (Buchanan et al., 2000). Today there is a wide range of different techniques (artificial insemination, IVF and embryo transfer, gamete intrafallopian tube transfer, and surrogacy) that are being used; all, however, have one common characteristic: "the separation of human reproduction from the act of coitus" (Mastroianni, 2004, p. 2261).

Although the primary motivation for developing new reproductive technologies was medical infertility,[7] the current ethical debates surrounding reproductive technologies are opening questions related to abortion, PGD of embryos,

---

5. http://www.swedishmedcare.com/en/services/smc-ivf-in-vitro-fertilisation/ivf-history.

6. The future could potentially bring us some kind of over-the-counter sex-selection kit (Remaley, 2000).

7. It is however interesting to note that there are different approaches to infertility: today it is accepted that infertility is not a disease (in a strict sense), but a way of disability; it is a kind of bodily "malfunction" and people suffering from it should get appropriate help (Warnock report, 1984): "As a disability, infertility can sometimes be prevented, sometimes cured, or must otherwise be accommodated." (Beauchamp et al., 2014).

sperm sorting, moral status of embryo, autonomy, sex selection, sex ratio at birth,[8] and sexism.

In the following chapter, our aim is to present some historical and modern issues associated with sex selection (with an emphasis on countries such as China and India), to highlight the diverse legal and ethical questions raised by sex selection in these countries, and to point out potential long-term social consequences generated by these techniques.

2. THE ROOTS AND REASONS OF SEX SELECTION

An ancient Chinese statement claims that "eighteen goddess-like daughters are not equal to one son with a hump" (Cook et al., 2012, p. 363). Although the commitment to sex selection has a long history, going as far back as to ancient times (Singer and Wells, 1985), it is only in the last few decades that sex selection has become an important issue in the social, medical, and ethical literature. Prior to modern sex-selection techniques (prenatal diagnosis and abortion, PGD and embryo selection, sperm sorting), the most radical traditional technique of sex selection was infanticide performed by midwives. The strong intention to assure the birth of at least one son led to a number of recommendations and customs to be used before conception, also known as "preconception folklore prescriptions," including different religious methods for conceiving a (usually male) child (Remaley, 2000, p. 249).

According to the Talmud, to have a male heir, the marriage bed should be positioned in the north-south direction. Also, it is crucial that, during the moment of copulation, the woman must emit her semen before the man to ensure that the child will be a boy (Leviticus 12:2; Schenker, 2002, pp. 401–401). In ancient Greece, Golden noted that men would lie on their right side during sex in order to guarantee a boy. In 18th century France, men desiring the same result would tie their left testicle. Medieval Germans would place a hammer under the bed to produce a boy, while their Danish cousins would put a pair of scissors under the bed in order to ensure a girl is conceived (Golden, 1998, p. 82). Pogrebim reported that a male child would be easier to conceive if the weather was dry, there was a full moon, the wind was coming from the north, and the nut harvest was plentiful. Women were not spared from pertinent advice, either. Women were to have a special diet and pinch their husband's right testicle before intercourse; they were also advised to dress in men's clothing on their wedding night (Pogrebin, 1981, p. 82).

One of the best-known nonprofessional tests for selecting the gender of the offspring is known under the name Chinese Gender Predictor. The test dates back to the 13th century, and was rediscovered in Taiwan in 1972. It uses mother's date of birth, the lunar month and the date of conception, and the lunar

---

8. "Sex ratio at birth" is defined as the number of male live births for every 100 female births (Hesketh and Wei Xing, 2006, p. 13271).

age at the time of pregnancy. Another approach, developed by Dr. Landrum Shettles in 1960, suggests using the timing of the intercourse, due to different speeds of X and Y chromosome in relation to their viability. Currently, similar questionable options are widely offered as "safe-at-home" tests, being found on the internet over 101 ways of determining the gender prenatally, depending on the timing of the intercourse. An article published more than 20 years ago in the *The New England Journal of Medicine,* showing that timing of copulation in relation to ovulation has no effect in determining the gender of the future child, did not change in any substantial way the popularity of such methods (Wilcox et al., 1995).

How to have a child of a particular gender is a challenge from a biological, medical, and practical point of view. While couples try various unpleasant and uncertain tactics and procedures, one question still needs to be asked: why does a couple needs to have a child of a certain gender? Medical reasons for the selection of an embryo of a specific sex when aimed to avoid the transmission of sex-linked diseases have been advanced. The most common examples of such diseases are Hemophilia (A and B), fragile X syndrome, Fabry disease, color blindness, and Duchenne muscular dystrophy. The use of sex selection for medical reasons is now possible due to discoveries of genetic linkage and differences of X and Y chromosomes, followed by the rapid development of PGD techniques.

Beyond medical reasons, the demand for sex selection these days has more to do with *social reasons* than with the aim of avoiding any health-related conditions. These reasons reflect various and variable culture, society, economy, demography, and ethical norms. Less problematic, but worth noting, is the scenario of preferring a specific gender for family balancing, when there are already one or more children of a specific gender in a family, and the parents wish a child of the opposite gender (Mcmillan, 2004).

According to statistics, the average biological probability of having a boy is 51.2%, i.e., 105 boys are born per 100 girls, making the probability of having boys over girls approximately 2.4% higher (Holland, 2002).[9] However, this initial advantage in favor of boys is counter-balanced by a higher mortality rate of boys in every stage of life, resulting in girls having a longer life expectancy. In developed countries such as the United States, Canada, and the United Kingdom, women outnumber men at around the age of 55. The largest discrepancy in favor of females is in Soviet Union, Armenia, Ukraine, and Latvia (where 80% of the suicide attempts are by men).[10] The Caribbean Islands have the

---

9. Although male to female ratio is higher in Arab countries, Norther Africa, and some parts of Asia, recent trends confirm a similar male surplus in Norway (2011), Sweden (2015), and possibly Germany, Denmark, the United Kingdom and Switzerland. This novelty still urge for more complex research, but the reason could be hidden in recent decades of constant migration from men surplus cultures (https://www.washingtonpost.com/news/worldviews/wp/2015/08/19/see-where-women-outnumber-men-around-the-world-and-why/?utm_term=.df79b499dec9).

10. http://www.npr.org/sections/goatsandsoda/2015/08/26/434616512/selecting-boys-over-girls-is-a-trend-in-more-and-more-countries.

highest ratio of women (84.5 men per 100 women), while United Arab Emirates is at the opposite spectrum (274 men per 100 women; Qatar 265.5; Oman 197; Bahrain 163).[11]

Traditionally, parents preferred to have boys, who were considered more important due to family recognition and economic utility (especially in agrarian economies); also, the sons were seen as more reliable for providing support for their parents in their final years (Hesketh et al., 2011, p. 1374). Boys often got increased maternal support (maternal gender preference), which was shown to be crucial for child development. This optimization of resources addressed to the preferred child/children allowed him/them to have increased body mass, increased weight-for-age, improved overall health, a decreased number of sick-days, and a decreased risk of malnutrition (Palloni, 2017, p. 20). Said differently, if you were a boy, born into a relation of maternal male preferences (which were dominant), your chances for a healthier life were maximized.

Most arguments suggesting a male gender preference for the child are rooted in studies traditionally performed in less developed countries, showing prevailing preferences toward sons and/or, in the case of more children, demanding for a balanced number of daughters and sons (Hank and Kohler, 2002, p. 4). As resources in a variety of settings (access to technology, health care, education, and values) are becoming widely available, more recent studies suggest that many developing countries do not have, nowadays, a dominant sex-preference (Palloni, 2017), at least not as an inimitable or unique cultural expectation. The fading of cultural conditioning pressures for a child of a specific sex, gradually shifted from population-based preferences to individual-based preferences,[12] and the economical reasons for wishing a child of a specific gender were replaced by emotional reasons. As one observer puts it, "The children become socialized, not in terms of what they do for the nation or for their economic utility, but because of the opportunities they provide for the individual to experience affection and to give it" (Wirtberg, 1992, p. 21, according to Westerlund, 2005, p. 126). Such shift did not, however, dramatically change the prevailing gender preferences. Daughters are usually preferred for family balancing, and an absolute preference for daughters is seen in very few cases, most often in developed countries, with low reproduction rates.[13]

---

11. https://www.washingtonpost.com/news/worldviews/wp/2015/08/19/see-where-women-outnumber-men-around-the-world-and-why/?utm_term=.78f5e6b425c2.
12. The so-called individual eugenics, claiming individuals freedom in decisions including reproduction was already well elaborated by Darko Polšek (Polšek, 1999).
13. In usual instances, the sex ratio of birth is very constant, with 105–107 male births per every 100 female births. There are however numerous indicators with different impact on the sex ratio at birth, "including family size, parental age, parental occupation, birth order, race, coital rate, hormonal treatments, exposure to environmental toxins, stress, several diseases, and even war (it was confirmed in Europe and the United Staes during First and Second War, as well as in the United States for the Korean and Vietnam Wars; but similar finding were no confirmed during Balkan Wars and Balkan Wars and Iran-Iraq war" (Hesketh and Wei Xing, 2006, p. 13271).

3. FROM POLITICS ON REPRODUCTION TO BIOPOLICY

During the 20th century there were two main issues appertaining to the family planning movement. The first was represented by the birth control movement, which focused on the individual rights of women, autonomy, and unwanted childbearing. This faction was deeply rooted in the rise of feminism and the works of Margaret Sanger and Marie Slopes. A second faction originated during the Industrial Revolution and stemmed from the ideas of Thomas Malthus and his *Essay on the Principle of Population* (1826). Malthus pointed to the discrepancy between fertility rates and long-term available resources to support life. Today, we are aware that Malthus's predictions were wrong because it is an unequal distribution of resources we are facing, and not an absolute lack of resources (Szabo, 2016).

During the middle of the last century, the proponents of these various ideas started to cooperate, building consensus on common grounds for a family planning movement. Many feminists, however, considered the term itself to be a "distracting euphemism." The first programs directed at family planning at the national levels were launched in the 1960s in India, Pakistan, Sri Lanka, Hong Kong, the Republic of Korea, Singapore, and Taiwan (China), even before the new international consensus was fully reached (Sinding, 2007).

Due to political reasons, China remained outside to the international efforts in that duration. Starting just after the end of World War II, the new political structures in China were among the first in the world to face a rapid population growth. Although China faced a lack of women throughout its history (even in the royal family of the Qing Dynasty) (Lee et al., 1994), the phenomenon intensified featuring this phenomenon occurred in the second half of the 20th century. After the Communist revolution in 1949, China went forward in its attempts to reduce mortality in the population by improving nutrition and water supply, and introducing vaccination programs and public health disease prevention activities. This period was also characterized by strong a pronationalistic policy, including child subsidies, and the prohibition of contraception, abortion, and sterilization (Hemminki et al., 2005). Despite hunger and poverty, it is estimated that China gained 1.5 years of life expectancy for every calendar year during the 1950s (Daugherty and Kammeyer, 1995, p. 229). Soon after, due to a rapid population increase, the government started to interfere with population fertility by providing contraceptives for voluntary use, and later played a stronger and more restrictive role in controlling individual reproductive freedoms (Feng et al., 2012). In 1953, contraceptives and abortions became available under certain conditions. After 1957, legal access to abortion was even easier to achieve. Such a turn was easy to achieve due to the fact that abortion on request was less controversial in China than in Western countries. Some studies actually show that many Chinese people do not consider abortion an ethical issue at all, probably because they believe that life begins after birth (Zhang and Cong, 2014, p. 1000).

Although the Chinese government claimed gender equality, male children were strongly preferred by the Chinese society, modeled by the example of the late imperial family, which emphasized the importance of men (Mann, 2011, p. 60). Traditionally, the Chinese society had a strong preference toward men because it was considered an honor to have a son who would bear and perpetuate the family name. According to Coale and Banister, the male preference is deeply rooted in the Daoist and Confucian systems (Lai-Wan et al., 2004; Coale and Banister, 1994). Thus couples often feel obliged to have a son, and the likelihood of making this possible was by having more children or even marrying multiple times (Sudbeck, 2012). Not having a son was often seen as infertility, or a "curse" (Chan et al., 2002).

After population peaked during the 1960s, China decided to introduce one of the strongest population control policies in the world—the one-child policy program in 1979. This policy allowed families to have only one child.[14,15] Indeed, where desired lower birth rates could not be achieved, the government used less than voluntary methods to impose this rule, often seen as coercion (Sinding, 2007, p. 8). The policy set a national target, namely a population of less than 1.2 billion and zero growth in 2000. Soon, the country had to face the severe consequences of enacting this policy, such as illegal pregnancies, abortions, sex selection, and gender imbalance (Hemminki et al., 2005).

In the decades following the introduction of the one-child policy program, there were no official transparent data regarding a female deficit, and when such alerts appeared in foreign agencies, they were claimed to be false by the Chinese authorities. Despite this fact, the National Commission for Family Planning and the Ministry of Health introduced regulations banning prenatal diagnosis at the request of the mother, except when performed by hospitals in order to diagnose specific genetic conditions. Sex identification based on nonmedical grounds was strictly forbidden, as it could lead to the abortion of female fetuses and the generation of sex imbalance, which could threaten the prosperity of the nation (Nie, 2011).

Things rapidly changed after the 2000 census, which showed an alarmingly skewed sex ratio (100 girls to 120 boys nationwide). In 2004, speaking at a conference about population, resources, and the environment, president Hu Jintao announced strategic actions aimed at solving the sex imbalance within 3–5 years. In 2005, China started the national worldwide campaign on "Care for Girls" focusing on the promotion of the value of female children and the consequences of sex imbalance (Nie, 2011).

Although India's first national family planning policy dates back to 1952, there are numerous signs confirming population growth concern before that date (Pyare Kishan Wattal's book *The Population Problem in India* from

---

14. From 1973 to 1979, the policy was applied only in selected provinces.
15. The same year abortion law dimished previous restrictions, setting the time limit for legal abortion at 28 weeks.

1916; the first birth control clinic in Poona established in 1923[16]; and 1928—the neoMalthusian league in Madras). After the World War II, India started national plans of controlling fertility, including contraceptives, vasectomy (with monetary subsidies during 1970–1977), abortion (from 1971), sterilization (for families with three children in 1975, 8.25 millions procedures being performed), and the introduction of a minimum age for marriage (Harakavy and Roy, 2007).

Just like China, India had a long social tradition of discrimination toward women that has resulted in female infanticide and, in the last few decades, sex-selective abortion. The motivation was partially rooted in the economics of marriage (the amount of dowry is decided by the family of the groom, and women are not allowed to inherit family wealth, or to provide help to parents). According to the Hindu religion, men, preferably sons of the interested party, are able to do the final rites needed for the salvation of a person. Until the banning of the sex determination tests, abortion clinics were advertised as a short-term investment for the long-term savings of dowry payments ("Invest Rs. 500 now, save Rs. 50,000 later"). In 1976 the government of India has decided that performing ultrasound and amniocentesis for the purpose of sex determination was illegal. Despite this fact India, as many other Asian countries, is still facing sex disparities as a consequence of previous prenatal sex selection. Like sex-selective abortion, female infanticide is another problem female new borns are facing in India. In addition, researchers have recognized several other phenomena, such as the "parity effect." This phenomenon appears when a woman having more than one child reluctantly takes lesser care of the female children, who are less likely to survive as a consequence. Another phenomenon is the "intensification effect" where, as total fertility in the country is falling, female mortality is rising, or "as fertility declines, the total number of children couples desire falls more rapidly than the total number of desired sons" (Myers, 2012, p. 9; Das Gupta and Mari Bhat, 1997). However, unlike other countries, urbanization and education in India did not have an impact on son preferences—the desire to have a smaller family increases sex-selective abortions (Jha et al., 2011).

4. MISSING GIRLS: WHY AND HOW MANY?

Although the phenomenon of missing girls is widely present in many Asian countries (with the exception of Japan), most authors agree that the greatest problem was in China and India (Das Gupta and Shuzhuo, 1990). We have already elaborated on the main cultural reasons related to India and China, but to understand the complexity of the missing girls phenomena, sex selection procedures have to be correctly ascertained.

---

16. Later on closed due to Gandhian resigment toward artificial methods of birth control.

In the last few decades, there have been numerous researches and studies aimed at identifying the reasons, conditions, and consequences associated with the phenomenon of missing girls, but very few directly targeted the development and use of modern prenatal diagnosis technologies. Pioneers of PGD technologies used them to detect genetic carriers of disorders mainly affecting boys, making it possible for parents to choose girls (Cyranoski, 2017, p. 273). In countries free of a dominant sex matrix, such opportunities had little or low influence on the sex or gender ratio, and were mainly used for health-related issues. At the same time, in countries with sex preferences deeply rooted in cultural traditions (such as China and India), PGD technologies became recognized as efficient tools to perpetuate an already existing practice. An interesting example of a country that does not fit the dominant Asian framework today is South Korea despite the fact that it shares certain similarities with other Asian countries. South Korea, in last few decades, has not followed prevailing statistics. Primarily, South Korea was a leader in the early introduction of sex-selective technologies. The first sex-ratio birth-indicating imbalance was reported in mid-1980s (Hesketh et al., 2011). Several decades later, after taking a series of measures including a law from 1988 that made it illegal to reveal the sex of a child to future parents, South Korea managed to bring its population back into balance. According to these studies, the main reasons were not public campaigns and legal regulation, but rather urbanization and rapid industrialization (Chung and Das Gupta, 2007).

Reports on missing girls in the Chinese society documented the use of infanticide and abandonment of female children during the last few centuries, confirming that today's gender imbalance is not just a direct consequence of sex selection technologies. According to Jiang, in the late 19th century, a missionary questioned 40 women aged 50 or above. Together, the women had given birth to 183 sons and 175 daughters, but while 126 of the sons survived to age 10, only 53 of the daughters did so. The women admitted to killing 78 of their female children (Jiang et al., 2012, p. 2). In 2000, it was estimated a total number of 40.9 million missing girls in China representing 6.7% of the expected female population; in India, the percentage was even higher (7.9%)[23].

Current literature recognizes two types of missing women: truly and nominally missing women. Truly missing refers to females who were made to disappear before or after birth mainly by the use of sex-selective abortion, but also infanticide or denial of care, especially at an early age. Nominally missing includes not only truly missing but also those omitted from statistics (falsely missing) (Jiang et al., 2005). From the perspective of sex-selective technologies' use, we relate to truly missing girls, aborted due to information about the sex of a child obtained during pregnancy. The gender intolerance, caused by sex relative technologies, has been diminished due to an increase in women's lifespan in the late 20th century China.

Just recently, girls missing from official record poped-up as a topic, putting a new light on the overall numbers of missing women in China. This phenomenon of unreported births and delayed registrations,[17] especially present in the countryside, might falsely reduce the number of (truly) missing women in China. This does indicate the inherent problems of a top-down methodology, as well as how cultural and political propaganda can be misleading (Shi and Kennedy, 2016).[18]

5. LEGAL AND ETHICAL QUESTIONS

Laws and regulations concerning PGD techniques widely vary all over the world. In Europe, most countries restrict the use of sex-selection techniques, although with great differences. Social use of PGD (including sex selection) is prohibited in Italy, being allowed only for the purpose of protecting the health and development of the embryo itself; however, there are specific regulations about which conditions meet those aims. In France, each request for PGD is evaluated by the *Centre Pluridisciplinaire de Diagnostic Prénatal*, an entity in charge of making the determination about conditions sufficiently severe to warrant such treatments. French law is only one of the few regulating fertility specialists who are allowed to perform PGD. In the United Kingdom, PGD is regulated by the Human Fertilization and Embryology Act (2008). The law allows the use of PGD only in cases of significant risks for the embryo (serious physical or mental disability, serious illness, and any other serious medical conditions). Unlike other Western countries, the United States lacks a specific institution that can specifically implement state or federal laws regarding PGD; thus the current state of the art is confusing in terms of regulatory clarity between assisted reproduction and genetic testing (Shelby Deeney, 2013), opening a landscape for fertility tourism (Bayefsky, 2017), also known as "cross-border reproductive care" (CBRC), or "reproductive exile."[19] Both China and India have a tradition of restricting nonmedical sex selection. The two countries started introducing a series of prohibitive laws, policies, and regulations on such procedures during the 1980s. The Chinese National Commission for Family Planning and the Ministry of Health prohibited, in September 1986, prenatal diagnosis at the request of the mother, except if performed by authorized hospitals to diagnose certain hereditary conditions. Breaches of this regulation led to penalties. The use of medical techniques for non medical reasons such as the

---

17. According to UNICEF 2002 report, during 1990s there were 6 millions of unregistered children (Pais, 2002).
18. According to some estimation, there are millions of orphans in China, nearly all of them being girls (Saeph, 1996).
19. Most famous form of such tourism is commercial surrogacy (renting wombs)—this industry is estimated to be worth $US6 billion annually (Deonandan, 2015). Due to the high cost in Western countries, most popular countries for surrogacy are Thailand, Ukraine, and Russia, while other countries with no or free regulation (such as Mexico, Nepal, Poland, and Georgia) are rapidly approaching to same reputation.

identification of the fetus' sex was also strictly prohibited through the Law on Maternal and Infant Health Care (1994) and the Law on Population and Family Planning of 2001 (Nie, 2011, p. 13). Another policy statement, the Action Plan to Reverse the Distorted Sex Ratio of Newborns, drafted after the national symposium, "The Distorted Sex Ratio of Newborns: Ethical, Legal and Social Issues" (2005) confirmed an academic consensus regarding the official position: "nonmedical sex-selective abortion, along with non-medical prenatal sex diagnosis, is morally wrong and should be legally prohibited" (Nie, 2011, p. 14). However, the dominant clinical practice (especially in China) remained resistant to changes for a long time, and often practiced something known under the name "the two illegalities"—mandatory (sometimes even forced) termination of pregnancies, generated by the national population policy (OCP), and the strict prohibition on sex-selective abortion (Nie, 2011).

Although India has faced gender disparities for a long time, public awareness increased after the 2001 National census, revealing rapidly increasing gender disparities after 1981 and 1991, especially in more urban and developed regions such as Haryana (861 F: 1000 M), and Punjab (876 F: 1000 M). Pioneering efforts regarding the national regulation of the population in India started during Indira Gandhi's government, which enacted measures such as granting employees promotions, loans, and housing permits based on proof of sterilization. In the following decades, India's population control model implicitly started to encourage sex-selective abortion, which was a side effect of the greater accessibility to prenatal sex identification technologies, and especially amniocentesis. In 1976, the use of ultrasound and amniocentesis to reveal the sex of the fetus to the mother became illegal, but a few years later the government decided to make expensive tests cheaper and more accessible (Patel, 2007). Ultrasound technologies became commonly misused, and are currently being replaced by PGD methods. Access to PGD in India is regulated by national law and its practice was allowed only with a specific license; this is also the case in countries such as France, the United Kingdom, and the Netherlands (Stern, 2014). The banning of abusing sex-selection techniques, which was introduced in 1994,[20] practically had no results, their usage remaining unabated (Cook et al., 2012, p. 364).

Although the official Chinese and Indian positions regarding nonmedical sex selection are in many ways identical to those found in most other countries, morally condemning and often legally restricting the practice, in reality things tend to be significantly different. Although many ethicists tried to explain this paradox by simply extending arguments of reproductive liberty and individual autonomy to the cases of China and India, arguing at the same time for the nonbanning of social sex selection, numerous arguments were rejected because they were deeply rooted in Western tradition. When asked Chinese women

---

20. The Prenatal Diagnostic Techniques (Regulation and Prevention of Misuse) Act, 1994. Indian Parliament Act No. 57.

regarding the moral acceptance of sex-selective abortion, most of them considered it morally wrong. However, the 2000 survey results confirmed that 36% of the interviewed women admitted to have performed one (Chu, 2011). This result suggests that sex selection is still performed on a large scale both because breaching the legal regulation does not have legal repercussions and because there is a significant gap between individual attitudes and values. As a consequence, it seems that the risks associated with the procedure are acceptable when compared with the benefit of achieving a socially desirable aim, namely to have a male child. It is not surprising that ethicists are not harmonized in their positions, argumentations, and recommendations, although most of them based their claims on women's autonomy and freedom, especially regarding their reproductive rights. This is justifiable as strict limitations of women's rights continue to exist in some parts of the world (Macklin, 1999). Feminist oriented ethicists like Wendy Rogers, Angela Ballantyne, and Heather Draper recommended caution about emphasizing freedom, responsibility, and choice in specific cultural frameworks, including where the decision to have a son, in a society that strongly prefers this gender, is never an autonomous one—women do not "choose" to sex-select in a social vacuum. The German bioethicist Otto Döring has an even stronger view regarding the cultural dimension of this issue, by considering the current trends of transferring western standards to China as "culturally insensitive and 'utterly inappropriate'" (Döring, 2008). Or, as Adorno would summarize "There is no right way with wrongness" (Adorno, 1951).

Keeping in mind the "serious drawbacks entailed in state intervention in sex-selective abortion in the Chinese context",[21] but also trying not to jeopardize reproductive (especially women's) freedom, we would summarize the main points and recommendations that should be implemented by physicians, and also by the general population, about limiting the use of PGD in nonmedical purposes, as:

1. A clear regulatory framework of the activities that institutions and individual specialists should be allowed to provide through PGD.
2. All institutions providing PGD, both public and private, should be involved in the continuous reporting of the issues raised during their activities in this field to relevant administrative structures.
3. Systematic early education on family planning, gender selection, and social consequences of sexism are needed to limit sex selection, especially in high-risk areas, such as India and China.

---

21. "Neglecting reproductive liberty and reproductive rights; overlooking the hidden dangers of state power; inconsistency with existing abortion policies; practical ineffectiveness; underestimating the costs and resistance involved; simplifying and misrepresenting the key issues; a lack of public discussion; and ignoring the moral and political principles established in traditional Chinese thought." (Nie, 2011, pp. 17–18).

4. Restrictions could be relaxed if the couples intend to have a balanced sex ratio in their offsprings (Riley, 2004).
5. Physicians providing sex-related abortion should have their license to practice medicine revoked.
6. The marketing of medical equipment used for PGD should be limited and controlled.
7. Virtue ethics should be taught to physicians.
8. Introducing alternative procedures (sperm sorting or dual fluorescence in situ hybridization for simultaneous detection of X- and Y-chromosome-specific sequences) might be less physically and emotionally harmful for women than sex-related abortion.

6.   CONCLUSION

The reproductive revolution that we are facing today is to a large extent a result of advances in medical technologies that support families in their pursuit of having children, with a special emphasis on preimplementation genetic diagnosis and other procedures able to detect medical conditions of the future child. There are many reasons, primarily social (cultural, religious, and economic) and characterial, which allowed modern medical technologies to be widely used for nonmedical purposes, such as detecting the gender of the embryo/fetus as a prerequisite for sex-selective abortion. Despite legal and moral restrictions, the practice of sex selection is especially present in Asian societies, headed by India and China, stemming from a dominant cultural preferences toward male off-springs. Combined with national family planning programs designed to control rise in population, decades of sex-selection practices have led to millions of missing women and gender imbalance.

Historically, the right to reproduce did not entail the right to choose the characteristics of the future offspring; parents had hopes that the future child will have certain characteristics, and the partners select themselves based on certain characteristics. Medical resources were scarce, and access to them was often dependent upon the socio economic status. A shift from this traditional health-care system, generated by the introduction of new methods available to a broader population became a reality in recent times. Usually, medical procedures generate positive outcomes for the patients; however, sometimes their reckless use for nonmedical purposes (e.g., selecting the gender of the child), might generate negative social and potentially discriminatory results.

Sex-selective abortion can be seen as method that blurs the line between women's right to have a safe abortion and, potentially, discrimination against (female) babies (Eklund and Purewal, 2017). Although abortion is seen today as a standard reproductive right in many developing world countries, in India and China its discriminate use cannot be indiscriminately defended purely based on the right to respect individual choices. Another important aspect of allowing sex-selective abortion is that it is a slippery slope to increased acceptance of eugenic techniques for other traits of future babies—if we allow sex selection today, will we also allow tomorrow the selecting of eye color, height, IQ? If moral standards and legal regulation do not work on sex selection, why would they work for other issues?

To conclude, discussions on reproductive rights today represent some of the most important ways of responding to intriguing questions in modern bioethics, introducing not only medical, moral, and legal perspectives, but also a broader social context, and cultural norms and values. Trying to understand these issues, without leaving our own cultural heritage behind, could lead to doubtful and concerning opinions and recommendations. It would be wrong to say that the current situation of sex and gender selection in countries such as China and India is not without concern and should be preserved, but any plan and action for change should be gradual, inclusive, and culturally sensitive. Otherwise, we would only be trapped in repeating gender stereotypes and deepening already existing and troubling social consequences.

REFERENCES

Adorno, T.W., 1951. Minima Moralia. Reflection from the Damaged Life. Sage, London.

Bayefsky, M. 2017. Comparative preimplantation genetic diagnosis policy in Europe and the USA and its implications for reproductive tourism Reprod. BioMed. Soc. *Online*, 3, 41–47.

Beauchamp, T., Walters, L., Kahn, J.P., Mastroianni, A.C., 2014. Contemporary Issues in Bioethics. Wadsworth Cengage Learning.

Biggers, J.D., 2012. IVF and embryo transfer: historical origin and development. Reprod. Biomed. Online 25, 118–127.

Buchanan, A., Brock, D.W., Daniels, N., Wickler. D., 2000. From Chance to Choice: Genetics and Justice. Cambridge University Press, Cambridge.

Chan, C.L.W., Yip, P.S.F., Ng, E.H.Y., Ho, P.C., Chan, C.H.Y., Au, J.S.K., 2002. Gender selection in china: its meanings and implications. J. Assist. Reprod. Genet. 19, 426–430.

Chu, J., 2011. Prenatal sex determination and sex-selective abortion in rural central China. Popul. Dev. Rev. 27, 259–281.

Chung, W. & Das Gupta, M. 2007. Why is son preference declining in South Korea? The role of development and public policy, and the implications for China and India. Policy Research Working Paper, 4373.

Churchill, F., 1846. On the Theory and Practice of Midwifery, second American ed. Lea and Blanchard, Philadelphia.

Coale, A.J., Banister, J., 1994. Five decades of missing females in China. Demography 3, 459–479.

Cook, R.J., Dickens, B.M., Fathalla, M., 2012. Reproductive health and human rights. In: Integrating Medicine, Ethics, and Law. Clarendon Press, Oxford.

Cyranoski, D., 2017. China's push for better babies. Nature 548, 272–274.

Das Gupta, M., Mari Bhat, P.N., 1997. Fertility decline and increased manifestation of sex bias in India. Population investigation committee. Popul. Stud. 513, 307–315.

Das Gupta, M., Shuzhuo, L., 1990. Gender bias in China, South Korea and India 1920–1990: the effects of war, famine, and fertility decline. Dev. Chang. 30, 619–652.

Daugherty, H.G., Kammeyer, K.C.W., 1995. An Introduction to Population. The Guilford Press.

Deonandan, R., 2015. Recent trends in reproductive tourism and international surrogacy: ethical considerations and challenges for policy. Risk Manage. Healthcare Policy 8, 111–119.

Döring, O., 2008. What's in a choice? Ethical, cultural and social dimensions of sex selection in China. Hum. Ontog. 2, 1–14.

Eklund, L. & Purewal, N. 2017. The bio-politics of population control and sex-selective abortion in China and India, Feminism Psychol., doi: https://doi.org/10.1177/095935351668226.

Feng, W., Cai, Y., Gu, B., 2012. Population, policy, and politics: how will history judge China's one-child policy? Popul. Dev. Rev. 38 (Supplement), 115–129.

Fugger, E., Black, S., Keyvanfar, K., Schulman, J.D., 1998. Births of normal daughters after Micro-Sort sperm separation and intrauterine insemination, in-vitro fertilization, or intracytoplasmic sperm injection. Hum. Reprod. 13, 2367–2370.

Golden, F., 1998. Boy? Girl? Up to you. Time 82–83. 21 September.

Hank, K., Kohler, H.-P., 2002. Gender Preferences for Children Revisited: New Evidence from Germany. MPIDR Working Paper WP 2002-017. Max-Planck-Institut für demografische Forschung. pp. 1–24. https://www.demogr.mpg.de/papers/working/wp-2002-017.pdf.

Harakavy, O., Roy, K., 2007. Emergence of the Indian National Family Planning Program. The Global Family Planning Revolution: Three Decades of Population Policies and Programs The International Bank for Reconstruction and Development/The World Bank. pp. 301–321.

Hemminki, E., Wu, Z., Cao, G., Viisainen, K., 2005. Illegal births and legal abortions—the case of China. Reprod. Health. 2. https://doi.org/10.1186/1742-4755-2-5.

Hesketh, T., Lu, L., Wei Xing, Z., 2011. The consequences of son preference and sex-selective abortion in China and other Asian countries. Can. Med. Assoc. J. 183, 1374–1377.

Hesketh, T., Wei Xing, Z., 2006. Abnormal sex ratios in human populations: causes and consequences. Proc. Natl. Acad. Sci. U. S. A. 36, 13271–13275.

Holland, B.K., 2002. What Are the Chances? Woodoo Deaths, Office Gossip, and Other Adventures in Probability. The John Hopkins University Press.

Ingle, D.J., 1971. Gregory Goodwin Pincus. Biographical Memories. Columbia University Press, National Academy of Sciences, New York, pp. 229–270.

Jha, P., Kesler, M. A., Kumar, R., Ram, F., Ram, U., Aleksandrowicz, L., Bassani, D. G., Chandra, S., Banthia, J. K. 2011. Trends in selective abortions of girls in India: analysis of nationally representative birth histories from 1990 to 2005 and census data from 1991 to 2011. Lancet, 9781, 1921–1928. https://doi.org/10.1016/S0140-6736(11)60649-1 (November 25, 2017).

Jiang, Q., Feldmann, M. W. & Jin, X. 2005. Estimation of the Number of Missing Females in China: 1900–2000. http://citeseerx.ist.psu.edu/viewdoc/download;jsessionid=2689629CC60AFEC13B1510D5195B61C0?doi=10.1.1.372.3037&rep=rep1&type=pdf (November 22, 2017).

Jiang, Q., Shuzhu, L., Feldmann, M.W., Sánchez-Barricarte, J.J., 2012. Estimates of missing women in twentieth century China. Contin. Chang. 3, 1–16. https://www.ncbi.nlm.nih.gov/pmc/articles/PMC3830941/pdf/nihms513861.pdf.

Lai-Wan, C.C., Blyth, E., Hoi-Yan, C.C., 2004. Attitudes to and practices regarding sex selection in China. Prenat. Diagn. 26, 610–613.

Lee, J., Wang, F., Cambell, C., 1994. Infant and child mortality among the Qing nobility: implications for two types of positive check. Popul. Stud. 3, 395–411.

Macklin, R., 1999. Against Relativism: Cultural Diversity and the Search for Ethical Universals in Medicine. Oxford University Press, New York.

Malmquist, E., 2008. Good Parents. Better Babies. An Argument about Reproductive Technologies Enhancement and Ethics. In: Linköping Studies in Arts and Science. 447. Dissertations on Health and Society No. 14. Linköping University, Department of Medical and

Health Sciences–Linköping http://liu.diva-portal.org/smash/get/diva2:1757/FULLTEXT01. pdf (August 4, 2017).

Mann, S.L., 2011. Gender and Sexuality in Modern Chinese History. Cambridge University Press.

Mastroianni, L., 2004. Reproductive Technologies. I. Introduction, third ed. Encyclopedia of Bioethics, Gale Cengage Learning. pp. 2261–2267.

Matthews, L., 2013. World Religions. Cengage Learning.

Mcmillan, J., 2004. Reproductive Technologies. II. Sex Selection, third ed. Encyclopedia of Bioethics, Gale Cengage Learning. pp. 2267–2271.

Myers, C. 2012. Sex selective abortion in India. Global Tides, 3, 1–19. http://digitalcommons. pepperdine.edu/globaltides/vol6/iss1/3 (August 4, 2017).

Nie, J.B., 2011. Non-medical sex-selective abortion in China: ethical and public policy issues in the context of 40 million missing females. Br. Med. Bull. 98, 7–20.

Nisker, J.A., Jones, M., 1997. The Ethics of Sex Selection. Ethical Dilemmas in Assisted Reproduction. Parthenon, pp. 41–50.

Pais, M.S., 2002. Birth registration: right from the start. UNICEF Innoc. Digest. 9, 1–32.

Palloni, G., 2017. Childhood health and the wantedness of male and female children. J. Dev. Econ. 126, 19–32.

Patel, T., 2007. Sex-Selective Abortion in India: Gender, Society and New Reproductive Technologies. Sage Publications, New Delhi.

Pogrebin, L.C., 1981. Growing Up Free: Raising Your Child in the 80's. Bantam Books.

Polšek, D., 1999. Laissez-faire eugenika: Strategije druge geneze *Homo sapiens*a [Laissez faire eugenics. Strategy of second *Homo sapiens* genesis]. Društveni značaj genske tehnologije [Social meaning of gene technology], Institut društvenih znanosti Ivo Pilar. pp. 259–285.

Remaley, R.E., 2000. The original sexist sin: regulating preconception sex selection technology. Health Matrix J. Law Med. 102, 248–298.

Riley, N.E., 2004. China's population: new trends and challenges. Populat. Bull. 59, 2. http://www. prb.org/Source/59.2ChinasPopNewTrends.pdf.

Schenk, S.L., 1887. Das Sägethieri künstlich befruchter ausserhalb des Mutterthieres. Mitteilungen aus dem Embryologischen Institute der K.K. Unitersität Wein. 2, 107–118.

Schenker, J.G., 2002. Gender selection: cultural and religious perspectives. J. Assist. Reprod. Genet. 19 (9), 400–410.

Shelby Deeney, M., 2013. Bioethical considerations of preimplantation genetic diagnosis for sex selection. Washington Univ. Jurisprudence Rev. 5 (2), 333–360.

Shi, Y., Kennedy, J., 2016. Delayed registration and identifying the "missing girls" in China. China Quarter. 228, 1018–1038.

Sinding, S.W., 2007. Overview and Perspective. The Global Family Planning Revolution: Three Decades of Population Policies and Programs, The International Bank for Reconstruction and Development/The World Bank. pp. 1–12.

Singer, P., Wells, D., 1985. Making babies: the new science and ethics of conception. Scribner's Sons.

Speroff, L., 2009. A Good Man. Gregory Goodwin Pincus. Amica Publishing Inc., Portland, Oregon.

Stern, H.J., 2014. Preimplantation genetic diagnosis: prenatal testing for embryos finally achieving its potential. J. Clin. Med. 3 (1), 280–309.

Sudbeck, K., 2012. The Effects of China's One-Child Policy: The Significance for Chinese Women. DigitalCommons@University of Nebraska—Lincoln.

Szabo, S., 2016. Urbanization and Inequalities in a Post-Malthusian Context: Challenges for the Sustainable Development Agenda. Springer.

Westerlund, K. 2005. Cultural aspects on reproductive technology and genetic diagnosis. PGD and Embryo Selection Report from an International Conference on Preimplantation Genetic Diagnosis and Embryo Selection, Nordic Council of Ministers, p. 124–135.

Wilcox, A.J., Weinberg, C.R., Baird, D.D., 1995. Timing of sexual intercourse in relation to ovulation. Effects on the probability of conception, survival of the pregnancy, and sex of the baby. N. Engl. J. Med. 333, 1517–1521.

Wirtberg, I., 1992. His and her childlessness. Dissertation Karolinska Institute, Stockholm.

Zhang, H. & Cong, Y. 2014. China. Handbook of Global Bioethics. Springer, pp. 993–1009.

FURTHER READING

ask n.d. http://ask.rks-gov.net/media/2065/demografske-promene-kosova-u-periodu-1948-2006.pdf (September 5, 2017).

Bennett, N.G., 1983. Sex Selection of Children: An Overview. Bennett NG: Sex Selection of Children. New York, pp. 1–12.

chinesegenderchart. n.d. https://www.chinesegenderchart.info/101-gender-predictors.php (November 19, 2017).

Cohen, J., Trounson, I., Dawson, K., Jones, H., Hazekamp, J., Nygren, K.-G., Hamberger, L., 2005. The early days of IVF outside the UK. Hum. Reprod. Update 5, 439–459.

Openknowledge. n.d. https://openknowledge.worldbank.org/bitstream/handle/10986/7367/wps4373.pdf?sequence=1&isAllowed=y (November 23, 2017).

researchgate n.d. https://www.researchgate.net/profile/Monica_Das_Gupta/publication/233115102_Fertility_Decline_and_Increased_Manifestation_of_Sex_Bias_in_India/links/58ad844f92851cf7ae85798d/Fertility-Decline-and-Increased-Manifestation-of-Sex-Bias-in-India.pdf (November 27, 2017).

Spaeth, A., 1996. A: Life and Death in Shanghai: China Lashes out at a Human-Rights Report Alleging Widespread Abuse of Orphans. Time Magazine (International Edition), p. 147.

Wolf, A.P., Huang, C., 1845–1945. Marriage and Adoption in China. Stanford.

Yang, J., Song, Y., Qu, Z., et al., 2009. Shenyu Zhengce yu Chusheng Xingbiebi [Fertility Policy and Sex Ratio at Birth]. Social Sciences Academic Press, Beijing.

Chapter 7

The Impact of Big Data on Beginning of Life Issues

Dario Sacchini and Antonio G. Spagnolo

Institute of Bioethics and Medical Humanities, Fondazione Policlinico Universitario "A. Gemelli" – Università Cattolica del Sacro Cuore, Rome, Italy

1. INTRODUCTION: FIRST OF ALL, WHAT IS REALLY "BIG DATA"?

"Big Data" (BD) is a term that defines an intriguing reality highlighted by complexity: from its definition to the technical meaning and to the economic, organizational, ethical, legal, and social consequences, without omitting public awareness. In general, BD is an extremely wide, complex, either structured or unstructured amount of information, so large that routine data processing software/procedure are not able or inadequate to address them properly. From a historical point of view, the origin of the term is uncertain, becoming a buzz key concept in 2011, especially in Biology/Life Sciences/Health. Before addressing the focus of the chapter—the impact of BD in beginning-of-life (BOL) issues—it is useful an overview, firstly about BD as a whole, and secondly about BD in health care.

There are several definitions of BD. Among them we have chosen one that in our opinion better encompass its nature: "a cultural, technological, and scholarly phenomenon that rests on the interplay of: (1) Technology: maximizing computation power and algorithmic accuracy together, analyze, link, and compare large data sets. (2) Analysis: drawing on large data sets to identify patterns in order to make economic, social, technical, and legal claims. (3) Mythology: the widespread belief that large data sets offer a higher form of intelligence and knowledge that can generate insights that were previously impossible, with the aura of truth, objectivity, and accuracy." (Boyd and Crawford, 2012, p. 663).

Therefore, BD is the result of the increasing generation of information due to massive digitization allowed by information technologies (IT). The per-capita

Clinical Ethics at the Crossroads of Genetic and Reproductive Technologies.
https://doi.org/10.1016/B978-0-12-813764-2.00007-6

131

capacity to store information has doubled every 40 months since the 1980s (Hilbert and López, 2011) and by 2025, the forecasts indicate 163 zettabytes of data (Reinsel et al., 2017). Conventionally, BD has data sets in the petabyte (1024 terabytes) or exabite (1024 petabytes) range.

Currently, the main BD sources are: media industry, entertainment industry, health care, life sciences, video surveillance, and transportation, logistics, retail, utilities, environment, and telecommunications (Villars et al., 2011). In the last years, many BD inputs were obtained from wearable/connected/IT devices, including remote sensors, software logs, microphones, radio-frequency identification (RFID) readers, wireless sensor networks, computers, tablets, smartphones, embedded sensors, cameras, all of which allow us not only to become more connected, networked, and traceable, but also increased opportunities for gathering digital information from wide variety of large granular data sets, for very different fields like human genomics, health care, oil and gas, search, surveillance, and finance (Villars et al., 2011).

These large data sets cannot be properly analyzed using standard statistical software; therefore BD environment requests "ad hoc" skills and procedures. BD analytics is the process of examining very large variety of data types to discover/uncover hidden/unknown patterns/correlations in different fields (market, science, etc.); "data scientists" are experts who have skills in handling BD, while "data mining" is the process of finding useful patterns from databases by using different theories/technologies (Zhang et al., 2016). Moreover, BD has three properties that have to be contextually expanded: *volume* (the amount of data), *variety* (types of data), and *velocity* (speed of data processing), coined as the "3Vs" model by the Gartner analyst Doug Laney (2001).

In short, BD leads to potentially great opportunities for us to better understand the world we are living in; however, at the same time, it causes significant concerns.

The first one is about the "BD paradigm" and the debate around the deep epistemological shift caused by BD. As Kitchin said: "There is little doubt that the development of Big Data and new data analytics offers the possibility of reframing the epistemology of science, social science, and humanities, and such a reframing is already actively taking place across disciplines. Big Data and new data analytics enable new approaches to data generation and analyses to be implemented that make it possible to ask and answer questions in new ways .... There is an urgent need for wider critical reflection on the epistemological implications of Big Data and data analytics, a task that has barely begun despite the speed of change in the data landscape" (2014, p. 10).

In other words, BD would change/reframe the definition, the methods, and the objects of knowledge and our understanding of human networks, community, and social life (Boyd and Crawford, 2012) as well as of the research processes, because the term encompasses large data sets but also the tools to manage them (Burkholder, 1992), and even a particular way to manage information, the nature, and the categorization of reality and individual/social behavior (Snijders et al., 2012).

Another issue is that unparalleled accuracy and objectivity, which is sustained by BD "supporters," has to face unquestionable realities including: data volume, velocity, and a variety of data storage and transmission; the issue of transparency of data collection methods; the complexity of data processing; and the human element when it comes to data gathered, especially in social media. Even if these risks and complexities are to be taken into account, one should not forget that BD offers unprecedented levels of insights about the analyzed issues and it tends to deeply change the way in which enterprises are run and studies are being performed, which are now based on hard data rather than intuition (Merelli et al., 2014).

As Kimble et al. said, "there is clearly a need to de-bunk some of the myths that surround it… First, big data does not provide easy answers… Second, there is a need to de-bunk the belief in the supposed objectivity of big data. There are several technologies and methodological reasons why bigdata may not be as complete and objective as itseems, which should be apparent to even a casualuser of social media" (2015, p. 32).

Moreover, the claims to BD objectivity and accuracy are misleading. As literature points out: "Big Data offers the humanistic disciplines a new way to claim the status of quantitative science and objective method. It makes many more social spaces quantifiable. In reality, working with Big Data is still subjective, and what it quantifies does not necessarily has a closer claim on objective truth—particularly when considering messages from social media sites" (Boyd and Crawford, 2012, p. 667). A similar idea was supported by Bollier, who said that, "As a large mass of raw information, Big Data is not self-explanatory. And yet the specific methodologies for interpreting the data are open to all sorts of philosophical debate. Can the data represent an 'objective truth' or is any interpretation necessarily biased by some subjective filter or the way that data is 'cleaned?'" (Bollier, 2010, p. 13). And again, Boyd concluded that, "Interpretation is at the center of data analysis. Regardless of the size of a data, it is subject to limitation and bias. Without those biases and limitations being understood and outlined, misinterpretation is the result. Data analysis is most effective when researchers take account of the complex methodological processes that underlie the analysis of that data" (2012, p. 668). In brief, assuming that, through the use of BD, numbers speak for themselves, sidelining other methodologies for the analysis of issue at hand is a methodological and epistemological bias.

The scientific literature also suggests "bigger data are not always better data" (Boyd and Crawford, 2012). In fact, BD and whole data are not synonymous. Sometimes, a small sample can bring out more information about a particular issue than BD, which can be rendered meaningless; also, the sample can be difficult to understand when its source is unknown. In some circumstances, small data size, or a proven scientific methodology can yield better results, as clearly shown by the failure of Google Flu Trends, a poster child for BD, to predict the actual prevalence of the flu. BD often tends to focus on the data as a whole rather than on a representative sample, potentially causing biases due to logistical and analytical challenges such as the synthesis of heterogeneous data resources (Jones et al., 2006).

We should also take into account that data context is essential for generating meaning from data, and without knowing it we are only doing math and not applied science. The management of the context in Big Data is another unmet challenge (Boyd and Crawford, 2012) that has to be properly addressed, especially in health care, where any mishaps could increase morbidity or mortality.

Another general concern about BD is represented by the difficulty to provide reliable guarantees for assuring subjects' privacy and civil liberties (such as autonomy or anonymity) in a strongly connected society such as ours (Nayef Al-Rodhan, 2014).

As of today, we have a poor understanding of the ethical, legal, and social implications (ELSI) underpinning BD, being still many unanswered questions, such as: "Should someone be included as a part of a large aggregate of data? What if someone's 'public' blog post is taken out of context and analyzed in a way that the author never imagined? What does it mean for someone to be spotlighted or to be analyzed without knowing it? Who is responsible for making certain that individuals and communities are not hurt by the research process? What does informed consent look like?" (Boyd and Crawford, 2012, p. 672). Moreover, the ethics of BD research methodologies implies and requests a deep debate, because new methods complicate long-standing principles of research ethics [e.g., the "data (human) subject" considered as a corollary] (Buchanan, 2017).

We need to address specific ethical issues about BD research, collection, analysis, and use as well as to guarantee a rigorous accountability—conceived as a multidirectional relationship (Dourish and Bell, 2011)—both of researchers/professionals and research subjects, for protecting the rights of the people involved, with an emphasis on privacy issues. Moreover, ethical issues differ depending on the type of BD usage (Fairfield and Shtein, 2014).

Finally, great inequalities can rise in case of limited access to social media data (Boyd and Crawford, 2012) for those who do not have money (e.g., not well resourced universities) or those who do not work in a company that owns data, potentially raising justice-related ethical issues (such as equitable use of data, resource allocation, and so on).

2.   BIG DATA AND HEALTH CARE: AN EXPANDING UNIVERSE

The BD "saga" in Biology/Life Sciences/Health care explicitly starts from 2011 onwards (Villars et al., 2011; Chute et al., 2013; Luo et al., 2016). It is generated by both academics under the well-known neologism "omics" on the one side, and by medical devices and applications, health-care professionals reports, social health services, utility providers, mobile devices, apps, and sensors set up by medical/social teams, patients, biomedical companies, payers, and governments, on the other side.

BD in health care is coined as biomedical BD (BBD) (Luo et al., 2016; Ulfenborg et al., 2017). As early as 2012, as shown in different research papers, translational bioinformatics was able to disrupt human health and health care using BD by discovering patterns and making clinically relevant predictions (Shah, 2012; Phan et al., 2012).

Since 2013, the scientific literature highlighted the "inevitability" of application/utilization of BD in health care (Murdoch and Detsky, 2013; Milton, 2017; Brennan and Bakken, 2015). Currently, BBD includes (but is not limited to): bioinformatics, clinical informatics, imaging informatics, and public health informatics. Each field has specific characteristics, methodologies, tools, and aims. According to Luo et al. the definitions of these terms are: "in *bioinformatics*, high-throughput experiments facilitate the research of new genome-wide association studies of diseases, and with *clinical informatics*, the clinical field benefits from the vast amount of collected patient data for making intelligent decisions. *Imaging informatics* is now more rapidly integrated with cloud platforms to share medical image data and workflows, and *public health informatics* leverages big data techniques for predicting and monitoring infectious disease outbreaks." (2016)

The main tools/technologies used in *bioinformatics* are: data storage, retrieval and analysis, error identification, and platform integration deployment while *clinical informatics* pays particular attention to data sharing and security issues, because it works with heterogeneous (structured and unstructured) data. *Imaging Informatics* uses data storage, retrieval, sharing, and analysis as well, even if it works with particular heterogeneous data produced by different modalities of medical images. Finally, *public health informatics* deals with infectious disease surveillance; population health, mental health and chronic disease management, but it has not introduced new approaches, as of 2016 (Luo et al., 2016).

According to Murdoch et al. there are four ways for improving quality and efficiency of health care through BD (2013):

1. expand the capacity to produce new knowledge, overcoming the difficulties and costs to answer clinical questions/doubts prospectively/retrospectively, in terms of generalizability;
2. promote knowledge dissemination, especially for physicians who struggle to stay up-to-date with the latest evidence aimed at guiding their best clinical practice (e.g., clinical management of comorbidities for chronic patients) as well as to get resource savings and appropriate standardization of care;

3. help the translation of data gathered from BD into clinical practice, through the integration of systems biology (e.g., genomics) with electronic health-care records data sets;
4. the transformation of health care by more direct access to information "by" and "on" patients, allowing a useful merging of traditional health care-related data (e.g., medical record) with other personal data (e.g., education info).

It should also be added that BD are expected to optimize the coverage of healthcare policies and to develop innovative economic models able to address such a complex matter as well (Phillips et al., 2014). The wide usage or BBD poses a series of significant barriers, which include (Murdoch and Detsky, 2013):

1. a lack of incentives for using BD by physicians/health-care organizations (e.g., hospitals);
2. health-care stakeholders that support BD (e.g., health service researchers, pharmaceutical companies, public health organizations) do not deliver direct individual patient care;
3. the need for adequate requirements for ensuring privacy. Early further literature on the matter (Gehrke, 2012; Schadt, 2012) highlighted the need for specific differential privacy/disclosure control procedures as well as policy recommendations aimed at adequating practice to privacy expectations.
4. the fragmentation of BD platforms [e.g., electronic health records (EHR)] and, consequently, limited interoperability;
5. BD safety issues (e.g., inadvertent selection of incorrect medications in EHR management);
6. overreliance on electronic systems as whatever automated systems can require human override in case of system failure.

Other challenges regarding BBD include: issues generated by data and information quality (O'Donoghue and Herbert, 2012), the need for intelligent tools to control the accuracy, believability and handling of missing data (Mirkes et al., 2016), the difficulty of using a large amount of heterogeneous, unstructured electronic information in health care (Murdoch and Detsky, 2013; Costa, 2014); the need to set up new types of consent forms and procedures, which would allow study subjects to openly share the data generated about themselves (Vayena et al., 2013), the costs for BD storage, security, transfer, and analysis of big data are high, posing accessibility issues to small-medium or not so well-funded academic/research institutions (Costa, 2014), the fact that the "sequence first, think later" approach is vastly overcoming the hypothesis-driven approach in computational biology (Merelli et al., 2014), the need for an open data sharing environment, because there are incentives to restrict data distribution specially from an industry perspective, aimed at maintaining competitive advantages. The latter is a necessary prerequisite for setting up effective predictive disease models, tools, and software for the integration of large-scale, distinct data collected on whole populations (Schadt et al., 2010), their large-scale data storage, management and analysis, requiring specific computational solutions. In this regard, Schadt et al. (2010) concluded that "sequencing DNA, RNA, the epigenome, the metabolome, and the proteome from numerous cells in millions of individuals, and sequencing environmentally collected samples routinely from thousands of locations a day, will take us into the exabyte scales of data in the next 5–10 years. Integrating these data will demand unprecedented high-performance computational environments." (Schadt et al., 2010, p. 13). Also, because the application of data mining in clinical medicine is currently in its infancy, it needs further, deep research, because it deals with critical elements from both patients' and health-care professionals' perspectives: assessing disease risks, supporting clinical decisions, predicting disease developments, practical drug use guidance, strengthening medical management, benefiting evidence-based medicine (Zhang et al., 2016).

BD in health care also raises important ELSI, especially regarding informed consent, privacy, confidentiality, ownership of data, research, epistemology/objectivity, oversight about information collected "omics" database and in electronic medical record (EMR), Big Data Divides' due to economic constraints for conducting studies on increasingly larger BD, or public health issues related to digital disease detection (context sensitivity, methodology, legitimacy requirements) (Knoppers et al., 2012; Rothstein, 2015; Mittelstadt and Floridi, 2016; Vayena et al., 2015).

Mittelstadt and Floridi (2016) underlined six further areas of ethical concern needed for a wider debate: "the dangers of ignoring group-level ethical harms; the importance of epistemology in assessing the ethics of Big Data; the changing nature of fiduciary relationships that became increasingly data saturated; the need to distinguish between 'academic' and 'commercial' Big Data practices in terms of potential harm to data subjects; future problems with ownership of intellectual property generated from analysis of aggregated datasets; and the difficulty of providing meaningful access rights to individual data subjects that lack necessary resources." (Mittelstadt and Floridi, 2016).

To address all these issues, there are a few needed steps, including (1) to overcome a one-size-fits-all approach, because the specific characteristics of different BD sets and (2) to verify whether current governance approaches are adequate to handle BD, especially for managing communication with subjects involving in biobanking that utilize next-generation sequencing technologies (NGS) (Knoppers et al., 2012).

Some current uses of BBD, as revealed by a systematic analysis of the PubMed database (last access on January 21, 2018), will be presented below, after which we will analyze the particularities of BBD in BOL issues.

According to Costa, BBD should be driven by the premises of personalized medicine (PM) (2014). In fact, Vicini et al. state that, "The sheer volume of patient data generation is rapidly outgrowing the scientific and computational resources that are needed for accurate interpretation and subsequent reporting of test results to support clinical decision-making. As such, the entire current diagnostic instrumentation and processing infrastructure may need to change so that its workflow better supports." (2016, p. 198). Moreover, Chawla and Davis (2013) proposed a patient-centered approach for fulfilling a personalized health care.

Many authors analyzed the difficulties of integrating big biomedical data (Phan et al., 2012; Costa, 2014), because they are gathered, and should be analyzed on different levels (molecular, cellular, tissue, organ, system, and patient, population, society as a whole): "data heterogeneity, community standards in data acquisition, and computational complexity are big challenges for such decision making" (Phan et al., 2012). In other words, how to adequately manage and gather big health care-related data (such as the one generated by NGS), and translating them in an optimized personalized "bedside" care is a real challenge (Chute et al., 2013).

Moreover, progress in PM/Genomics are both influencing health-care delivery for the patient and the reshaping of biomedical discovery (Issa et al., 2014). Large information datasets can figure out a personalized biochemical fingerprint that could allow therapies tailored specifically on particular patients. Merelli et al. (2014) argued that bioinformatics can be "the glue" to address and win all the open issues about BD. Bibault et al. underlined the need of a better integration between the usage of big data in radiation oncology and PM (2016).

The selection and drug discovery is already influenced by large data sets (Hu et al., 2014). The real challenge here is how to actually use big data to transform the discovery processes (Chen and Butte, 2016; Leyens et al., 2017). Currently, we can harness big data for Systems Pharmacology modeling (SPM) aimed at holistically understand mechanisms of drug actions at multiple levels: genetic, molecular, cellular, organismal, and environmental, in order to support drug discovery and clinical practice (Xie et al., 2017). Since 2014, scientific literature has highlighted that "Big data" could "allow for networks-driven systems pharmacodynamics whereby drug information can be coupled to cellular- and organ-level physiology for determining whole-body outcomes. Patient '-omics' data can be integrated for ontology-based data-mining for the discovery of new biological associations and drug targets" (Issa et al., 2014).

To optimize BD analysis aimed at pharmaceutical and bioinformatics, some authors suggested a potential use of deep artificial neural networks and neuromorphic chips (Pastur-Romay et al., 2016). Moreover, BD and computational biology can be useful to better address complex chronic diseases (Ren and Krawetz, 2015). The "Omics revolution" would allow to better address topics such as human male infertility by the integration of BD in a systems biology approach, even if research is in its infancy: "The accumulation of genome sequence data from carefully phenotyped infertile men in the coming years will invariably shed new light on the complexities of male infertility, and will hopefully pave the way for improved diagnostic and treatment strategies." (Carrell et al., 2016, p. 298).

In Dermatology, there are some early contributions suggesting a potentially useful role of BD (Anbunathan and Bowcock, 2017), while in Psychiatry and Mental health research we have a real "conundrum" about the ability of BD to clarify etiology of psychiatric diseases because the complex interaction among biological and social factors. BD simply does not come with useful explanations, particularly in this field (Mewes, 2013; Stewart and Davis, 2016). Experiences have also been carried out in Obstetrics and Gynecology (Altmäe et al., 2014), in Cardiology and Telecardiology, even if several issues, such as cloud data confidentiality, data interoperability among health-care institutions, network latency and accessibility—are still to be solved (Hsieh et al., 2013; Mayer-Schönberger, 2016). BD is going to be addressed in Pathology practice (Gu and Taylor, 2014), with the need for a strong integration of different pathology data. Also in Neuroendocrine tumors (NET) we have early reports (Pyo et al., 2016), showing that BD can help for a better identification of the individual and family risk factors for these disorders. In Allergology, Belgrave et al. highlighted the need for a multidisciplinary approach for a synergic coupling between data-driven methods (BD logics) and hypothesis-driven one (clinical interpretation, understanding, and external validation of information), with the final goal of a real PM (2017).

BD was also analyzed in gastroenterology (Fabijanić and Vlahoviček, 2016), where it could be used both to classify human gut microbiome samples according to the pathological condition diagnosed in the human host, and "as a powerful analytical tool to predict for community lifestyle-specific metabolism" applying "codon usage adaptation on entire metagenomic data sets" (Fabijanić and Vlahoviček, 2016).

BD perspectives are also investigated on early contributions in radiation therapy (Guihard et al., 2017; McNutt et al., 2016), in general surgical practice as well as in plastic and reconstructive surgery (Mathias et al., 2016; Smith and Chaiet, 2017), nephrology (Sutherland et al., 2017), radiology (Aphinyanaphongs, 2017), pediatrics (Asante-Korang and Jacobs, 2016), hepatology (Wooden et al., 2017), hematology (Westin et al., 2016), microbiology (Yang and Li,

2015), otology (Lei et al., 2016), ophthalmology (Clark et al., 2016), anesthesiology (Levin et al., 2015), transfusion medicine (Pendry, 2015), neurology (Liebeskind et al., 2015).

Another promising application of BD is in public health, namely in disease surveillance systems, especially for infectious diseases, through the development of new tools based on digital and social data streams (Simonsen et al., 2016) and spatial BD (as well Lee et al., 2016).

In clinical experimentation, BD data can allow to set up new tools as, for example, the Patient Generated Health Data (PGHD), which offer opportunities to improve the efficiency and output of clinical trials, as was shown in oncological (Wood et al., 2015) or in critical care research (Docherty and Lone, 2015).

As a conclusion to this subchapter, we will quote Luo et al. stating, "bioinformatics is the primary field in which big data analytics are currently being applied, largely due to the massive volume and complexity of bioinformatics data. Big data application in bioinformatics is relatively mature, with sophisticated platforms and tools already in use to help analyze biological data, such as gene sequencing mapping tools. However, in other biomedical research fields, such as clinical informatics, medical imaging informatics, and public health informatics, there is enormous, untapped potential for big data applications." (Luo et al., 2016, p. 8).

3. BIG DATA AND BEGINNING-OF-LIFE ISSUES: A FIELD IN ITS INFANCY

The last quote of previous paragraph introduces on the state of the art regarding scientific contributions about BD and BOL in Biomedicine. The literature is still surprisingly poor, considering BOL in the first stages of a human being's existence, that is to say when oocyte activation (i.e., the mechanism that blocks polyspermy) occurs after the first sperm has penetrated the egg (Mio et al., 2012). Though we want to extend BOL stage considering neonatal time (up to 28 days after the birth) or human environmental aspects in addition with particular attention to the development of self-awareness the result is the same: the scientific literature is scarce in the area. A preliminary explanation is that "BD era" in general and in BOL time as well, is in its infancy, even if public debate is growing.

Anyway, a synthetic review of available references shows the following: Flint et al. (2017) have used BD for drug effect evaluation (doxapram, a respiratory stimulant), on the oxygenation in preterm infants. The authors showed the detailed effects over time of pharmacotherapy using high-frequency monitoring data, aimed at a better personalized therapy.

Luo et al. (2017) proposed machine learning as key technology for conducting prediction/classification using large clinical datasets aimed at improving patients' health outcomes and saving economic resources.

Van Dijk et al. analyzed the opportunities of mobile health (mHealth) in preconception care (PCC), by carrying out a qualitative study for patients who participated in a randomized controlled trial (RCT), named "Smarter Pregnancy," and professionals involved in PCC (2017). "Pros" and "cons" of BD collection by mHealth were discussed in focus group sessions. The authors conclude that: "The health care providers and professionals unanimously agreed that big data can be of great medical and scientific importance. By obtaining detailed information on target groups and populations, interventions can be designed and clinical care can be tailored at specific behaviors, needs, or risk factors of specific patient groups. Although the health care providers and professionals were aware of the perception of patients toward the use of big data, they believed that commercial use could also be beneficial in creating large-scale awareness." (Van Dijk et al., 2017).

BD can also be used to decrease neonatal death. In fact, Singh et al. (2017) showed the experience of iNICUs (integrated Neonatology Intensive Care Unit) using Beaglebone and Intel Edison based IoT integration with biomedical devices in NICU. iNICU allows millions of data points per day per child. "It acts as first Big Data hub (of both structured and unstructured data) of neonates across India offering temporal (longitudinal) data of their stay in NICU and allow clinicians in evaluating efficacy of their interventions" (Singh et al., 2017, p. 1).

Also in the NICU, Spitzer et al. (2015) showed "The Pediatrix Baby-Steps® Data Warehouse," "an electronic data capture system for the assessment of outcomes, the management of quality improvement (QI) initiatives, and the resolution of important research questions… for novel observations that could not be made otherwise." (Spitzer et al., 2015). BD was also shown to be useful in the investigation of the incidence of brain injuries of intraventricular hemorrhage (IVH) and periventricular leukomalacia (PVL), in premature infants in China (Chen et al., 2013). Homaira et al. (2016) demonstrated, through a BD linkage study, a high burden of respiratory syncytial virus (RSV) hospitalization in Australia, particularly in children aged <3 months.

Moreover, BD analysis can produce more reliable information on population trends results and live birth rates associated with common artificial reproductive technologies (ART) treatment strategies as showed by Chambers et al. (2016) in Australia. In a recent contribution a framework for BD analysis was used to study early differentiation of human pluripotent embryonic stem cells (hPSCs) toward the mesoderm and cardiac lineages (Ulfenborg et al., 2017).

The majority of scientific and ELSI issues that we listed before, both in the Introduction and in the paragraph on BD in health care, can be somewhat applied to BOL field. Therefore, further wide research is requested for addressing more specifically those issues.

4.   CONCLUSION: A WORK IN PROGRESS

The future uses of BD, including in BOL, is an open and rapidly growing universe, where not all factors can be identified in advance and we need further research, analyses, reflections. According to Chute et al.: "The safest and most promising application of big data in health care will be driven by clinical and genomic data that aregenerated with or transformed into standards-based representations, to ensure comparabilityand consistency. There is convergence in both the genomic and the clinical phenotypingworlds, driven by the application of genomics to clinical practice and Meaningful Use. Together, these promise to make good on the promise of higher quality and lower costhealth care in an information-driven industry" (2013). Also, we agree with Kranzberg who stated that: "Technology is neither good nor bad; nor is it neutral (...) technology's interaction with the social ecology is such that technical developments frequently have environmental, social, and human consequences that go far beyond the immediate purposes of the technical devices and practices themselves." (1986, p. 545).

A recent expert panel on clinical genomics in clinical routine (Delaney et al., 2016) proposed some recommendations on the matter, particularly referred to: strengthening evidence aimed at a better assessment of medical benefit/utility and cost-effectiveness, customizing regulation, promoting new payment models to match more research and development with clinical care, promoting adequate "learning" data systems, and educating all stakeholders about both strengths and weaknesses of clinical genomics.

Something like this, in the light of the potentialities and critical issues related to big data, seems undoubtedly desirable, adding that the ELSI should be more explicitly addressed both in the scientific community and in civil society, paying particular attention to an adequate, affordable social communication aimed at an appropriate public and political debate.

So, the use of BD can be meaningful not only regarding its clinical high-quality for patients, cost-saving use for health-care services, and methodological/technical reliability for BD experts, but firstly referring to the main actor in Biomedicine: whatever ill person in all stages of existence, from the very beginning of his/her individual life.

REFERENCES

Al-Rodhan, N 2014, The Social Contract 2.0. Big Data and the Need to Guarantee Privacy and Civil Liberties, viewed 17 January 2018, http://hir.harvard.edu/article/?a=7327.

Altmäe, S., Esteban, F.J., Stavreus-Evers, et al., 2014. Guidelines for the design, analysis and interpretation of 'omics' data: focus on human endometrium. Hum. Reprod. Update 20 (1), 12–28.

Anbunathan, H., Bowcock, A.M., 2017. The molecular revolution in cutaneous biology: the era of genome-wide association studies and statistical, big data, and computational topics. J. Invest. Dermatol. 137 (5), e113–e118.

Aphinyanaphongs, Y., 2017. Big data analyses in health and opportunities for research in radiology. Semin. Musculoskelet. Radiol. 21 (1), 32–36.

Asante-Korang, A., Jacobs, J.P., 2016. Big data and paediatric cardiovascular disease in the era of transparency in healthcare. Cardiol. Young 26 (8), 1597–1602.

Boyd, D., Crawford, K., 2012. Critical questions for Big Data. Inform. Commun. Soc. 15 (5), 662–679.

Bollier, D., 2010. The Promise and Peril of Big Data. The Aspen Institute, Washington, DC. https://www.emc.com/collateral/analyst-reports/10334-ar-promise-peril-of-big-data.pdf (last access 17 January 2017).

Brennan, P.F., Bakken, S., 2015. Nursing needs big data and big data needs nursing. J. Nurs. Scholarship 47 (5), 477–484.

Buchanan, E., 2017. Considering the ethics of big data research: a case of Twitter and ISIS/ISIL. PLoS One 12(12).

Burkholder, L. (Ed.), 1992. Philosophy and the Computer. Westview Press, Boulder, San Francisco, and Oxford.

Carrell, D.T., Aston, K.I., Oliva, R., et al., 2016. The "omics" of human male infertility: integrating big data in a systems biology approach. Cell Tissue Res. 363 (1), 295–312.

Chambers, G.M., Wand, H., Macaldowie, A., et al., 2016. Population trends and live birth rates associated with common ART treatment strategies. Hum. Reprod. 31 (11), 2632–2641.

Chawla, N.V., Davis, D.A., 2013. Bringing big data to personalized healthcare: a patient-centered framework. J. Gen. Internal Med. 28 (Suppl. 3), S660–S665.

Chen, B., Butte, A.J., 2016. Leveraging big data to transform target selection and drug discovery. Clin. Pharmacol. Ther. 99 (3), 285–297.

Chen, H.J., Wei, K.L., Zhou, C.L., et al., 2013. Incidence of brain injuries in premature infants with gestational age $\leq$ 34 weeks in ten urban hospitals in China. World J. Pediatr. 9 (1), 17–24.

Chute, CG, Ullman-Cullere, M, Wood, GM, & al 2013, 'Some experiences and opportunities for big data in translational research', Genet. Med., vol. 15, no. 10, pp. 802–809.

Clark, A., Ng, J.Q., Morlet, N., Semmens, J.B., 2016. Big data and ophthalmic research. Surv. Ophthalmol. 61 (4), 443–465.

Costa, F.F., 2014. Big data in biomedicine. Drug Disc. Today 19 (4), 433–440.

Delaney, S.K., Hultner, M.L., Jacob, H.J., et al., 2016. Toward clinical genomics in everyday medicine: perspectives and recommendations. Expert Rev. Mol. Diagn. 16 (5), 521–532.

Docherty, A.B., Lone, N.I., 2015. Exploiting big data for critical care research. Curr. Opin. Crit. Care 21 (5), 467–472.

Dourish, P., Bell, G., 2011. Divining a Digital Future: Mess and Mythology in Ubiquitous Computing. MIT Press, Cambridge, MA.

Fabijanić, M., Vlahoviček, K., 2016. Big data, evolution, and metagenomes: predicting disease from gut microbiota codon usage profiles. Methods Mol. Biol. 1415, 509–531.

Fairfield, J., Shtein, H., 2014. Big data, big problems: emerging issues in the ethics of data science and journalism. J. Mass Media Ethics 29 (1), 38–51.

Flint, R., Weteringen, W.V., Völler, S., et al., 2017. Big data analyses for continuous evaluation of pharmacotherapy: a proof of principle with doxapram in preterm infants. Curr. Pharmaceut. Des. https://doi.org/10.2174/1381612823666170918121556 [Epub ahead of print].

Gehrke, J., 2012. Quo vadis, data privacy? Ann. N. Y. Acad. Sci. 1260, 45–54.

Gu, J., Taylor, C.R., 2014. Practicing pathology in the era of big data and personalized medicine. Appl. Immunohistochem. Mol. Morphol. 22 (1), 1–9.

Guihard, S., Thariat, J., Clavier, J.B., 2017. Big data and their perspectives in radiation therapy. Bull. Cancer 104 (2), 147–156.

Hilbert, M., López, P., 2011. The World's technological capacity to store, communicate, and compute information. Science 332 (6025), 60–65.

Homaira, N., Oei, J.L., Mallitt, K.A., et al., 2016. High burden of RSV hospitalization in very young children: a data linkage study. Epidemiol. Infect. 144 (8), 1612–1621.

Hsieh, J.C., Li, A.H., Yang, C.C., 2013. Mobile, cloud, and big data computing: contributions, challenges, and new directions in telecardiology. Int. J. Environ. Res. Public Health 10 (11), 6131–6153.

Issa, N.T., Byers, S.W., Dakshanamurthy, S., 2014. Big data: the next frontier for innovation in therapeutics and healthcare. Expert Rev. Clin. Pharmacol. 7 (3), 293–298.

Jones, M.B., Schildhauer, M.P., Reichman, O.J., Bowers, S., 2006. The new bioinformatics: integrating ecological data from the gene to the biosphere. Annu. Rev. Ecol. Evol. Syst. 37 (1), 519–544.

Knoppers, B.M., Zawati, M.H., Kirby, E.S., 2012. Sampling populations of humans across the world: ELSI issues. Annu. Rev. Genomics Hum. Genet. 13, 395–413.

Laney, D 2001, 3D Data Management: Controlling Data Volume, Velocity and Variety, viewed 17 January 2018, https://blogs.gartner.com/doug-laney/files/2012/01/ad949-3D-Data-Management-Controlling-Data-Volume-Velocity-and-Variety.pdf

Lee, E.C., Asher, J.M., Goldlust, S., et al., 2016. Mind the scales: harnessing spatial big data for infectious disease surveillance and inference. J. Infect. Dis. 214 (suppl 4), 409–413.

Lei, G., Li, J., Shen, W., Yang, S., 2016. Three applications and the challenge of the Big Data in otology. Zhonghua Er Bi Yan Hou Tou Jing Wai Ke Za Zhi 51 (3), 230–234.

Levin, M.A., Wanderer, J.P., Ehrenfeld, J.M., 2015. Data, big data, and metadata in anesthesiology. Anesth. Analg. 121 (6), 1661–1667.

Leyens, L., Reumann, M., Malats, N., Brand, A., 2017. Use of big data for drug development and for public and personal health and care. Genet. Epidemiol. 41 (1), 51–60.

Liebeskind, D.S., Albers, G.W., Crawford, K., 2015. Imaging in StrokeNet: realizing the potential of big data. Stroke 46 (7), 2000–2006.

Luo, G., Stone, B.L., Johnson, M.D., et al., 2017. Automating construction of machine learning models with clinical big data: proposal rationale and methods. JMIR Res. Protoc. 6 (8), e175.

Luo, J., Wu, M., Gopukumar, D., Zhao, Y., 2016. Big data application in biomedical research and health care: a literature review. Biomed. Inform. Insights 8, 1–10.

Mathias, B., Lipori, G., Moldawer, L.L., Efron, P.A., 2016. Integrating "big data" into surgical practice. Surgery 159 (2), 371–374.

Mayer-Schönberger, V., 2016. Big data for cardiology: novel discovery? Eur. Heart J. 37 (12), 996–1001.

McNutt, T.R., Moore, K.L., Quon, H., 2016. Needs and Challenges for big data in Radiation Oncology. Int. J. Radiat. Oncol. Biol. Phys. 95 (3), 909–915.

Merelli, I., Pérez-Sánchez, H., Gesing, S., D'Agostino, D., 2014. Managing, analysing, and integrating big data in medical bioinformatics: open problems and future perspectives. Biomed. Res. Int. 134023. https://doi.org/10.1155/2014/134023. Epub 2014 Sep 1.

Mewes, H.W., 2013. Perspectives of a systems biology of the brain: the big data conundrum understanding psychiatric diseases. Pharmacopsychiatry 46 (Suppl 1), S2–S9.

Milton, C.L., 2017. The ethics of big data and nursing science. Nurs. Sci. Q. 30 (4), 300–302.

Mio, Y., Iwata, K., Yumoto, K., et al., 2012. Possible mechanism of polyspermy block in human oocytes observed by time-lapse cinematography. J. Assist. Reprod. Genet. 29 (9), 951–956.

Mirkes, E.M., Coats, T.J., Levesley, J., Gorban, A.N., 2016. Handling missing data in large healthcare dataset: a case study of unknown trauma outcomes. Comput. Biol. Med. 75, 203–216.

Mittelstadt, B.D., Floridi, L., 2016. The ethics of big data: current and foreseeable issues in biomedical contexts. Sci. Eng. Ethics 22 (2), 303–341.

Murdoch, T.B., Detsky, A.S., 2013. The inevitable application of big data to health care. JAMA 309, 1351–1352.

O'Donoghue, J., Herbert, J., 2012. Data management within mhealth environments: patient sensors, mobile devices, and databases. J. Data Inform. Qual. 4 (1), 1–20.

Pastur-Romay, L.A., Cedrón, F., Pazos, A., Porto-Pazos, A.B., 2016. Deep artificial neural networks and neuromorphic chips for big data analysis: pharmaceutical and bioinformatics applications. Int. J. Mol. Sci. 17 (8). pii: E1313.

Pendry, K., 2015. The use of big data in transfusion medicine. Transfus. Med. 25 (3), 129–137.

Phan, J.H., Quo, C.F., Cheng, C., Wang, M.D., 2012. Multiscale integration of -omic, imaging, and clinical data in biomedical informatics. IEEE Rev. Biomed. Eng. 2 (5), 74–87.

Phillips, K.A., Trosman, J., Kelley, R.K., et al., 2014. Genomic sequencing: assessing the health care system, policy, and big data implications. Health Aff. 33 (7), 1246–1253.

Pyo, J.H., Hong, S.N., Min, B.H., et al., 2016. Evaluation of the risk factors associated with rectal neuroendocrine tumors: a big data analytic study from a health screening center. J. Gastroenterol. 51 (12), 1112–1121.

Reinsel, D., Gantz, J., Rydning, J., 2017. Data Age 2025: The Evolution of Data to Life-Critical. Seagate.com, Framingham, MA.

Ren, G., Krawetz, R., 2015. Applying computation biology and "big data" to develop multiplex diagnostics for complex chronic diseases such as osteoarthritis. Biomarkers 20 (8), 533–539.

Rothstein, M.A., 2015. Ethical issues in big data health research: currents in contemporary bioethics. J. Law Med. Ethics 43 (2), 425–429.

Schadt, E.E., 2012. The changing privacy landscape in the era of big data. Mol. Syst. Biol. 8, 612.

Schadt, E.E., Linderman, M.D., Sorenson, J., et al., 2010. Computational solutions to large-scale data management and analysis. Nat. Rev. Genet. 11 (9), 647–657.

Shah, N.H., 2012. Survey: translational bioinformatics embraces big data. Yearbook Med Informat. 7 (1), 130–134.

Simonsen, L., Gog, J.R., Olson, D., Viboud, C., 2016. Infectious disease surveillance in the big data era: towards faster and locally relevant systems. J. Infect. Dis. 214 (S4), S380–S385.

Singh, H., Yadav, G., Mallaiah, R., et al., 2017. iNICU—integrated neonatal care unit: capturing neonatal journey in an intelligent data way. J. Med. Syst. 41 (8), 132.

Smith, A.M., Chaiet, S.R., 2017. Big data in facial plastic and reconstructive surgery: from large databases to registries. Curr. Opin. Otolaryngol. Head Neck Surg. 25 (4), 273–279.

Spitzer, A.R., Ellsbury, D., Clark, R.H., 2015. The Pediatrix BabySteps® Data Warehouse—a unique national resource for improving outcomes for neonates. Indian J. Pediatr. 82 (1), 71–79.

Stewart, R., Davis, K., 2016. 'Big data' in mental health research: current status and emerging possibilities. Soc. Psychiatry Psychiatr. Epidemiol. 51 (8), 1055–1072.

Sutherland, S.M., Goldstein, S.L., Bagshaw, S.M., 2017. Leveraging big data and electronic health records to enhance novel approaches to acute kidney injury research and care. Blood Purif. 44 (1), 68–76.

Ulfenborg, B., Karlsson, A., Riveiro, M., et al., 2017. A data analysis framework for biomedical big data: application on mesoderm differentiation of human pluripotent stem cells. PLoS One 12 (6), e0179613.

Van Dijk, M.R., Koster, M.P.H., Rosman, A.N., RPM, S.-T., 2017. Opportunities of mhealth in preconception care: preferences and experiences of patients and health care providers and other involved professionals. JMIR mHealth uHealth 5 (8), e123.

Vayena, E., Mastroianni, A., Kahn, J., 2013. Caught in the web: informed consent for online health research. Sci. Transl. Med. 5 (173), fs6.

Vayena, E., Salathé, M., Madoff, L.C., Brownstein, J.S., 2015. Ethical challenges of big data in public health. PLoS Comput. Biol. 11 (2), e1003904.

Villars, R.L., Olofson, C.W., Eastwood, M., 2011. Big Data: What It Is and Why You Should Care. International Data Corporation (IDC). https://www.idc.com/ (last access Jan 17, 2017).

Westin, G.F., Dias, A.L., Go, R.S., 2016. Exploring big data in hematological malignancies: challenges and opportunities. Curr. Hematol. Malign. Rep. 11 (4), 271–279.

Wood, W.A., Bennett, A.V., Basch, E., 2015. Emerging uses of patient generated health data in clinical research. Mol. Oncol. 9 (5), 1018–1024.

Wooden, B., Goossens, N., Hoshida, Y., Friedman, S.L., 2017. Using big data to discover diagnostics and therapeutics for gastrointestinal and liver diseases. Gastroenterology 152 (1), 53–67.

Xie, L., Draizen, E.J., Bourne, P.E., 2017. Harnessing big data for systems pharmacology. Annu. Rev. Pharmacol. Toxicol. 6 (57), 245–262.

Yang, K., Li, S.Z., 2015. Application of big data mining technology in monitoring and early-warning of schistosomiasis. Zhongguo Ji Sheng Chong Xue Yu Ji Sheng Chong Bing Za Zhi 33 (6), 461–465.

Zhang, Y., Guo, S.L., Han, L.N., Li, T.L., 2016. Application and exploration of big data mining in clinical medicine. Chin. Med. J. (Engl.) 129 (6), 731–738.

FURTHER READING

Belgrave, D., Henderson, J., Simpson, A., et al., 2017. Disaggregating asthma: big investigation versus big data. J. Allergy Clin. Immunol. 139 (2), 400–407.

Bibault, J.E., Giraud, P., Burgun, A., 2016. Big data and machine learning in radiation oncology: state of the art and future prospects. Cancer Lett. 382 (1), 110–117.

Hu, Y., Bajorath, J., 2014. Learning from 'big data': compounds and targets. Drug Discov. Today 19 (4), 357–360.

Jacofsky, D.J., 2017. The myths of 'big data' in health care. Bone Joint J. 99-B (12), 1571–1576.

Kimble, C., Milolidakis, G., 2015. Big data and business intelligence: debunking the myths. Glob. Business Organiz. Excell. 35 (1), 23–34.

Kitchin, R., 2014. Big data, new epistemologies and paradigm shifts. Big Data Soc. 1–12.

Kranzberg, M., 1986. Technology and History: Kranzberg's laws. Technol. Cult. 27 (3): 544–560.

Mayer-Schönberger, V., Cukier, K., 2013. Big Data: A Revolution That Will Transform How We Live, Work, and Think. Houghton Mifflin Harcourt, Boston, MA.

Snijders, C., Matzat, U., Reips, U.D., 2012. 'Big data': big gaps of knowledge in the field of internet. Int. J. Internet Sci. 7, 1–5.

Vicini, P., Fields, O., Lai, E., et al., 2016. Precision medicine in the age of big data: the present and future role of large-scale unbiased sequencing in drug discovery and development. Clin. Pharmacol. Ther. 99 (2), 198–207.

Chapter 8

The Moral Status of the Embryo From the Standpoint of Social Perceptions

Rafael Pardo
BBVA Foundation, Spain

1. INTRODUCTION

The dominant view at a given time of the human embryo and its moral status stands as a central cultural and symbolic construct, with implications on multiple planes, especially on the regulatory framework for its protection, demarcating what are deemed to constitute acceptable and unacceptable practices in the embryo's creation and use. How this status is defined is, accordingly, a factor of weight in the conduct of scientific research in the area, conditioning the ends pursued and imposing strong operational constraints (signally, the provenance of the embryo and developmental limits for its use in research). The advance of science, in turn, contributes to shaping the dominant social perception and, indirectly, the regulatory treatment of the embryo's condition, initially by refining the categorization of its successive development stages with their corresponding biological properties, but also by updating the biomedical benefits potentially achievable and, in so doing, redefining the trade-off between use of the human embryo and its rights.

The conceptualization of the embryo's attributes and moral status is a key formal object of sustained academic interest in the bioethics field. It is also a cause of intense public debate, which periodically resurfaces in one or other guise in the light of scientific advances in developmental biology or clinical practice, and as a response to regulatory milestones. As well as the academic community specializing in bioethics, numerous and, above all, influential social agents participate in a strongly polarized debate that at times tips over into outright confrontation through recourse to instruments of collective action. Among the most important of these agents is the scientific community itself (with its institutions and publications), churches or organized religions, the feminist movement, pro-life and pro-choice associations and movements, associations of patients suffering conditions that

Clinical Ethics at the Crossroads of Genetic and Reproductive Technologies.
https://doi.org/10.1016/B978-0-12-813764-2.00008-8

might benefit from the progress of embryonic research and, last but not least, political parties. The mass media, furthermore, give salience to the embryo as a cultural and symbolic construct at times of heightened public controversy, amplifying the debate and, frequently, taking sides with one or other view.

All these actors are both influenced by and help to shape a variable that is more latent than explicit, but no less potent for that: social perceptions of the embryo among the public at large. In other words, they engage in a two-way interaction with the cultural schemata surrounding the embryo and the values associated with its moral condition. This cultural matrix is the result of the concurrence of plural forces that differ in presence and strength from one society to another; a fact that goes some way to explaining the commonalities and differences exhibited in views of the embryo between societies. The predominant view in a given society and time conditions the degrees of freedom available to regulators, subject as they are to pressures of varying intensity and valence, to establish and enforce the legal framework for the creation and use of human embryos.

The contexts where the conceptualization of and controversy over the moral status of the embryo have gained most traction are in vitro fertilization (IVF) research and clinical practice, abortion (at times distinguishing between pre-embryo, embryo, and fetus, at others viewing them as a continuum) and, most recently, research with embryonic stem cells, created ad hoc or left over from other processes (like IVF treatments). It also bears mention that some of the facets of embryo use being argued over today overlap with the discussion initiated in the 1970s regarding fetal research, with the scientific community on one side and the public on the other or, in abstract terms, with the specific culture of the scientific community counterposed to the moral values and principles prevailing in society:

> *"Researchers stress the scientific merit of the research; nonscientists stress its moral implications [...] Although scientists tend to discuss the issues in technical terms, the controversy is rooted in basic value questions: Is the fetus fully human? What are its rights and its legal and social status?".*

(Maynard-Moody, 1992, pp. 3–4)

These same questions come up again in discussions on the moral status of the embryo.

The contexts of IVF and abortion have already given rise to widely known conceptual frames and an ample literature. But it is worth devoting a few paragraphs to the public's attitudes on these two issues, using data from the *European Mindset Study*, a survey on European beliefs and values carried out in 2009 in 12 European Union member states.[1] IVF is now accepted as a

---

1. The *European Mindset Study* was carried out in 2009 and designed by the Public Opinion and Social Studies Department at the BBVA Foundation. The fieldwork was conducted by IPSOS between November and December 2009. The sample size in each country was 1500 cases, level of confidence 95.5%, sampling error ±2.6%. The countries included were the following: Belgium, Bulgaria, Denmark, France, Germany, Greece, Italy, Poland, Portugal, Spain, Sweden, and the United Kingdom.

routine procedure in most developed countries, albeit with significant differences both between and within countries. The level of acceptance is positively associated with the educational level and, conversely, negatively associated with religiosity (the greater a person's religious sentiment and practice, the lower their acceptance of IVF), with approval levels being highest among non-believers[2] (see Table 1). Regarding abortion, the levels of acceptance are similar to those recorded for IVF, although salience and litigiousness are greater in this case due to the existence of a single-issue group, the pro-life movement. Acceptance reaches a mean value of 5.2, on a scale from 0 (totally unacceptable) to 10 (totally acceptable), for the 12 European Union countries included in the survey, with scores ranging from practically unanimous approval in the Scandinavian countries to strong rejection, notably in Poland and, less so, Italy.[3] As in the case of IVF, higher educational level is associated with a more liberal position on abortion (acceptance score of 6 points among those with university studies against 4.3 among those with primary school studies only). But the most marked differences emerged in regard to a person's declared level of religiosity, over and above the particular religion he or she belongs to. Hence for the same 12 countries, approval averaged 6.8 among those who say they are not at all or not very religious, but less than half that (just 3.2) among those at the other extreme, defining themselves as very religious. Among the countries most opposed to abortion, the difference between both groups, that is, "not at all/not very religious" vs "very religious" is close to 4 points (6.1 vs 2.4) in the case of Poland, stretching to almost 5 points (8 vs 3.4) in Italy.

The embryo as considered in the context of abortion and, especially IVF has lost most of its currency in the public opinion landscape since the turn of the century, in which time regulation of these two practices has remained, for the most part, stable. Both have been replaced in terms of level of salience and narrative by research programs using embryonic stem cells, with potential to form the hub of the future regenerative medicine. The pages that follow will focus on this more recent domain from the standpoint of public perceptions of the embryo, employing empirical evidence from the widest international survey available to date, with proven analytical power to detect patterns of perceptions

---

2. According to data from the *European Mindset Study*, in the majority of the 12 European societies included in the survey, the IVF procedure meets with average acceptance of over 5.4 (on a scale from 0 to 10), with scores highest in Bulgaria (8.0), Denmark (7.9), Sweden (7.7), Spain (6.6), and Belgium (6.3), lower but still above the halfway mark in Germany (5.8), France (5.5), and Poland (5.4), below the approval threshold in Portugal (4.8), the United Kingdom (4.5), and Italy (4.2), and firmly in the rejection zone in Greece (2.7), which stands here as an outlier in the distribution.

3. Approval of abortion as such (i.e., with no mention of special cases or circumstances) averages 5.2 on a scale from 0 (totally unacceptable) to 10 (totally acceptable) in the 12 countries referred to above. Denmark registers the highest acceptance (8.6), followed by Sweden (8.2), Belgium (6.5), France (6.3), Bulgaria (5.8), Spain (5.1), Portugal (5.1), Germany (5.0), the United Kingdom (4.9), Italy (4.7), Greece (3.7), and Poland (3.6).

TABLE 1 IVF Acceptance

Can You Tell Me Whether You Think the Things I Am Going to Read Out Are Acceptable or Unacceptable? Please Use a Scale From 0 to 10, Where 0 Means You Think It Is Totally Unacceptable and 10 Means You Think It Is Totally Acceptable. In Vitro Fertilization

		All EU Countries	Bulgaria	Denmark	Sweden	Spain	Belgium	Germany	France	Poland	Portugal	UK	Italy	Greece
Total		5.4	8.0	7.9	7.7	6.6	6.3	5.8	5.5	5.4	4.8	4.5	4.2	2.7
Sex	Male	5.4	7.7	7.8	7.3	6.5	6.1	5.9	5.3	5.5	4.8	4.4	4.3	2.7
	Female	5.4	8.2	8.0	7.9	6.8	6.3	5.7	5.7	5.3	4.8	4.5	4.1	2.6
Age finished studying	15 or under	4.7	7.2	7.1	6.7	6.2	5.6	5.4	4.9	4.3	4.2	4.0	3.9	2.4
	16–19	5.5	8.0	7.4	7.7	6.8	6.0	5.9	5.4	5.4	5.1	4.5	4.5	2.3
	20 or older	6.1	8.3	8.0	7.8	7.2	6.8	6.1	5.9	5.8	5.8	5.2	4.5	3.3
	Still studying	5.8	7.7	8.0	7.7	6.9	6.5	6.4	6.0	6.0	5.7	4.8	4.3	3.1

Religion	Nonreligious	5.9	7.6	7.9	7.8	7.1	6.3	6.1	5.9	6.8	6.1	4.8	5.2	2.6
	Catholic	5.1				6.6	6.3	5.3	5.4	5.2	4.6	4.2	4.0	
	Protestant	5.7	8.0	7.9				5.9				4.3		
	Christian unspecified	4.6	8.0	7.8									4.1	
	Orthodox Christian	5.0	8.1											2.6
Religiosity	0–2	6.2	7.8	8.2	7.9	7.3	6.7	6.1	6.3	7.3	5.9	4.6	5.8	3.4
	3–4	5.8	8.2	8.1	7.8	6.7	6.1	5.9	5.5	6.1	5.3	4.5	5.2	2.7
	5	5.6	8.0	7.8	7.7	6.6	6.4	6.2	4.9	5.9	4.7	4.5	4.6	3.2
	6–7	5.2	7.9	8.0	7.4	6.7	6.1	5.8	4.9	5.7	4.7	4.6	4.4	2.7
	8–10	4.5	7.8	6.8	6.4	5.7	5.6	4.7	5.2	4.5	4.3	4.0	3.5	2.4

European Mindset Study (2009), BBVA Foundation. Number of cases: 1500 in each of the 12 countries.

and attitudes. The standpoint adopted here counsels a brief digression to introduce the intersection of the empirical perspective in bioethics with the interdisciplinary field engaging with public perceptions of science.

2. THE EMPIRICAL PERSPECTIVE IN BIOETHICS AND PUBLIC PERCEPTIONS OF SCIENCE STUDIES

One of the main arguments justifying an interest in bioethical analysis, beyond the immediately concerned academic circle, is that for approximately the last 20 years it has been generally held that any regulation (or nonregulation) of new developments in the life sciences and biomedicine should be preceded by the corresponding bioethics report (Kuhse and Singer, 1990, p. 37). The formula of bioethics committees formally appointed by parliaments or other regulatory agencies, and charged with preparing official reports, has become the institutional mechanism of choice to guide regulators in their mission, an arrangement that some authors criticize on the grounds that it has supplanted the use of proper democratic mechanisms based on agent representativeness and accountability, and others welcome precisely because it gives a solid conceptual grounding to the work of regulators and policy-makers (Friele, 2003). As a rule, the reports these committees issue adhere to the traditions of philosophical and ethical analysis, applying them to the formal objects and issues of genetics and the life sciences, and the body of medical ethics and ethics of human experimentation that developed over time.

The large majority of bioethical analyses have paid little or no heed to any views or ethical judgments that the public might form, despite the availability of empirically grounded social science analyses of values and ethical views (Pardo and Hagen, 2016). But in the last 15 years, the empirical perspective, associated in part with the social sciences, has increasingly found a place in the ethics field, particularly in bioethics, though not without opposition from those who advocate a strict adherence to the conventional philosophical and ethical framework (Birnbacher, 1999; Molewijk et al., 2004; Sugarman, 2004; Lassen et al., 2006; Thiele, 2009).

Regardless of bioethical analysis' greater or lesser receptivity to the "datum" of the ethical values informing public preferences, attitudes and conduct, the legislators and regulators of pluralist, democratic societies undoubtedly keep close track of public opinion in sensitive areas like those that concern us here; among other reasons, because they know that regulations must be aligned with public opinion if they are to be complied with rather than widely contested or ignored. One powerful crystallization of public opinion is undoubtedly the salience and the treatment that "issues" are accorded in the mass media, but beneath this layer of published opinion lies the public's actual opinion, shaped by the intervention of forces, groups, channels, conventional and digital social networks, and personal experience, which together far outweigh the influence of conventional media. While the public's opinion can be easily swayed in matters peripheral to the individual mindset and social culture, it is resistant to change in regard to issues and objects such as the embryo that overlap with or impinge on core values.

Knowledge of this deeper, more stable stratum of "opinion" regarding sensitive issues at the science-society interface is based on the theoretical and methodological tradition known as "public perceptions and attitudes to science," built upon contributions from the social sciences, from sociology to politology by way of social psychology and the interdisciplinary field of communication and public opinion studies (Miller et al., 1980; Shapin, 1990; Wynne, 1995; Gregory and Miller, 1998; Knight, 2006; Hallman, 2017). Its interest for bioethics lies not only in the field's instrumental function in getting public views to count as an input variable for policy-makers and, increasingly, the science community and its institutions, but also because its analyses and measurements rest on a robust theoretical and methodological repertoire capable of generating results of substantive interest to bioethics. Indeed, it is in this domain, more markedly so than in the general ethics field, that a growing number of authors have championed and made room in their analyses for empirical studies of the ethical values and standards associated with scientific and technological change. This is particularly true of the life sciences, where the issues addressed include cloning, the embryo and its biomedical uses, PGD and human reproductive technologies, and, more recently, embryonic stem cells (see Bayertz, 2006).

The explosive growth in public perception of science studies since the field emerged over 30 years ago has been linked with a shift away from science at large as its central object toward the subsets of "green" (agriculture, food) and "red" (biomedicine) biotechnology; areas that are highly dynamic from a purely scientific standpoint, with a very significant economic and health-care potential, but also received by the public with unease, fear, and controversy for upsetting the perception of risk and the moral and environmental values proper to the culture of the late 20th and early 21st centuries. In contrast to the cultural narrative of progress linked to the onward march of science and technology observable from late 19th century to the end of the 1960s, with only brief interruptions (the most serious to do with nuclear weapons), the rapid advance of the life sciences and biomedicine in the closing third of the past century has met with critical or, at least, ambivalent feelings, controversies, and expressions of resistance in areas impacting on the natural environment, the moral landscape, and widely shared symbolic constructs (concerning, chiefly, the embryo, the gene, and so-called "naturalness") (Roy et al., 1991; Nelkin, 1992; Bauer, 1995; Bauer and Gaskell, 2002; Gaskell and Bauer, 2006; Atkinson et al., 2009; Siipi, 2008, 2011).

This tradition of public perceptions of science analyses tended to progress in parallel to bioethics, with little explicit contact between them. Recently, however, some cross-fertilization has been apparent, with mutual borrowings, the enlargement of their respective conceptual repertoires, and an exploratory sharing of analytical tools and results (Birnbacher, 1999, 2016; Thiele, 2009). Still, this approach open to the association of bioethics and public perceptions of (bio) science studies tends to operate under an implicit division of labor between the two, which consigns the latter to a rather limited role: the description of people's values and moral standards, that is, the provision of "raw" data for bioethical analysis. But social science studies of values and perceptions of science not only describe (measure) people's ethical mindset, they also analyze its formal properties such as depth and structure (scope, interrelationships between views on different facets of an area or issue, degree of consistency or inconsistency among values, and ethical criteria) and its variability in the population (showing areas of consensus and of disagreement and polarization) and, frequently, offer explanatory schemas on its underlying whys or "reasons" (Pardo and Hagen, 2016, pp. 130–131).

Some leading bioethical specialists have observed that although the ethical angle is essential to regulation and the latter, as we have said, generally has to wait upon a bioethical analysis, the public's understanding of the nature of ethical reasoning is seriously flawed. On the one hand, according to those authors, ethics tend to be seen as equating to a purely subjective opinion, such that all ethical standards and judgments would have a similar or rationally indiscernible validity. On the other, they postulate that it is widely held that ethical answers are laid down clearly and for all time in each religion's sacred texts (Kuhse and Singer, 1990, p. 37), relieving believers of any effort beyond applying the proper hermeneutics, usually provided by the institutional apparatus of churches. Having likewise rejected the view that ethical standards are somehow anchored in or derived from "human nature," along the lines posited by sociobiology, these bioethicists propose an approach that puts the accent on the "standards" of reasoning, i.e., cognitive or discursive elements required for rational debate: respect for "relevant facts"—especially the scientific facts of a field or object affecting ethical criteria—and the principle of logical consistency in the chain of argument (Kuhse and Singer, 1990, 39). An additional criterion, no less demanding than the previous two, is that "an ethical judgment must be impartial or (...) universal," meaning that it must start from a perspective that declines to assign greater weight to the person's own interests and attitudes than those of other agents (Kuhse and Singer, 1990, p. 39).

From an analytical standpoint tested against empirical evidence, the first point to be made about the public's supposed vision of morality is that it is by no means clear that a majority adhere to the "ethics = subjective judgment" position, or would agree that "ethics = doctrine set out in the sacred texts" of the corresponding religious belief. There is evidence, indeed, that a large subset of society believes that ethical standards or principles are not individual or merely subjective but, rather, communal or shared, de facto or, at least, as a desideratum by those who live within the framework of a given culture and society.

According to data from a wide-ranging empirical study into values in Europe,[4] a large majority believe that "there are clear principles about what is right and what is wrong," with 72% concurring against the 25% affirming that no such principles exist (Fig. 1). At the other extreme, that of the subset of the public who regard ethics as flowing directly from church doctrines, this opinion is primarily ascribable—with qualifications—to the section of the population that recurrently "activates" the religious doctrinal frame through their customary exposure to and participation in religious ceremonies and rites. For the majority of the population of secular democratic societies, including a significant percentage of less committed believers who participate only occasionally in religious rites, there are considerable degrees of freedom between the corresponding doctrinal teachings and ethical criteria. In 8 of the 10 countries included in the just-mentioned survey, a clear majority do not believe there is a

Which of the following two opinions do you tend to agree with more?

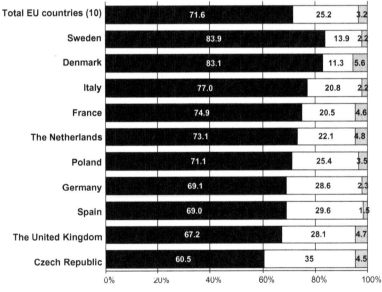

FIG. 1 Beliefs about the existence of clear ethical principles. *(Values and Worldviews (2013), BBVA Foundation. Number of cases: 1500 in each of the 10 countries.)*

4. The *Values and Worldviews Study* was designed by the BBVA Foundation and the fieldwork carried out by IPSOS from November 2012 through January 2013 in 10 countries: Germany, Denmark, Spain, France, Italy, the Netherlands, Poland, the United Kingdom, Sweden, and the Czech Republic. Sample size in each country: 1500; sampling error ±2.6%.

necessary relationship between religious and ethical beliefs: on a scale running from 0 to 10, the statement "it is necessary to believe in a religion to have values and act in an ethical way" meets with below-average agreement in six countries (Denmark 2.4; France 2.7; Sweden 2.9; Spain 3.2; Czech Republic 3.3; Netherlands 3.3; the United Kingdom 3.6; Germany 4.7). Only in Italy and, above all, Poland, two countries where Catholicism is widely embraced and practiced, do a majority believe in this link between religiosity and ethics (Fig. 2).

Another relevant facet of the general framework of social perceptions about the relationships between ethics and religion is how each is perceived as interacting with science. According to data from a systematic examination of society's scientific culture,[5] the dominant view in the majority of the 10 European countries surveyed was that "ethics should set limits on scientific advances," with support for this idea especially pronounced in Denmark, Austria, Germany, the Czech Republic, and France. A majority were also in agreement, though with more division of opinion, in Poland, the United Kingdom, Italy, and the United States. Citizens of the Netherlands and Spain were the least inclined to believe that ethics should set limits on the advances

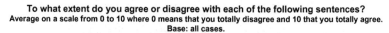

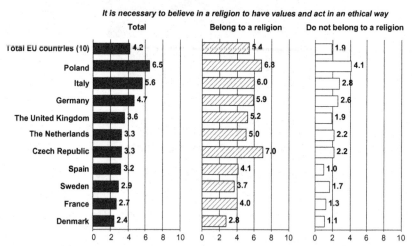

FIG. 2   Ethical principles and religious beliefs. *(Values and Worldviews (2013), BBVA Foundation. Number of cases: 1500 in each of the 10 countries.)*

---

5. The *Scientific Culture Study* was designed by the BBVA Foundation and the fieldwork carried out by TNS-Opinion from October through November 2011 in 10 European Union countries (Germany, Austria, Denmark, Spain, France, Italy, the Netherlands, Poland, the United Kingdom, and the Czech Republic) and the United States. Sample size in each country: 1,500; sampling error ±2.6%.

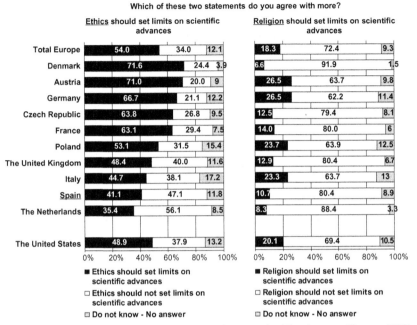

Which of these two statements do you agree with more?

	Ethics should set limits on scientific advances				Religion should set limits on scientific advances		
Total Europe	54.0	34.0	12.1		18.3	72.4	9.3
Denmark	71.6	24.4	3.9		6.6	91.9	1.5
Austria	71.0	20.0	9		26.5	63.7	9.8
Germany	66.7	21.1	12.2		26.5	62.2	11.4
Czech Republic	63.8	26.8	9.5		12.5	79.4	8.1
France	63.1	29.4	7.5		14.0	80.0	6
Poland	53.1	31.5	15.4		23.7	63.9	12.5
The United Kingdom	48.4	40.0	11.6		12.9	80.4	6.7
Italy	44.7	38.1	17.2		23.3	63.7	13
Spain	41.1	47.1	11.8		10.7	80.4	8.9
The Netherlands	35.4	56.1	8.5		8.3	88.4	3.3
The United States	48.9	37.9	13.2		20.1	69.4	10.5

- ■ Ethics should set limits on scientific advances
- □ Ethics should not set limits on scientific advances
- □ Do not know - No answer

- ■ Religion should set limits on scientific advances
- □ Religion should not set limits on scientific advances
- □ Do not know - No answer

FIG. 3 (A and B) Ethics and religion as constraints on scientific advances. *(Source: BBVA Foundation International Study on Scientific Culture (2011). Number of cases: 1500 in each of the 11 countries.)*

of science (see Fig. 3A). The relations between religion and science are viewed very differently: in all the countries surveyed the prevailing view was that religion should not impose limits on science, with agreement reaching almost 90% in Denmark and the Netherlands, and more moderate levels, bordering on 60%, in Poland, Italy, Germany, and Austria (see Fig. 3B).

This is not to say that religious beliefs have run out of influence; simply that their sway is weaker and, above all, indirect and obliged to compete with other traditions and institutions—science, among them—of the social culture of the time. Their role in the formation and content of central beliefs and attitudes has not disappeared entirely, even in mainly secular societies, but plays out along a lengthy path running through the general culture of society.

While acknowledging the importance of abiding by the criteria of good moral reasoning in any "enlightened" debate, advocated for by a number of influential bioethicists, it must be said that the first requisite—to know and respect the facts or findings established by scientific research—is a tough proposition for the public at large in a domain like the life sciences, with the high demands it places on scientific literacy. The abundant empirical evidence of more than two decades (chiefly, the several surveys of the European Commission known as Eurobarometers, and similar ones carried out in the United

States; see George Gaskell et al., 2003)[6] shows that the majority of the population in socioeconomically advanced societies fail to reach even the basic knowledge threshold in biology and genetics, relying instead to make sense of complex issues involving scientific knowledge on other types of conceptual "devices" such as schemas, frames, and heuristics of mixed origins supplied by the overarching culture. As regards the principle of maintaining a logical consistency in statements of an ethical nature, it presupposes a cognitive mode of operation seldom observed in people's day-to-day thought processes as they grapple with a multifaceted object or issue, which tend to follow a pattern bearing little resemblance to the proof of a theorem or formal reconstruction of a theory. Cognitive science and social science studies show that, when confronting complex evaluative issues, from political to moral, most individuals, except the minority operating with a tightly integrated ideological structure—usually belonging to organizations with an explicit culture or credo—(see Philip E. Converse's influential 1992 contribution) do not start from a fundamental ethical or normative principle and proceed from there in a deductive reasoning mode, but tend to apply fuzzy holistic templates derived from worldviews of disparate origin (some inspired by science, others by religion or popular culture) and more bounded frames, sensitive to the particularities and salient attributes of the issue at hand (see Gigerenzer et al., 1999; Gigerenzer, 2008). And certainly, public opinion formation and change do not progress according to an elegant and formally consistent train of thought (Zaller, 1992). What from a bioethical perspective may seem to be inconsistent values held by an individual, from a social science angle may be understood as the result of individuals, for varied reasons, giving different weights to concurrent heterogeneous frames and worldviews when evaluating a specific situation; distinct from those they would apply in another context with different salient attributes (Pardo, 2009, p. 129). Further, in some contexts a degree of inconsistency is not necessarily a bad thing in evaluative and ethical terms, as argued by Polish philosopher Leszek Kolakowski:

> *"[...] the race of inconsistent people continues to be one of the greatest sources of hope that possibly the human species will somehow manage to survive. [...] Total consistency is tantamount in practice to fanaticism; while inconsistency is the source of tolerance". The totally consistent individual is defined thus by Kolakowski: "He who is in possession of a series of non-contradictory universal principles and strives to abide by them strictly in all he does and in all his opinions about what is right".*

(Kolakowski, 1967, p. 214)

---

6. According to the Eurobarometer data, the mean score in biology knowledge among the European public in 15 EU member countries, measured on an scale from 0 to 9, was 4.77 in 1996, 4.78 in 1999, and 4.93 in 2002, a very modest gain in a 6-year period (George Gaskell et al., "Europeans and Biotechnology in 2002. A report to the EC Directorate General for Research from the project 'Life Sciences in European Society'", March 2003, 22, available at http://ec.europa.eu/commfrontoffice/publicopinion/archives/ebs/ebs_177_en.pdf).

In the universe of lived values, as opposed to the domain of scientific theory, a degree of flexibility, fragmentation, and looseness between beliefs and values is not only a facet of observed conduct, but also a potentially desirable one at that. This vision in fact corresponds with the general mindset in some advanced societies, where a majority believe that "Ethical principles should be applied in accordance with the circumstances of the moment." This is the case in Sweden (74%), the Netherlands (73%), France (68%), and Denmark (66%) and, by a smaller margin, in the United Kingdom (54%). Conversely, in countries where the Catholic Church is in a hegemonic position or enjoys a significant role in the continuous activation of ethical values and criteria following from that faith, people believe that "Ethical principles should be applied regardless of the circumstances of the moment": a statement meeting with widespread agreement in Poland (67%) and Italy (66%), and more limited support in Spain (56%) and Germany (51%) (see Fig. 4). These general differences are reflected in part in the dissimilar patterns existing between countries in connection with specific ethical and bioethical issues, among them the moral status of the embryo.

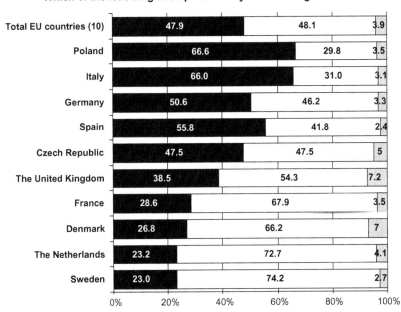

Which of the following two opinions do you tend to agree with more?

Country	Regardless	In accordance	Do not know
Total EU countries (10)	47.9	48.1	3.9
Poland	66.6	29.8	3.5
Italy	66.0	31.0	3.1
Germany	50.6	46.2	3.3
Spain	55.8	41.8	2.4
Czech Republic	47.5	47.5	5
The United Kingdom	38.5	54.3	7.2
France	28.6	67.9	3.5
Denmark	26.8	66.2	7
The Netherlands	23.2	72.7	4.1
Sweden	23.0	74.2	2.7

■ Ethical principles should be applied regardless of the circumstances of the moment
☐ Ethical principles should be applied in accordance with the circumstances of the moment
☐ Do not know – No answer

FIG. 4 Applicability of ethical principles. *(Values and Worldviews (2013), BBVA Foundation.* Number of cases: *1500 in each of the 10 countries.)*

Studies of public perceptions and attitudes to biomedical science and bio-technology show not so much the prevalence of universal viewpoints about central objects and issues posed by those life sciences, as views and attitudes that exhibit large variability within and between societies. This is not only relevant in order to understand the factors and sources of moral concern accounting for that variability, but also informs about the level of consensus and dissensus in each society and across regions, and about the opportunities for rational debate and regulation in pluralistic societies, as well as the diverse obstacles they must face.

In sum, public perceptions of life science studies offer descriptive but also explanatory models, complementing conventional approaches in bioethics, to explore the varied facets of the evaluation of scientific developments that impinge on values and symbolic constructs operating in the social mindscape. These analyses are certainly not going to resolve the key question of the indispensable "ought to" in the bioethics domain, but they may shed useful light on the "is," that is, the ethical values and their interaction with technologies, a component of the "impact value assessment in applied ethics" perspective championed by Ruth Chadwick:

> *"Value impact assessment would be an interdisciplinary activity on three levels. The first level would involve empirical research on social values and attitudes concerning new technologies and their applications".*

(Chadwick, 2000, p. XXI)

In turn, these social science studies benefit from the concepts and explanatory schemas constructed by bioethics. One such bioethical concept is precisely that of the moral status of the embryo, discussed here from the standpoint of public understanding and perceptions.

3.   THE NOTION OF MORAL STATUS

Moral status is a foundational concept in bioethics, in the sense that the attribution or otherwise of this condition to a given entity and, as the case may be, the degree of such attribution, has major consequences for the treatment it is accorded. One systematic contribution on this notion characterizes it elegantly as follows:

> *"To have moral status is to be morally considerable, or to have moral standing. It is to be an entity towards which moral agents have, or can have moral obligations. If an entity has moral status, then we may not treat it in just any way we please; we are morally obliged to give weight in our deliberation to its needs, interests, or well-being".*

(Warren, 1997, p. 3)

Intuitively most people nowadays—it certainly was not always so—are readily disposed to attribute a moral status to individual humans, regardless of age, race, or gender, precisely because they belong to a group (human beings) with certain acknowledged properties. The borderline appears when, for diverse reasons, an individual's mental capacity is seriously and irreversibly impaired (this capacity being generally considered an essential attribute for an individual's full inclusion in the human group). Still, the large majority of society would recognize individual as a human being whose rights should be preserved. There is less social and doctrinal consensus regarding the moral status of animals, including "higher" nonhuman animals, and a lot less still about the objects belonging to the set of inanimate entities lacking sentience, present throughout nature. Even so, some philosophers and other specialists (ethologists, animal welfare experts, bioethicists) have launched a formidable conceptual battle, paralleled and reinforced by collective action in the case of animal welfare and animal rights movements, to have this moral status extended fully or in part to nonhuman animals. Further, philosophers seeking to provide a conceptual foundation for nature preservation, as well as some conservationist movements, reject out of hand the notion of moral status, which they condemn as anthropocentric, postulating in its stead an ethics anchored on a "biocentric" or even "ecocentric" vision, giving an abstract and systematic foundation to the beliefs and feelings of cultures whose view of nature differs from that prevailing in the Western world. Such cultures, historically less indebted than the West to rationality and scientific information, readily ascribe a sacred nature to places or objects that the scientific gaze and the social mindscape it has shaped would immediately class as inanimate. Indeed they regard these objects as, at least, endowed with spiritual life and therefore deserving of a special respect, akin to the attribution of moral status (for this non-Western cultural vision, see, for instance, Bruun and Kalland, 1996).

The ontologies and classifications used to characterize the types of observable entities (human beings, nonhuman animals, inanimate natural entities, machines) have varied over time, with the dividing lines between them significantly redrawn, at times in a piecemeal or reformist fashion, at others holistically or revolutionarily, joining or aggregating into a single set groups previously seen as separate or, conversely, splitting some of their number into other, differentiated sets, perhaps with some overlap between them, or treating them as disjunct entities on the grounds that they have no fundamental properties in common (see Sheehan and Sosna, 1991). This shifting "ontology," a product of both formally arrived at conceptualizations, perceptions embedded in the culture of a society and time, and also, collective action and regulatory decisions, has been accompanied by disparate ethical views about the moral obligations of the human group toward all other classes of entities or individuals. Ontology and ethical views tend to be in flux and strongly correlated, influencing each other.

If this ontology, its boundaries and interactions are subject to reconceptualization and controversy, the debate becomes even more pointed when it involves the inclusion or otherwise of the few-day-old human embryo in the set of individual human beings and, where applicable, the extent to which it can be considered to have the attributes of an adult human. In turn, the vision held of the moral status of the embryo does much to shape perceptions of the legitimacy of its creation and uses, increasing or reducing the degrees of freedom permitted in the regulation of the corresponding basic research and clinical domain. Hence the intense and highly polarized academic and social debate raging since the 1970s on the subject of the embryo and the criteriology that assigns it to the human group or, alternatively, sets it fully or partially apart.

This modeling of the view of the embryo refers back to a large body of philosophical thought, religious beliefs, theological studies, and scientific theory, as well as incorporating elements from the ideology of more or less recent social movements, including the feminist movement, pro-choice and pro-life organizations, and patient associations. Of all the social actors involved, beyond the professional community of bioethicists, the two wielding the greatest doctrinal influence in the matter are, without doubt, the churches (especially the Catholic Church) and scientific institutions. Given the enormous diversity of traditions shaping the key theoretical constructs, styles of thought, and modus operandi of these multiple agents, the quest to find an analytical and parsimonious space for public discussion and eventual social consensus on the issue is fraught with major difficulties, starting with the communication problems between the scientific community and members of the public, who tend to understand and use key terms in radically different ways than scientists (Illies, 2016, p. 115), the more so in the case of terms of a highly symbolic character, deeply embedded in the culture of a given society and period.

4. TWO CONTRASTING NARRATIVES ON THE MORAL STATUS OF THE EMBRYO

Unlike the case of abortion, where analysis and debate about its regulation has had as one of its main axes the status of the embryo and the fetus versus women's rights (see Condit, 1990; Marx Ferree et al., 2002, on the cultural shaping of public discourse on abortion in the contrasting cases of Germany and the United States),[7] in the case of the few-day-old embryo, the discussion has centered on the biological nature of the embryo as defined by science vs the vision active in the culture and social mindscape. And, as a corollary of this characterization, on the existence or absence of a moral status; in effect counterposing the moral condition of the embryo to the scientific community's right to do research and, particularly, the rights of society at large and especially patients to the potentially attainable benefits in biomedical knowledge and their translation to the clinic.

This instrumental, trade-off, perspective "activates" an attitude of rejection in the public opinion sphere and each individual's moral conscience toward

---

7. One of the most significant frames on abortion has been the characterization of "fetal life," a frame that could take the values "*pro*-abortion" (associated to the belief that "life begins at birth"), "*anti*-abortion" (supported by the belief that "life begins at conception"), and *neutral* ("the real issue is when life begins") (see Marx Ferree et al., 2002, p. 106 and following).

practices that recall highly salient cases of the use of humans in biomedical experiments without their consent, whether the goals pursued were legitimate or otherwise. Obviously, many of the most brutal experiments of this nature could offer no such justification, as well as selecting most of their subjects from society's most vulnerable groups, and these cases have had a major impact in shaping society's moral awareness about the compulsory enrollment of human subjects in biomedical research (see Rubenfeld and Benedict, 2014; Resnik, 2018). Even if these highly publicized and widely condemned cases have no resemblance to current research involving early human embryos, their "ripple effects" (indirect costs of a behavior or practice affecting parties "minimally related to the initial event," see Kasperson et al., 1988) are unavoidable. If the embryo is perceived as a human individual, or more fuzzily, as a living being in the human class, its use in research is considered an illegitimate and totally unacceptable practice, even though this could mean giving up potentially valuable and uncontroversial goals—medical benefits for many people. The general rejection of research with human subjects without their consent becomes particularly intense if early human embryos are believed to be "not mere biological tissues or clusters of cells," but "the tiniest of human beings," to use the characterization of the Center for Bioethics and Human Dignity. The critical step from which highly significant consequences follow is then how to characterize a human embryo.

Mary Warnock has revisited the moral status of the embryo issue many times and contributed to providing the conceptual underpinning for the United Kingdom's pioneering regulation of human embryology and fertilization. Her perspective, as such, is of particular relevance here. From her applied ethics standpoint, Warnock states that in cases like the use of human embryos, which are matters of public morality, i.e., moral issues that society at large is called to decide on through its representatives, the standard practice in pluralistic societies is to apply an instrumental or "consequentialist" approach: an assessment of the cost/benefit trade-off of a given regulatory treatment, or, more specifically, who and how many will benefit set against those that will suffer the undesired effects. But this approach falls down in the presence of a human biological entity that must inevitably be destroyed in the exchange; an unacceptable outcome from the point of view of those who straightforwardly equate a human embryo to a human individual. This precisely was the stumbling block encountered when drawing up the recommendations of the influential *Warnock Report on Human Fertilisation & Embryology* (Warnock, 1985), prepared in the wake of the boom in embryological research that followed the proof of the effectiveness of IVF treatments:

> *"The central question, What status should be accorded to the early embryo?, had to be settled by different means. And the only means available was the degree to which people felt that the early embryo was in fact, to all intents and purposes, a child".*
> (Warnock, 1998a, p. 49)

The reference to people's feelings is not just rhetorical, but a very significant component in Warnock's way of thinking about the regulation of important public morality issues, as she clarifies here:

> *"The ethical issue that arose from the use of embryos was at once seen to be a matter of public rather than private morality. [...] Any new law would have to reflect, as far as possible, the moral views not merely of Parliament but of the country as a whole".*

(Warnock, 1998b, p. 392)

To determine the degree of correspondence between the view of the legislator and that of society as a whole, centrally involves taking as an input to bioethical analysis the public's perceptions of the embryo, measured and analyzed with the help of the interdisciplinary field known as public perceptions of science studies, devoted to understanding the areas of culture shaped by science-society interactions.

The approach taken by Warnock coincides with the arguments of Kuhse and Singer, mentioned previously, as regards the imperative or, more leniently, the desirability of the public's ethical judgment being anchored on a proper grasp of the strictly factual basis of the moral issue in question as defined by state-of-the art scientific knowledge:

> *"No one can make a sound moral judgment without knowledge of the facts about which they are to judge".*

(Warnock, 1998a, p. 46)

But this view by no means postulates that moral divergences about the human embryo can be resolved by sole reference to the knowledge provided by embryology, which the public would consequently have to acquire and accept as the true and only vision. Knowledge of the embryo's development process gleaned from the results of biological and biomedical research is undoubtedly a variable that plays or could play a role in shaping the view held of its status (see Dawson, 1990; Trounson, 1990). Some authors have stressed the importance of the scientific community making a decided effort to equip society with the terms needed to describe biological entities like the embryo, which would mean educating the public in basic biological notions, particularly the distinction between the blastula and gastrula:

> *"If the public decides to protect embryos, we had better know whether we mean by that fertilized eggs, blastocysts, gastrulas, or only post-implantation stages".*

(Maienschein, 2006, p. 51)

Yet this scientific enlightenment of the public is by no means a determining factor, because the embryo is not just a formal object in the science domain, but also a cultural construct derived from other narratives, particularly the religious matrix, which rather than operating, as science does, in a minimalist, analytical and evaluative neutral manner, works holistically, assigning to it properties and values that elude empirical falsification in Popper's terms. It is, in short, a belief-mediated construct, with a powerful symbolic dimension and emotions attached to it, drawing on widely dissimilar doctrinal foundations, rather than a stylized concept forming part of a theory and open to intersubjective operationalization and proof. If we add to this the public's scant knowledge of biology and genetics, it becomes readily apparent that a wide subset of society holds a view of the embryo that owes more to society's culture, influenced in turn by religious doctrine and associated beliefs, than to the scientific narrative provided by developmental biology. Relying on the methodology of qualitative interviews conducted in the United States, two sociologists specialized in the study of religion in public life have shown that believers, when assessing the legitimacy of using embryos for biomedical research, apply two conceptual templates: the creation narrative (God as the only creator, humans should abstain from intervening in life domains that would implying a "playing God" attitude) and/or the sacredness of human life (according to the teachings of the Church or religious authority). The result of these two correlated perspectives is that "protecting embryos takes precedence over potential cures that could save human lives" (Howard Ecklund and Scheitle, 2018, p. 127).

This confluence and clash of perspectives regarding the human embryo came to light during the research and subsequent translation to the clinic of IVF treatments in the period running from 1978, when the first child was born using the technique, to the end of the 1980s, and reemerged in 1998 with the successful creation of human embryonic stem cells. Michael Mulkay has offered a rich and useful account of how the IVF technique was received in the United Kingdom, some of whose conclusions are of particular relevance here (Mulkay, 1997).

The two main actors in the social debate, as played out in Parliament, were religious faiths and the scientific community. Then as now, at the heart of the discussion was the issue of

"when the developing human embryo becomes an individual and, accordingly, when the moral principle requiring protection of the individual comes into effect".
(Mulkay, 1997, p. 101)

The scientific community deployed a persuasive narrative based not just on specific knowledge about the development of the embryo, but also on general stereotypes of high rhetorical value, whereby "reason" or scientific rationality was opposed to "faith" or religious dogma, "science-based facts" to "traditional beliefs" (Mulkay, 1997, p. 104). The vision of the human embryo conveyed by the research community was then—and still is—that of a biological entity in gradual development. According to this view, the moment of fertilization (actually a process, not a moment) has no particular significance that might grant that "cluster of cells" the status of a human individual, with the respect due to that condition. This biological view of the embryo has since been supplemented by a second, completely different line of argument wielded also by prominent scientists and organizations; that of the "immorality" of halting research that could produce medical benefits for many people:

> *"[An analysis of] the rights and demands of the sick, makes one wonder if one group in society ever has the moral authority to impose its religiously based beliefs on other groups in society".*

(Ruse and Pynes, 2006, p. 225)

The result is that for four decades, the minimalist, gradualist view of the few-day-old embryo as a mere cluster of cells, allied with the promise of a vast range of potentially attainable clinical benefits and a homogeneous public voice articulated by scientific institutions have posed a formidable challenge to the other main actor competing to win public opinion and the regulator to a radically different vision of the human embryo. On the whole, the scientific community has had more success in shaping the embryo frame used by political elites than in changing the more general perceptions and culture of the public at large.

Not all religious faiths think alike on this issue, and while they may exhibit some commonalities such as style of thinking or conceptualizing, what stands out most are the marked differences in how the human embryo is viewed. This relative disunity has been a limiting factor as regards their influence on legislators. The Catholic Church has long upheld a doctrinal tradition of total equivalence between the embryo and the human individual, both of which it considers to be endowed with a soul from the moment of the union of egg and sperm; a line of theological thought whose main elements some trace back to early Christian theologians like Tertullian[8] and Thomas Aquinas or the school of Scholasticism, but not to Augustine: "Augustine remained uncertain whether a human soul was present in a fetus that dies in its mother's womb, whereas Thomas Aquinas asserted that a soul is present at an unspecified point following conception" (Waters, 2003, p. 67). Conversely, Umberto Eco believes that, for Aquinas, God does not endow the fetus with a rational soul until it is a formed body, and Giovanni Sartori takes up this same argument, affirming that the thesis that "the soul appears with the embryo" [...] "was never the doctrine of the Catholic Church" until the last century (Sartori, 2016, pp. 86–87). Whether or not the stance taken by the Catholic Church refers back to theological sources like those just mentioned, most commentators locate firmly in the 20th century the view unequivocally stated in encyclicals and other doctrinal texts to the effect that:

---

8. Tertullian (c.155/160-c.220) argued that "the embryo, after conception, has a soul, and it is man (*homo*) when it attains its final form. [...] The prevailing Christian understanding (...) seems to have followed the Septuagint in distinguishing between an unformed and formed stage. [...] It is abundantly clear from these discussions [among Christians, Augustine and others] that the most anyone contends is that ensoulment occurs at conception; the dominant view is that the fetus becomes a man only when 'formed'" (Noonan Jr., 1986, p. 90).

"The use of human embryos or fetuses as an object of experimentation constitutes a crime against their dignity as human beings who have a right to the same respect owed to a child once born, just as to every person" (Ioannes Paulus II, 1995, p. 50). "The number of embryos produced [in IVF treatments] is often greater than that needed for implantation in the woman's womb, and these so-called 'spare embryos' are then destroyed or used for research which, under the pretext of scientific or medical progress, in fact reduces human life to the level of simple 'biological material' to be freely disposed of" (Ioannes Paulus II, 1995, p. 11).
"Nothing and no one can in any way permit the killing of an innocent human being, whether a fetus or an embryo, an infant or an adult, an old person, or one suffering from an incurable disease, or a person who is dying".
(Congregation for the Doctrine of the Faith, 1980, p. 546)

Protestant traditions, meanwhile, do not have an officially enshrined, univocal doctrine in the manner of the Catholic Church, but are more pluralistic in their approach. They nonetheless tend to converge in a position of rejection toward the instrumental use of human embryos, favoring in its stead the reprogramming of adult tissues:

"An ecumenical group of Protestant and Orthodox (but not Roman Catholic) thinkers submitted a position paper [...] entitled 'Therapeutic Uses of Cloning and Embryonic Stem Cells' [...] The ecumenical statement ultimately rejects the use of human embryos as means to ends, even the lofty end of promoting human health. The creation of embryos through nuclear transfer is quite clearly condemned [...]".
(Walters, 2006, p. 277)

At the other extreme, the Jewish tradition accords no special moral status to an embryo formed in a Petri dish:

"Genetic materials outside the uterus have no legal status in Jewish law, for they are not part of a human being until implanted in a woman's womb" (Dorff, 2001, p. 91); "In the Jewish tradition, moral status is not ascribed to the human embryo at the time of fertilization".
(Walters, 2006, p. 267)

Although in the case of Islam there is no single doctrinal voice but rather a set of opinions or fatwas, most Muslim commentators, according to Walters, concur that it makes no sense to talk about moral status in the earlier stages of embryonic or even fetal development, until 4 months after conception; a vision consistent with the absence from the Koran of a criterion like the "ensoulment" of the Christian tradition. There should therefore be no principled objection to using embryos for biomedical research purposes (Walters, 2006, pp. 268–269). The position derived from the Buddhist tradition tends toward ambivalence: favorable toward the biomedical ends or purposes of embryo research—finding treatments to relieve suffering, but hostile inasmuch as an axiom of the Buddhist worldview is to refrain from harming or killing any being. Hinduism, in contrast, rejects such research on the grounds that it involves the destruction of the embryo, which this religion accords protection from the very moment of conception (Walters, 2006, pp. 270–271).

The advance of science and religious traditions, through their corresponding institutions and conceptual frameworks, shapes the culture of a society at any given time, influencing in turn the perceptions, views, and attitudes of the public. Other actors, such as pro-life and pro-choice movements and patients associations, also contribute their share to molding the vision of the human embryo, but unlike in the cases of abortion and IVF, their voice and influence have so far been fairly limited regarding the creation and use of embryonic stem cells.

The pages that follow offer a profile of the public's stated views across a broad range of countries on the moral status of the embryo and the legitimacy of its possible use to advance biomedical research, furnishing the corresponding empirical evidence.

5.   THE STATUS OF THE EMBRYO FRAME AND THE BEGINNING OF INDIVIDUAL HUMAN LIFE

In the past 15 years, the scientific community has reached out directly through the mass media to policy-makers and public opinion to convey to them the huge biomedical potential of embryonic stem cells, at the root of what is known as regenerative medicine, which offer the promise of new and powerful therapeutic tools for the treatment of some of the most devastating conditions prevalent in advanced societies. As has occurred with other areas of science and technology since the start of the 21st century, the hype generated around this potential translation from basic knowledge to the clinic has attained formidable proportions, before the technique is even out of the laboratory. The majority of people, as discussed earlier, have very limited knowledge of biology and genetics, but have appropriated a narrative of promise anchored on the proven instrumental value and cultural power of science. This appreciation of science is less enthusiastic than a little over 50 years ago, but significantly higher and more universal than that of expectations and worldviews of a very different origin and nature (among them, political ideologies, and religious beliefs). These promises of science—biomedical science in the case that concerns us here—are endorsed by the prestige and trust that society accords to the scientific and medical community, far higher than that vested in any other professional group, with trust understood in the dual sense of, first, an expectation of fair play and regard for the public interest and, secondly, confidence in practitioners' knowledge-based skills and capacity to deliver.

Judging by these two factors, i.e., the beneficial nature of the goals pursued and trust in the main agent of their delivery (scientists and clinicians), one might expect wide social acceptance of embryonic stem cell research, with very little dissent. But, as occurs with other emerging areas of science (Pardo et al., 2009), the field has collided in its development with deeply entrenched cultural reservations about the means to be employed; reservations linked in this case to social perceptions of the human embryo, shaped in fundamental measure by certain religious faiths and their institutions. The scientific narrative (minimalist, analytical, axiologically neutral) clashes head-on with the religious narrative (holistic, belief-based, with strong values and emotions attached) in the cultural projection and activation of two strongly differentiated "frames" regarding the status of the human embryo, impairing their comparability and impeding any fruitful debate.[9]

Frame or *schema* is a central construct in artificial intelligence and cognitive psychology to capture the way individuals organize their information about the world. A frame does not represent an object as a chain of propositions, logically connected, but as a chunk or group of properties that may include cognitive and evaluative elements. When the image of a particular object is recalled from memory, a whole cluster of its most characteristic or salient attributes ("terminals") are immediately retrieved (see Minsky, 1975) and these "terminals" to use Minsky's term act "as connection points to which we can attach other kinds of information. […]. As soon as you hear a word like 'person', 'frog', or 'chair', you assume the details of some 'typical' sort of person, frog, or chair. […] Frames are drawn from past experience and rarely fit new situations perfectly. We therefore have to learn how to adapt our frames to each particular experience" (Minsky, 1986, p. 245). The identification of the relevant frame has proved to be a powerful resource for understanding social attitudes to certain objects: the bioethical frame known as the "moral status of the embryo" is the most powerful predictor of social attitudes to embryonic stem cell research and, more generally, to embryo experimentation.

Social acceptance of research on embryonic stem cells depends on a wide range of variables, including, familiarity with science (biological scientific literacy), worldviews (general expectations regarding science, religious beliefs), perception of the trade-off between medical benefits and the rights of the embryo, trust in the scientific and medical communities and religious institutions, and the frame of the moral status of the embryo. The available empirical evidence, presented here synthetically, shows that for a majority of people in most societies attitudes to embryo experimentation for stem cell research are, at present, only weakly related to knowledge, and strongly related to their perceptions of the moral status and rights of the embryo and associated beliefs about the moment when individual life begins.

The results set out below come from a lengthy module on perceptions of stem cell research and the status of the embryo within a study based on a

---

9. This is an extreme example of the problem of "incommensurability" proposed by Thomas Kuhn and Paul K. Feyerabend with regard to scientific theories whose conceptual frameworks do not overlap significantly enough to permit a direct comparison.

representative survey of the public's awareness, knowledge, and attitudes toward biotechnology. The study was conducted in 10 European countries and the United States.[10]

The questionnaire used in our study offered four options on the status of the embryo representative of the categories that appear with different formulations in the debates among better informed population segments and stakeholders: "a human embryo that is a few days old . . ." [. . .] (1) "is a mere cluster of cells, and it makes no sense to discuss its moral condition"; (2) "has a moral condition halfway between that of a cluster of cells and that of a human being"; (3) "is closer in its moral condition to a human being than to a mere cluster of cells"; (4) "has the same moral condition as a human being." Because what interests us here is the degree of polarization in views of the embryo and also the regulatory options available based on the public's perceptions (although obviously the regulator is also open to other influences, like the view of the scientific community), the data will be analyzed from various successive angles. The most general angle provides the results map appearing in Fig. 5. In six of the nine European countries, and also in the United States, a relative majority believe that the human embryo has the same moral condition as a human being, although there is also significant support for the two less extreme positions ("halfway," "closer"). In six of the nine countries, the position that finds the greatest approval (although it falls well short of absolute majorities), is the one that ascribes to a human embryo a few days old the same moral condition as to a human being. These are countries with very large Catholic majorities, where the Roman Catholic Church plays an active cultural role (the exception being Germany, where Catholic and Protestant religions are comparable in number, so that there is a sizeable Catholic population, but well below the other societies in the group). The six countries are, in descending order: Poland 38%, Italy 37%, Austria 32%, Spain 30%, Germany 29%, and France 28%. Because the percentage of Catholics in all these societies is much higher than the subset of the population that chose "the same moral status" category, this suggest that people are either not well aware of the Catholic Church's opposition to embryo experimentation or the religious doctrine on this issue is not shared by broad segments of this religion, who opt for more nuanced positions, possibly as a cultural reflection of the scientific narrative.

Only in two of the nine countries studied the majority view of the embryo is that of "a cluster of cells." These are Denmark, with 37% and, at a significant

---

10. *BBVA Foundation Biotechnology Study* was designed by the Public Opinion and Social Studies Department of the BBVA Foundation under the direction of the author of this chapter and with the collaboration in the United States study of Jon D. Miller (University of Michigan Institute for Social Research). Sample size: 1500 interviews in Germany, France, the United Kingdom, Italy, Spain, the Netherlands, Denmark, Austria, and Poland. Sampling error ±2.6%. The fieldwork was carried out by TNS-Opinion from October 2002 through February 2003. A parallel study was directed by Jon D. Miller and conducted in the United States in 2003, with 2040 online questionnaires collected through Knowledge Networks.

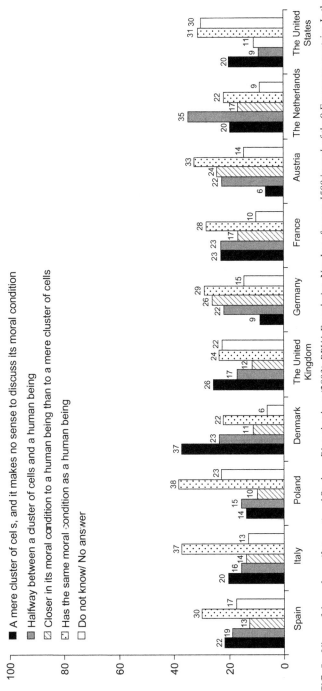

Which of the following four phrases do you agree with most?
"An embryo that is a few days old is...."

■ A mere cluster of cells, and it makes no sense to discuss its moral condition
▨ Halfway between a cluster of cells and a human being
▧ Closer in its moral condition to a human being than to a mere cluster of cells
⊡ Has the same moral condition as a human being
☐ Do not know/ No answer

FIG. 5 Views of the embryo. *(International Study on Biotechnology (2003), BBVA Foundation. Number of cases: 1500 in each of the 9 European countries. In the United States: Survey of Knowledge Networks (2003): 2040 cases.)*

distance, the United Kingdom with 25% (in this case, almost on an equal footing with the 24% who believe that a human embryo has the same moral condition as a human being). In Denmark, most of the population is Evangelical Lutheran (around 95%), while in the United Kingdom a large majority belongs to the Anglican church. Neither of these societies has a widespread organized religious faith that strongly opposes distinguishing between an early embryo and an individual human being. In the Netherlands, the intermediate option is the dominant perception with 35%. The majority of the Dutch are "unaffiliated" with any religion, with Catholics and Protestants weighing in at around 31% and 21%, respectively.

Table 2 shows the distribution of the population groups opting for the two extreme positions on the scale, clearly representative of the scientific and religious-driven view, respectively. These are in theory the two groups likeliest to turn to diverse forms of collective action, especially in the presence of organizations that have as a "cause" in their program either the defense of the embryo or the promotion of the corresponding scientific research. In one group of countries, for each individual who sees the embryo as a cluster of cells, there

TABLE 2 Ratio Views on the Embryo as a Human Being and as a Cluster of Cells

Ratio by Country between "Has the Same Moral Condition as a Human Being" and the Embryo as "a Mere Cluster of Cells"

	Ratio
Austria	5.1
Germany	3.2
Poland	2.8
Italy	1.8
The United States	1.5
Spain	1.4
France	1.2
The Netherlands	1.1
The United Kingdom	0.9
Denmark	0.6

International Study on Biotechnology (2003), BBVA Foundation. Number of cases: 1500 in each of the 9 European countries. In the United States: Survey of Knowledge Networks (2003): 2040 cases.

are from just under two to just over five who perceive it as having the same status as a human being. These are Austria, Germany, Poland, and Italy. In four other societies, there is greater equilibrium: from greatest to smallest, the Netherlands, France and, at a larger distance, Spain and the United States. Finally, the ratio is reversed—i.e., more people see the embryo as a cluster of cells than as a human being—in only two countries: the United Kingdom, where the difference is minimal, and Denmark, where the difference is marked.

A final exploratory angle consists of a size comparison between three segments: the *first* integrated by those who see the embryo as a mere cluster of cells, a *second* including those who see it as halfway between a cluster of cells and a human being, and a *third* that aggregates the semantically and eva-luatively close options of the human embryo as a human being and the embryo as closer in its moral condition to a human being than to a mere cluster of cells (see Table 3). To the extent that the regulation of embryo experimentation could be a function of each society's perception of the status of the embryo, one would expect a tendency toward restrictive rules in Austria, Germany, Italy, Poland, and the United States (where almost a third of the population was unable or unwilling to offer a response on the moral status of the embryo). Conversely,

TABLE 3 Perceptions of the Status of the Human Embryo (Three Groups)

Country	Cluster of Cells	Halfway Cluster Cells—Human Being	Closer to a Human Being+Human Being	Do Not Know—No Answer
Poland	14%	15%	48%	18%
Austria	6%	22%	57%	10%
The Netherlands	20%	35%	38%	7%
Britain	26%	17%	35%	10%
Spain	22%	20%	42%	16%
France	23%	23%	45%	7%
The United States	21%	8%	40%	31%
Denmark	37%	24%	34%	5%
Germany	9%	22%	55%	11%
Italy	20%	16%	51%	10%

International Study on Biotechnology (2003), BBVA Foundation. Number of cases: 1500 in each of the 9 European countries. In the United States: Survey of Knowledge Networks (2003): 2.040 cases.

a more permissive regulatory stance could be expected (in descending order) in Denmark, the United Kingdom, and the Netherlands. The middle ground, with more undefined positions susceptible to being swayed one way or another depending on the "voices" of organized stakeholders is represented by Spain and France.

The debate about the moral status of the embryo associated to stem cell research is a relatively new issue for public opinion as a whole. A more familiar "frame of reference" is that of the views on the "beginning of individual human life," which has influenced perspectives and regulations on abortion. Both frames—moral status, beginning of individual human life—are related from a logical standpoint but, depending on the nature of the debate or the decision to be taken (individual vs collective or regulatory), one or other frame can be "activated" and used as a conceptual framework. The main views about the beginnings of human life which have crystallized in the current map of societal beliefs are: (1) at the moment when the egg and the sperm unite, (2) 2 weeks after conception, when different tissues can be distinguished, (3) 3 months after conception, when growth of the fetus begins, and (4) at the time of birth. The response distribution in the nine European countries and the United States can be seen in Fig. 6. "Individual human life begins with conception": this idea is most widely held in Italy and Poland (over 50%), and by smaller relative majorities in the United States, Spain, Germany, France, Austria, and the United Kingdom. "Individual human life begins three months after conception" is the dominant view in Denmark and the Netherlands (40%).

As in the case of the status of the embryo, the map of views on human life is highly differentiated and plural in Europe and the United States, which implies that standardizing regulations across countries is a very difficult task. In some countries beliefs are so evenly divided between the two dominant options ("when the egg and the sperm unite" and "three months after conception") that the regulation of embryo research would demand extensive dialog in order to achieve a high level of legitimacy in a question affecting deep strata of the ontological and associated moral landscape of society.[11]

---

11. Although there is no strict logical dependence between the two frames (status of the embryo, beliefs about the beginning of a new human being), there are strong statistical associations between them (a Pearson Chi-square test was found to be highly significant in all countries). For instance, in the case of Denmark, among those responding that a new human life does not start until birth, 68% believe that a human embryo is simply a cluster of cells and only 7% that it has the same moral condition as a human being. Conversely, of the group responding that "a new human being can be considered to exist at the moment when the egg and the sperm unite," 57% believe that "a few days old human embryo has the same moral condition as a human being" and an additional 15% that "a human embryo is closer in its moral condition to a human being than to a mere cluster of cells." When an individual or society chooses the "intermediate" option in one of the variables, association with the options provided for the other variable will show a wider dispersion, suggesting either that attitudes are less well-defined or that they are more open to other influences. For the most part then both frames tend to form a cluster of beliefs.

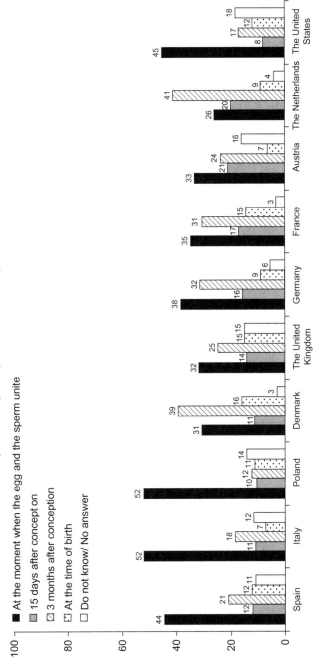

FIG. 6 Beliefs on the beginning of individual human life. *(International Study on Biotechnology (2003), BBVA Foundation. Number of cases: 1500 in each of the 9 European countries. In the United States: Survey of Knowledge Networks (2003): 2040 cases.)*

6. VIEWS ON THE STATUS OF THE EMBRYO, RELIGIOUS BELIEFS, SCIENTIFIC LITERACY, AND GENDER

As we have argued in the previous pages, religious beliefs (activated to varying extents by certain churches) and knowledge of biology (assimilated more or less effectively by the public) are the two main cultural forces shaping views of the embryo's status and, through them, a conviction of the legitimacy or illegitimacy of its uses. The evidence of the international study presented in the previous section allows us to calibrate with some precision the degree of influence of both variables.

The data show that the variation in perceptions of the embryo is significantly associated with religious beliefs. Across all countries, the unaffiliated tend to hold a view of the embryo as a cluster of cells with no moral condition, and to see human life as beginning at a later stage after conception, with the Catholic population at the opposite extreme (see Tables 4 and 5). While a religious divide (religious vs. nonreligious) exists to some extent in all countries, it is possible to identify three large groups: (1) Catholic countries: the religious divide (believers vs nonbelievers) is very deep, with Catholics strongly believing that individual human life begins with conception and that the embryo has the same moral condition as a human being. (2) Austria and Germany are statistical "outliers." The religious cleavage exists, but very few position themselves in the category "a human embryo is a mere cluster of cells" (i.e., this view is not prevalent even in the nonbelievers subset). The most plausible explanation is that the view on the embryo and its experimental uses is a highly sensitive issue in German culture due to the "ripple effects" of the brutal eugenics program applied during the Nazi regime. (3) Denmark, the United Kingdom, and the Netherlands: perceptions of the embryo are not so significantly differentiated by religious beliefs as in other countries, due to the presence of churches clearly less active in advocating a specific doctrine on the embryo.

The intensity of religious practice further differentiates perceptions of the embryo: individuals of sporadic religious practice follow a "middle road" between those who regularly practice their religion and the group of nonbelievers and nonchurchgoers (see Table 6).

The other variable with major power to shape a drastically different view of the embryo is the biological understanding acquired by the public. In effect, the data show that there is a significant association between level of biological literacy and the view held of the embryo's status, such that the propensity to see a few-days-old embryo as "a cluster of cells" increases with a person's scientific knowledge (see Table 7).[12] Notwithstanding this general pattern, there are other

---

12. In order to assess the public's familiarity with basic concepts of biology and, more specifically, biotechnology, a battery of 22 items was constructed. Dichotomizing responses into correct/incorrect we get a summated scale with a Cronbach's Alpha Reliability coefficient = 0.81 for the nine European countries together, and similar values in each individually. The scores obtained are then grouped into the following three segments: high biological literacy (16–22 points), making up 23.3% of the total European sample; medium literacy (9–15 points), making up 54.2%; and low literacy (0–8 points), making up the remaining 22.4%.

TABLE 4 Beliefs About the Embryo by Country and Religion

An Embryo That Is a Few Days Old Is....

		A Cluster of Cells Without Any Moral Condition	Halfway Between a Cluster of Cells and a Human Being	Closer to a Human Being than a Cluster of Cells	Same Moral Condition as a Human Being
Catholic countries (strong religious distinction)					
Italy	Catholic	17.7	15.6	14.5	40.2
	Nonreligious	33.0	17.6	9.5	22.0
Poland	Catholic	13.4	14.2	9.9	39.5
	Nonreligious	19.7	26.2	9.1	19.9
France	Catholic	14.6	20.7	18.8	35.6
	Nonreligious	32.1	23.7	14.8	20.1
Spain	Catholic	16.8	18.6	11.8	33.5
	Nonreligious	35.4	18.8	14.3	19.6
Countries with a highly specific pattern					
Austria	Catholic	4.9	21.1	23.4	33.8
	Nonreligious	8.4	24.8	25.1	27.5
Germany	Catholic	9.6	21.6	25.2	34.9
	Protestant	6.2	20.4	23.7	32.0
	Nonreligious	9.4	23.1	26.5	24.7
Countries with more complex religious distinction					
The Netherlands	Protestant	9.4	25.4	21.6	35.1
	Catholic	19.3	29.0	19.0	24.3
	Nonreligious	26.0	36.5	11.9	16.2
The United Kingdom	Anglican	28.0	14.9	16.2	25.3
	Evangelical	20.6	23.5	13.3	22.7
	Nonreligious	26.9	16.7	9.6	21.5
Denmark	Protestant	35.6	23.2	13.1	22.9
	Nonreligious	44.2	25.1	8.7	14.6
The United States	Protestant	18.6	7.9	11.4	34.2
	Catholic	14.8	9.2	11.9	37.1
	Nonreligious	38.2	9.6	7.6	10.1

International Study on Biotechnology (2003), BBVA Foundation. Number of cases: 1500 in each of the 9 European countries. In the United States: Survey of Knowledge Networks (2003): 2.040 cases. In the United States responses from "evangelists" and "other religions" excluded from chart for reasons of space. These two groups are important minorities but amount to a smaller proportion of the population than the three groups displayed.

TABLE 5 Beliefs About the Embryo by Religious Beliefs

Which of the Following Statements Do You Agree With Most?

"A Human Embryo That Is a Few Days Old Is...

Percentage Over Total Responses (Europe)

	Nonreligious	Protestant	Catholic	Total
...cluster of cells, no sense discuss moral condition	29.6%	19.7%	17.6%	22.1%
...halfway between cluster of cells and human being	26.9%	23.9%	20.9%	23.4%
...closer to a human being than a mere cluster of cells	18.3%	23.4%	17.9%	18.9%
...has the same moral condition as a human being	25.2%	33.0%	43.5%	35.6%
Chi-Square Test	Value	df	Asymp. Sig. (2-sided)	
Pearson chi-square	397,162(a)	6	.000	

International Study on Biotechnology (2003), BBVA Foundation. Number of cases: 1500 in each of the 9 European countries.

TABLE 6 Beliefs About the Moral Status of the Embryo by Religiosity

Percentage Over Total Responses (Europe)

	Cluster of Cells	Halfway Cluster Cells—Human Being	Closer to a Human Being	Same Moral Status as a Human Being
Never prays or practices +Nonbelievers	30.0	28.2	17.0	24.7
Believes and practices from time to time	21.6	26.5	19.6	32.4
Believes and weekly practice	15.0	22.5	19.5	43.0
Believes and daily practice	14.0	15.9	16.3	53.9

$\chi^2 = 728.3$, df= 9, Signif. = 0.000.
International Study on Biotechnology (2003), BBVA Foundation. Number of cases: 1500 in each of the 9 European countries.

TABLE 7 Beliefs About the Moral Status of the Embryo by Biological Literacy

		Cluster of Cells	Halfway Cluster Cells–Human Being	Closer to a Human Being	Same Moral Status as a Human Being	Do Not Know–No Answer	Difference Same Moral–Cluster of Cells
Denmark	Low biological literacy	22.4%	12.2%	7.3%	27.8%	30.3%	5.4%
	High biological literacy	42.2%	28.0%	10.5%	17.8%	1.5%	−24.4%
The United Kingdom	Low biological literacy	13.9%	10.8%	9.1%	28.3%	37.8%	14.4%
	High biological literacy	38.8%	16.9%	15.4%	18.6%	10.2%	−20.2%
Spain	Low biological literacy	15.3%	13.8%	10.9%	32.3%	27.7%	17.0%
	High biological literacy	31.8%	21.2%	8.9%	32.4%	5.6%	0.6%
France	Low biological literacy	14.4%	15.1%	7.3%	33.0%	30.2%	18.6%
	High biological literacy	30.8%	27.9%	19.1%	19.7%	2.6%	−11.1%
Italy	Low biological literacy	19.7%	13.0%	10.6%	31.3%	25.4%	11.6%
	High biological literacy	22.5%	15.4%	18.3%	38.8%	4.9%	16.3%

Continued

TABLE 7 Beliefs About the Moral Status of the Embryo by Biological Literacy—cont'd

		Cluster of Cells	Halfway Cluster Cells-Human Being	Closer to a Human Being	Same Moral Status as a Human Being	Do Not Know— No Answer	Difference Same Moral— Cluster of Cells
The Netherlands	Low biological literacy	17.3%	12.5%	11.5%	26.3%	32.4%	9.0%
	High biological literacy	22.0%	43.5%	17.8%	14.2%	2.4%	−7.8%
Poland	Low biological literacy	10.2%	9.5%	7.5%	36.5%	36.3%	26.3%
	High biological literacy	18.0%	26.9%	14.9%	35.1%	5.1%	17.1%
Germany	Low biological literacy	7.3%	7.7%	24.9%	33.8%	27.0%	26.5%
	High biological literacy	9.7%	26.0%	25.4%	28.3%	10.6%	18.6%
Austria	Low biological literacy	5.6%	14.7%	20.3%	33.8%	25.6%	28.2%
	High biological literacy	7.8%	32.2%	28.3%	26.1%	5.7%	18.3%

International Study on Biotechnology (2003), BBVA Foundation. Number of cases: 1500 in each of the 9 European countries.

more specific profiles of interest. First, the differences between high and low literacy groups are attenuated in the cases of Germany and Austria, which here as in regard to the aforementioned religiosity variable, stand apart from remaining societies in their vision of the embryo and its use in research. Second, responses in the Netherlands are concentrated in the "halfway" category, with a strong differentiation between high and low biological literacy groups (43.5% and 12.5%, respectively). Finally, Italy and Poland show a specific distribution, with the majority in both countries opting for a view of the embryo similar to that of the Catholic Church, evidencing the strong influence of its doctrine even among population sectors with greater biological knowledge.

Although the two variables of religious belief and scientific literacy are the two most powerful predictors of views on the embryo, gender is another primary interest variable in the study of perceptions of the embryo and attitudes toward embryo experimentation for obtaining stem cells. Some authors claim that it is wrong to see the embryo as a mere biological entity in isolation from the mother (see Krones et al., 2006), while others have explored the ethical consequences of giving one or other answer—the woman or the embryo—to the question, "Who is the subject of research?" in the IVF context (Gaze and Dawson, 1990). But in the case of stem cells, it is possible for an embryo to be created in the laboratory, in the absence of a mother and with the woman intervening solely as an egg donor, and it is this possibility precisely that is at the heart of the debate. At the same time, it is clear that the degree of personal and biological involvement with the embryo varies in intensity and nature from women to men, so we might expect to find women holding a view of the embryo beyond a purely biological entity and also more opposed to the use of embryos for purposes other than their own development and well-being. The data show that, indeed, in all countries, women are likelier than men to attribute a full human condition to the embryo: 33% consider that the embryo has the same moral condition as a human, compared to 27% of men, and 45% consider that life begins at the moment the sperm and egg unite, compared to 35% of men. However, parts of these differences are due to the variation in religiosity between men and women: while 36% of all men across the nine European countries do not hold religious beliefs, this percentage drops to 26% among women. In a multiple regression model to account for attitudes to embryo research, the gender variable (codified as a dummy variable, with 0 corresponding to male and 1 to female) was found to be very low and a nonsignificant explanatory variable in most of the countries analyzed (Pardo and Calvo, 2008, p. 36).

7. MORAL STATUS AND ATTITUDES TO EMBRYO RESEARCH

Awareness of the public's perceptions and beliefs concerning the embryo and its moral status, and an understanding of the main narratives that sustain this concept, are of intrinsic interest to elucidate the social foundations of the value ascribed to the embryo as one of the most potent constructs in society's ontological and moral landscape. But knowledge of this frame also serves to explain attitudes toward the creation and biological use of embryos, which in turn inform the regulatory framework (protective or permissive) deployed in each country. The following tables synthetically present the public's attitudes toward the use of human embryos by reference to differing levels of abstraction; i.e., general attitudes to their use without specifying either the goals or means, followed by attitudes modulated according to the origins of the embryo (created ad hoc or "spares" from IVF treatments) and the biomedical goals of finding possible treatments for serious diseases. When it comes to evaluate the trade-offs between goals and means

> *"it is interesting to gauge the sensitivity or elasticity of people's attitudes in response to the presence of more specific goals and means. A low response to these specifics would suggest we are up against evaluative structures of a holistic or ideological kind, resistant to change and favoring confrontation rather than dialogue with the public."*

(Pardo et al., 2009, p. 1070)

Table 8 shows that if the trade-off is left unspecified, "research with human embryos that are a few days old," the majority view is one of rejection, according to data from the *European Mindset Study* referred to previously. Average acceptance on a scale from 0 to 10 (totally unacceptable, totally acceptable) stands below the halfway mark in 9 out of the 12 European countries, with rejection particularly strong among the Catholic population, other (non-Protestant) Christians and, above all, Orthodox Christians. Rejection of this use also increases among those declaring a higher level of religiosity.

This general map acquires greater relief through the inclusion of two specific variables; one related to the goals pursued (to find treatments for serious diseases) and one to the means employed ("embryos left over from fertility treatments" and "embryos created specifically for research"), as occurred in the aforementioned *Study on Biotechnology* (see Table 9). Providing the biomedical goals of embryonic research are seen as genuine and of high social value, the debate revolves around two focal points or concrete scenarios: (1) the use of "spare embryos" (embryos left over from fertility treatments, i.e., embryos that may never develop) and (2) the use of embryos created specifically to further biomedical research. The pattern of acceptability exhibits a strong contrast: a majority in most countries accept the use of spare embryos, albeit with sizable groups against, while opinion is polarized in the case of Poland and firmly opposed to the practice in Austria. All societies, with the exception of Denmark, reject the creation of embryos to further biomedical research.

Acceptance of each scenario varies with people's beliefs about the moral status of the embryo. As the status of the embryo is perceived to be closer to the moral status of a human being, acceptability of the use of embryos in biomedical research declines significantly, both in the case of spare embryos and, particularly, embryos created specifically for research purposes (see Fig. 7). As

TABLE 8 Acceptance of Research With Human Embryos

Can You Tell Me Whether You Think the Things I Am Going to Read Out Are Acceptable or Unacceptable? Please Use a Scale From 0 to 10, Where 0 Means You Think It Is Totally Unacceptable and 10 Means You Think It Is Totally Acceptable. Research With Human Embryos That Are a Few Days Old

		All EU Countries	Denmark	Sweden	Spain	Belgium	Bulgaria	France	Portugal	Italy	Poland	The United Kingdom	Germany	Greece
Total		3.7	5.4	5.0	5.1	4.5	4.2	4.0	4.0	3.8	3.4	3.2	3.1	1.6
Sex	Male	3.9	5.7	5.1	5.1	4.6	4.1	4.1	4.1	4.0	3.7	3.4	3.3	1.6
	Female	3.6	5.4	4.8	5.1	4.3	4.4	3.9	3.9	3.7	3.2	2.9	2.9	1.5
Age finished studying	15 or under	3.4	4.5	3.1	4.8	3.8	3.3	3.8	3.4	3.6	2.7	2.8	2.8	1.4
	16 to 19	3.6	5.5	4.7	5.1	4.4	4.3	3.9	4.2	4.0	3.4	3.2	3.1	1.4
	20 or older	4.3	5.4	5.3	5.6	4.9	4.4	4.2	5.2	4.1	3.7	4.1	3.5	2.0
	Still studying	4.0	6.2	5.4	4.8	4.4	4.2	4.2	4.7	3.6	3.8	2.9	4.2	1.7
Religion	Nonreligious	4.2	5.7	5.2	5.6	4.8	3.5	4.2	5.4	4.8	5.2	3.4	3.6	2.1
	Catholic	3.6			4.9	4.4		3.9	3.7	3.6	3.2	2.6	2.4	
	Protestant	3.8	5.4	5.0								3.4	3.6	
	Christian unspecified	3.2	5.0	5.0								2.8		
	Orthodox Christian	2.7					4.5							1.6

Continued

TABLE 8 Acceptance of Research With Human Embryos—cont'd

Can You Tell Me Whether You Think the Things I Am Going to Read Out Are Acceptable or Unacceptable? Please Use a Scale From 0 to 10, Where 0 Means You Think It Is Totally Unacceptable and 10 Means You Think It Is Totally Acceptable. Research With Human Embryos That Are a Few Days Old

		All EU Countries	Denmark	Sweden	Spain	Belgium	Bulgaria	France	Portugal	Italy	Poland	The United Kingdom	Germany	Greece
Religiosity	0–2	4.4	6.2	5.3	5.9	5.1	3.9	4.6	5.3	5.5	5.2	3.3	3.7	1.9
	3–4	3.9	5.5	5.3	5.3	4.4	4.3	3.9	4.7	4.8	4.2	3.1	3.3	1.7
	5	3.7	5.2	4.9	4.9	4.2	4.9	3.6	3.8	4.2	3.7	3.2	3.0	1.8
	6–7	3.8	4.7	4.7	5.2	4.4	4.5	3.8	3.9	4.0	3.6	3.5	3.2	1.5
	8–10	3.0	3.6	3.5	4.0	3.7	3.6	3.5	3.3	3.2	2.7	2.7	1.7	1.5

European Mindset Study (2009), BBVA Foundation. Number of cases: 1500 in each of the 12 countries.

TABLE 9 Attitudes to Embryo Experimentation

To What Extent Do You Find It Morally Acceptable to Experiment With Human Embryos That Are a Few Days Old to Obtain Stem Cells so as to Find Treatments for Serious Diseases?

	In the Case of Embryos Left Over From Fertility Treatments				In the Case of Embryos Created Specifically for Research			
	Against 0–3	"Neutral" 4–6	In Favor 7–10	Do Not Know—No Answer	Against 0–3	"Neutral" 4–6	In Favor 7–10	Do Not Know—No Answer
Denmark	18.4	16.8	62.0	2.7	27.8	21.7	47.9	2.6
The Netherlands	22.6	22.1	46.6	8.7	46.4	23.3	22.0	8.4
France	20.5	21.5	50.1	7.8	43.5	21.0	26.5	9.1
Germany	30.8	24.5	37.7	7.0	61.0	18.3	16.5	4.3
The United Kingdom	22.4	20.0	37.5	20.1	32.2	21.4	24.7	21.7
Italy	20.9	25.4	36.6	17.2	36.7	22.8	23.2	17.3
Spain	21.5	21.7	35.3	21.5	29.7	23.3	25.3	21.7
The United States	21.5	12.0	38.3	28.2	36.8	10.9	24.0	28.3
Poland	28.9	15.7	28.2	27.1	36.5	14.8	20.3	28.4
Austria	41.4	21.8	15.8	20.9	50.4	19.1	10.5	20.0

International Study on Biotechnology (2003), BBVA Foundation. Number of cases: 1500 in each of the 9 European countries. In the United States: Survey of Knowledge Networks (2003): 2.040 cases.

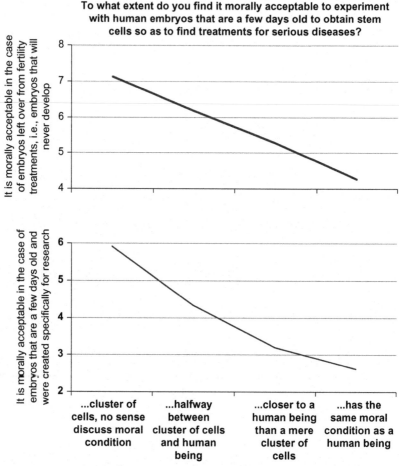

FIG. 7   Attitudes to embryonic stem cell research by beliefs about the moral status of the embryo. *(International Study on Biotechnology (2003), BBVA Foundation.* Number of cases: *1500 in each of the 9 European countries.)*

was to be expected, there is a statistically significant relationship between these two variables, the frame of the embryo and attitudes to its utilization. This result confirms the importance for all the main actors of strengthening the narratives they employ to shape such as powerful construct as the human embryo.

8. CONCLUSIONS

The moral status of the embryo is at once a core concept in the bioethical repertoire and a cultural construct of great symbolic weight, whose formal characterization and social perception have far-reaching consequences on the regulatory plane. Around four decades ago, it was the subject of intense social debate and academic analysis prompted by the successful introduction of IVF treatments. It has also been linked to the debate around abortion and its legislation (albeit with discussions focusing more on the fetus) and, more recently, has reemerged with considerable force in the context of research with embryonic stem cells. The cases of IVF and, more markedly, embryonic stem cells have stood out for the collision of two narratives concerning the embryo and its rights; one scientific and the other of a religious-moral nature. The scientific community has claimed that the promise of this research is immense, as the basis of what is becoming known as regenerative medicine with the ability to provide effective treatments for diseases rife in advanced societies. At the same time, some religious faiths have argued through their institutions that allowing embryos to be destroyed for the benefit of third parties is a morally reprehensible attack on the dignity of human life. Although science and religion nowadays inhabit their respective spheres without entering into moral or cultural clashes (except, in some few societies, over the acceptance of Darwinian evolutionary theory), in the case of the embryo, their perspectives or angles of approach, crystallized in incommensurable narratives (in the sense employed by science philosophers like Kuhn and Feyerabend), are in direct and outright conflict.

From the analytical, minimalist and value- and emotion-free standpoint of developmental biology, concerned solely with the biological properties emerging at each stage of the burgeoning human life from conception to the formation of a human individual, the few-days-old embryo, before the threshold of the 14th day, is merely a cluster of cells. Conversely, some bodies of doctrine of religious origin equate this same biological entity to a human being, in possession of the rights and obligations deriving from this status in the eyes of other agents.

Ontology (the properties, relationships, and boundaries between different entities) has vital consequences on the ethical plane (the treatment human individuals owe to all other entities). Specifically, the decision whether or not to include the human embryo in the set of human individuals and, if so, to what degree, immediately legitimizes or delegitimizes its use in biomedical research. The scientific community has developed a powerful narrative articulated on two axes, the first referring to the biological characteristics of the embryo (which concludes that there is a demarcating line before and after the 14th day following fertilization), the other concerning the instrumental value of embryo research for understanding and treating certain diseases. Some religious faiths (such as Catholicism) see the embryo as a human being in development from the very moment of conception, and emphatically reject the notion of it being an instrument for any end other than its own well-being. Other agents bring their own angles to the debate, although in the case of stem cell research, in contrast to the discussion on abortion, their voice and influence have so far been severely limited. These would include patients associations, and the pro-life and pro-choice movements.

As well as engaging in the corresponding conceptual elaborations and their submission to debate in institutional spaces like parliaments or bioethics committees, the scientific community and the churches compete to win public perceptions to their vision of the embryo, its status, and its rights. The available empirical evidence reveals that, generally speaking, the religious narrative has gained more influence than its scientific counterpart; a fact partly attributable to the public's low level of scientific literacy (especially as regards biology and genetics). The embryo frame and the associated frame regarding the beginning of human life bear the cultural imprint of the view promoted by religious beliefs more clearly than they do that emanating from the laboratory. And both frames influence public attitudes toward the use of embryos in biomedical research. From the public's standpoint, the importance of the ends (potential therapies for serious conditions) unquestionably dials down their moral reservations about the means, to some extent, but tends to do so only in the case of embryos left over from IVF treatments, and not in the case of their ad hoc creation for use in research.

The human embryo is a highly potent symbol of life embedded in a complex space, occupied by varied and deep-rooted evaluative structures and social imagery, shaped fundamentally by religious-moral creeds, in collision with the more recent and self-contained vision of contemporary biology and genetics. Although the moral landscape of today's society is clearly pluralistic regarding the nature or ontology of embryos and their instrumental use by science, the dominant valence is one of protection and respect. Improving the public's biological literacy could alter perceptions of the embryo and the legitimacy of its biomedical use, but would not be sufficient of itself. It seems logical that the current development of research with embryos for obtaining stem cells in countries with more permissive regulations, and a demonstration of the therapeutic potential it holds out, could help counteract the influence that religiously inspired views of the embryo currently exercise in public perceptions. Also, the majority of people could end up shifting toward the minimalist, biological end of the embryo frame as hitherto "weak voices," like those of patients and their associations and, particularly, those of the scientific community learn to better project the biological properties of the human embryo before 14 days of development and the many biomedical benefits that could be obtained. Knowledge would play its part, but also a dialog that includes values and worldviews.

REFERENCES

Atkinson, P., Glasner, P., Lock, M. (Eds.), 2009. Handbook of Genetics and Society. Routledge, London and New York.

Bauer, M., Gaskell, G., 2002. Biotechnology. The Making of a Global Controversy. Cambridge University Press, Cambridge.

Bauer, M.W. (Ed.), 1995. Resistance to New Technology. Cambridge University Press, Cambridge.

Bayertz, K., 2006. Struggling for consensus and living without it: the construction of a common European bioethics. In: Tristram, H., Engelhardt, H.T. (Eds.), Global Bioethics. The Collapse of Consensus (207-237). M&M Scrivener Press, Salem, MA.

Birnbacher, D., 1999. Ethics and social science: which kind of co-operation? Ethical Theo. Moral Pract. 2 (4), 319–336.

Birnbacher, D., 2016. Where and when ethics needs empirical facts. In: Brand, C. (Ed.), Dual-Process Theories in Moral Psychology. Springer, Wiesbaden, pp. 41–50.

Bruun, O., Kalland, A. (Eds.), 1996. Asian Perceptions of Nature. Curzon Press-Nordic Institute of Asian Studies, Richmond, Surrey.

Chadwick, R., 2000. Preface. In: Chadwick, R. (Ed.), The Concise Encyclopedia of the Ethics of New Technologies. Academic Press, San Diego-San Francisco.

Condit, C.M., 1990. Decoding Abortion Rhetoric. University of Illinois Press, Urbana and Chicago.

Congregation for the Doctrine of the Faith, 1980. Declaration on Euthanasia Iura et Bona (5 May 1980), II. AAS 72, 546.

Converse, P.E., 1992. The nature of belief systems in mass publics. In: Apter, D.E. (Ed.), Ideology and Discontent. The Free Press, New York, pp. 206–261.

Dawson, K., 1990. An outline of scientific aspects of human embryo research. In: Singer, P. et al. (Ed.), Embryo Experimentation. Ethical, Legal and Social Issues (3-13). Cambridge University Press, Cambridge.

Dorff, E.N., 2001. Stem cell research—a Jewish perspective. In: Holland, S., Lebacqz, K., Zoloth, L. et al. (Eds.), The Human Embryonic Stem Cell Debate. Science, Ethics, and Public Policy. The MIT Press, Cambridge, MA—London, England.

Friele, M.B., 2003. Do committees Ru(i)n the bio-political culture? On the democratic legitimacy of bioethics committees. Bioethics 17 (4), 301–318.

Gaskell, G., Bauer, M.W. (Eds.), 2006. Genomics and Society. Earthscan, London-Sterling, VA.

Gaskell, G., et al. 2003. Europeans and biotechnology in 2002. A report to the EC directorate general for research from the project 'Life Sciences in European Society', March 2003, 22. Available at: http://ec.europa.eu/commfrontoffice/publicopinion/archives/ebs/ebs_177_en.pdf.

Gaze, B., Dawson, K., 1990. Who is the subject of research. In: Singer, P., Kuhse, H., Buckle, S., Dawson, K., Kasimba, P. (Eds.), Embryo Experimentation. Ethical, Legal and Social Issues. Cambridge University Press, Cambridge, pp. 109–124.

Gigerenzer, G., 2008. Rationality for Mortals. How People Cope with Uncertainty. Oxford University Press, Oxford-New York.

Gigerenzer, G., Todd, P.M., the ABC Research Group, 1999. Simple Heuristics that Make Us Smart. Oxford-New York.

Gregory, J., Miller, S., 1998. Science in Public. Communication, Culture and Credibility. Plenum Press, New York-London.

Hallman, W.K., 2017. What the public thinks and knows about science and why it matters. In: Jamieson, K.H., Kahan, D., Scheufele, D.A. (Eds.), The Oxford Handbook of the Science of Science Communication (61-72). Oxford University Press, New York.

Howard Ecklund, E., Scheitle, C.P., 2018. Religion vs. Science. What Religious People Really Think. Oxford University Press, New York.

Illies, C., 2016. New debates in old ethical skins. In: Engelhard, M. (Ed.), Synthetic Biology Analysed. Springer International Publishing, Switzerland, pp. 89–126.

Ioannes Paulus II, P.P., 1995. Evangelium vitae. Available at: http://w2.vatican.va/content/john-paul-ii/en/encyclicals/documents/hf_jp-ii_enc_25031995_evangelium-vitae.html.

Kasperson, R.E., Renn, O., Slovic, P., Brown, H.S., Emel, J., Goble, R., Kasperson, J.X., Ratick, S., 1988. The social amplification of risk: a conceptual analysis. Risk Anal. 8 (2), 177–187.

Knight, D., 2006. Public Understanding of Science. Routledge, London-New York.

Kolakowski, L., 1967. Der Mensch ohne Alternative. R. Piper & Co Verlag, Munich.

Krones, T., Schlüter, E., Neuwohner, E., El Ansari, S., Wissner, T., Richter, G., 2006. What is the preimplantation embryo? Soc. Sci. Med. 63, 1–20.

Kuhse, H., Singer, P., 1990. The nature of the ethical argument. In: Singer, P. et al. (Eds.), Embryo Experimentation. Ethical, Legal and Social Issues (37-42). Cambridge University Press, Cambridge.

Lassen, J., Gjerris, M., Sandøe, P., 2006. After Dolly – Ethical limits to the use of biotechnology on farm animals. Theriogenology 65 (5), 992–1004.

Maienschein, J., 2006. The language really matters. In: Ruse, M., Pynes, C.A. (Eds.), The Stem Cell Controversy. Prometheus Books, Amherst, NY, pp. 37–54.

Marx Ferree, M., Gamson, W.A., Gerhards, J., Rucht, D., 2002. Shaping Abortion Discourse. Democracy and the Public Sphere in Germany and the United States. Cambridge University Press, Cambridge-New York.

Maynard-Moody, S., 1992. The fetal research dispute. In: Nelkin, D. (Ed.), Controversy. Politics of Technical Decisions. Sage, Newbury Park, pp. 3–4.

Miller, J.D., Suchner, R.W., Voelker, A.M., 1980. Citizenship in an Age of Science. Pergamon Press, New York.

Minsky, M., 1975. A framework for representing knowledge. In: Winston, P.H. (Ed.), The Psychology of Computer Vision. McGraw-Hill, New York, pp. 211–277.

Minsky, M., 1986. The Society of Mind. Simon and Schuster, New York.

Molewijk, B., Stiggelbout, A.M., Otten, W., Dupuis, H.M., Kievit, J., 2004. Empirical data and moral theory. A plea for integrated empirical ethics. Med. Health Care Philos. 7, 55–69.

Mulkay, M., 1997. The Embryo Research Debate. Science and the Politics of Reproduction. Cambridge University Press, Cambridge.

Nelkin, D. (Ed.), 1992. Controversy. Politics of Technical Decisions. Sage, Newbury Park.

Noonan Jr., J.T., 1986. Contraception. A History of Its Treatment by the Catholic Theologians and Canonists. Harvard University Press, Cambridge, MA.

Pardo, R. (2009). Public attitudes to pharming. In Rehbinder, E., Engelhard, M., Hagen, K., Jorgensen, R. B., Pardo-Avellaneda, R., and Thiele, F. et al. (2009). Pharming. Promises and Risks of Biopharmaceuticals Derived From Genetically Modified Plants and Animals (121-178). Berlin-Heidelberg: Springer.

Pardo, R., Calvo, F., 2008. Attitudes toward embryo research, worldviews, and the moral status of the embryo frame. Sci. Commun. 30 (1), 8–47.

Pardo, R., Engelhard, M., Hagen, K., Jørgensen, R.B., Rehbinder, E., Schnieke, A., Szmulewicz, M., Thiele, F., 2009. The role of means and goals in technology acceptance. A differentiated landscape of public perceptions of pharming. EMBO Rep. 10, 1069–1075.

Pardo, R., Hagen, K., 2016. Synthetic biology: public perceptions of an emergent field. In: Engelhard, M. (Ed.), Synthetic Biology Analysed. Springer International Publishing, Switzerland, pp. 127–170.

Resnik, D.B., 2018. The Ethics of Research With Human Subjects. Protecting People, Advancing Science, Promoting Trust. Springer International Publishing AG, Cham, Switzerland.

Roy, D.J., Wynne, B.E., Old, R.W. (Eds.), 1991. Bioscience and Society. John Wiley & Sons, Chichester-New York.

Rubenfeld, S., Benedict, S. (Eds.), 2014. Human Subjects Research After the Holocaust. Springer, Heidelberg-New York-London.

Ruse, M., Pynes, C.A., (Eds.), 2006. Introduction. In: The Stem Cell Controversy, Prometheus Books, Amherst, NY, pp. 225–226.

Sartori, G., 2016. La carrera hacia ningún lugar (Spanish translation of La corsa verso il nulla). Taurus, Madrid.

Shapin, S., Christie, J.R.R., 1990. Science and the public. In: Olby, R.C., Cantor, G.N., Hodge, M.J.S. (Eds.), Companion to the History of Modern Science. Routledge, London-New York, pp. 990–1007.

Sheehan, J.J., Sosna, M. (Eds.), 1991. The Boundaries of Humanity. Humans, Animals, Machines. University of California Press, Berkeley-Los Angeles-Oxford.

Siipi, H., 2008. Dimensions of Naturalness. Ethics Environ. 13 (1), 71–103.

Siipi, H., 2011. Is neuro-enhancement unnatural and does it morally matter? Trames 15 (65/60) (2), 188–203.

Sugarman, J., 2004. The future of empirical research in bioethics. J. Law Med. Ethics 32, 226–231.

Thiele, F., 2009. The ethical evaluation of pharming. In: Rehbinder, E., Engelhard, M., Hagen, K., Jorgensen, R.B., Pardo-Avellaneda, R., Thiele, F. et al. (Eds.), Pharming. Promises and Risks of Biopharmaceuticals Derived From Genetically Modified Plants and Animals. Springer, Berlin-Heidelberg, pp. 179–200.

Trounson, A., 1990. Why do research on human pre-embryos? In: Singer, P. et al. (Eds.), Embryo Experimentation. Ethical, Legal and Social Issues. Cambridge University Press, Cambridge, pp. 14–25.

Walters, L., 2006. Human embryonic stem cell research. An Intercultural perspective. In: Ruse, M., Pynes, C.A. (Eds.), The Stem Cell Controversy. Prometheus Books, Amherst, NY, pp. 259–282.

Warnock, M., 1985. A Question of Life. The Warnock Report on Human Fertilisation & Embryology. Basil Blackwell, Oxford.

Warnock, M., 1998a. An Intelligent Person's Guide to Ethics. Duckworth, London.

Warnock, M., 1998b. Experimentation on human embryos and fetuses. In: Kuhse, H., Singer, P. (Eds.), A Companion to Bioethics. Blackwell Publishers, Oxford, pp. 390–396.

Warren, M.A., 1997. Moral Status. Obligations to Persons and Other Living Things. Oxford University Press, Oxford-New York.

Waters, B., 2003. Does the human embryo have a moral status. In: Waters, B., Cole-Turner, R. (Eds.), God and the Embryo. Religious Voices on Stem Cells and Cloning. Georgetown University Press, Washington, DC, pp. 67–76.

Wynne, B., 1995. Public understanding of science. In: Jasanoff, S., Markle, G.E., Petersen, J.C., Pinch, T. (Eds.), Handbook of Science and Technology Studies. Sage Publications, Thousand Oaks-London-New Delhi, pp. 361–391.

Zaller, J.R., 1992. The Nature and Origins of Mass Opinion. Cambridge University Press, Cambridge.

Chapter 9

Fetal Reduction

Ana S. Carvalho*, Margarida Silvestre*,†,§, Susana Magalhães*,‡,¶
and Joana Araújo*

*Instituto de Bioética, Universidade Católica Portuguesa, Porto, Portugal, †Faculty of
Medicine, University of Coimbra, Coimbra, Portugal, ‡Universidade Fernando Pessoa, Porto,
Portugal, §UNESCO Chair in Bioethics, Institute of Bioethics, Universidade Católica Portuguesa,
Porto, Portugal, ¶CEGE, Research Centre in Management and Economics, Católica Porto Business
School, Universidade Católica Portuguesa, Porto, Portugal

1. BIOETHICAL ISSUES OF FETAL REDUCTION: WHY WORDS MATTER

Louise Brown was the first baby born through the fertility treatment in vitro
fertilization (IVF) in 1978. In the more than 30 years since that event, millions
of babies have been born through infertility therapies, including more than
2,000,000 IVF babies. These incredible success stories, however, have had a
corresponding "price to pay" (Evans et al., 2014); the increased use of assisted
reproduction technologies (ART), as well as ovulation induction treatments,
have led to an increased number of multifetal pregnancies in the last decades.
Almost 70% of all twins in the United States now come from infertility
treatments.

There is a large literature on the psychological distress experienced by couples who wish to have children but who cannot conceive naturally; long-term
infertility that is treated by ART superimposes cycles of hope and disappointment on the already depressed and vulnerable psyche of couples who are having
difficulty in conceiving. Following this cyclical "emotional roller coaster"
(Ring-Cassidy and Gentles, 2003), the fortunate couples may find themselves
pregnant. In increasing numbers, however, these pregnancies are "higher order"
with three or more implanted fetuses. Despite the advances in clinical care
that improved perinatal outcomes for multifetal gestation, twin pregnancies,
and specially triple and higher-order multiple pregnancies continue to experience elevated risks of adverse maternal and neonatal outcomes (Razaz et al.,
2017). Preterm delivery is the most common cause of morbidity and mortality
in triplet pregnancies (Zipori et al., 2017). Other adverse outcomes include
maternal medical complications, pregnancy loss, and intrauterine growth

Clinical Ethics at the Crossroads of Genetic and Reproductive Technologies.
https://doi.org/10.1016/B978-0-12-813764-2.00009-X

restriction (Tse et al., 2017). The risk of infant death in triplets is almost three times higher than in twins.

Taking these risks into account, one of the approaches that has been suggested is known as Fetal Reduction (FR)—in which the most accessible fetuses are terminated using different procedures and for different purposes and the overall pregnancy number is reduced to twins or a singleton (Ring-Cassidy and Gentles, 2003). Reduction of triplets or higher-order multiple pregnancies has been performed since the 1980s, aiming to reduce maternal and fetal adverse outcomes (Tse et al., 2017). In fact, in the majority of cases, authors use various terms to address the same procedure: "selective termination," "selective feticide," "multifetal pregnancy reduction," "embryo reduction," "fetal reduction," or "selective reduction"(Legendre et al., 2013). However, it seems important to highlight that this semantic vagueness is not, in our opinion, adequate to start our ethical reflection. Therefore taking into consideration that all terms deal with multiple pregnancy, fetus, and partial termination, we will use hereafter only the term fetal reduction. Meanwhile, we must recognize that the differences regarding the various uses, medical and nonmedical indication, and selective or nonselective (first trimester for nonselective and second trimester for selective reduction) purposes of the procedure, will have a different impact on the life of pregnant women and couples and have to be analyzed within specific ethical frames.

Improved outcomes of FR techniques, coupled with the growing use of ART, have led to new arena of ethical questions that must be considered. Early on, FR seemed warranted only in life and death situations—where the mother was of small stature and carrying four or more embryos, for example. As has been seen in numerous technologies, once the concepts are proven and the foundation for their use developed, the focus shifts from "life and death" to "quality of life" (Evans et al., 2014). This explains why there is currently a debate on whether or not it will be appropriate to offer FR routinely for twins, even for those created without ART intervention, with reasons other than medical ones.

This section focuses on the indications and contexts that differentiate the diverse uses of FR: the medical uses that can be (A) selective for cases of abnormal fetus and (B) nonselective of one or more fetus to decrease the medical risk for the mother and for the remaining fetuses; and nonmedical uses that, could be (C) selective, if aiming at a specific characteristic like sex without any medical justification and (D) nonselective if the motive of reduction is of social or economic nature.

1.1 Medical Use: Selective Fetal Reduction

With FR, it is possible to choose between fetuses, selecting only those without the relevant disorder; here the choice is between having a fetus with the serious and incurable disorder, or one without it. Generally, some combination of the four following criteria, all based on the concept of serious disorder could be seen as relevant for deciding whether the procedure of fetal reduction could be adequate: (a) the magnitude of the risk of developing the condition, (b) the seriousness of the condition, (c) the potential treatments for this condition, and (d) the combination of these factors to yield a high probability of a low quality of life for the child if born. However, "serious disorders" can be an extremely variable concept among physicians and widely differing answers could be given to the question whether the condition is "not effectively treatable" in the future. The absence of consensus could permit uses of FR that are likely to be interpreted very differently, and may create uncertainty for the participants in the process, introducing both flexibility and arbitrariness in decision-making.

1.2 Medical Use: Nonselective Fetal Reduction

The first type of medical use of fetal reduction is represented by the interruption of the development of one or more probably normal fetuses in a multiple pregnancy. This scenario lessens the maternal morbidity and fetal mortality by reducing the number of fetuses in utero in a high-risk multiple pregnancy (more than three fetuses), which could constitute a medically relevant risk to the pregnant woman and the remaining fetuses (Legendre et al., 2013).

Fetal reduction is often pointed as less risky than the pursuit of a multiple pregnancy, which presents many obstetrical risks (anemia, diabetes, eclampsia) and perinatal risks (intrauterine growth retardation, respiratory distress syndrome, miscarriage, preterm delivery). However, although there is evidence to support FR in quadruplet or higher order pregnancies, its use in triplets remains controversial, as the technique is not devoid of risks and potential complications.

1.3 Nonmedical Uses: Selective Fetal Reduction

In this case, FR will be used for the nonmedical selection of some characteristics, such as the gender of the offspring. Sex selection is the attempt to ensure the offsprings are of a specific sex, an aim that people have tried to pursue for hundreds of years. Methods have varied from special modes and timing of coitus to the practice of infanticide. But it was only recently that science has provided ways for people to select gender with any likelihood of success, before conception or birth (Silvestre, 2012). Whatever its methods or its reasons, sex selection has encountered significant ethical objections. The European Convention on Human Rights and Biomedicine, explicitly bans sex selection for other than health purposes by stating that: "The use of techniques of medically assisted procreation shall not be allowed for the purpose of choosing a future child's sex, except where "serious" hereditary sex related disease is to be avoided (Council of Europe, 1997)." The main ethical issues regarding the use of FR for nonmedical sex selection, are gender discrimination or stereotype, the appropriateness of expanding control over nonessential characteristics of the offspring, the relative importance of sex selection when weighed against medical burdens to parents and the future child and, commodification of human life (of both the child and the fetus). Like previously stated for preimplantation genetic diagnosis (Magalhães and Carvalho, 2010), the desired potential social benefits of sex selection may appear insufficiently significant when weighed against unnecessary burdens for all parties involved and above all, when weighed against its impact on the identity and self-understanding of the human species.

1.4   Nonmedical Use: Nonselective Fetal Reduction

It has been stated that "if it is right for a pluralistic society to curb a state's interference with the choice of abortion or other reproductive options, how could it be wrong for society to respect and protect the freedom of couples to choose to have one rather than two infants?" (Evans and Britt, 2010).

Here, the main ethical questions are: How can it be right or good to take the life of a twin for reasons other than a serious medical risk for woman or fetuses? Are reasons based on the family's quality of life morally acceptable?

Another issue that should be underlined is that the techniques used and the exact time of pregnancy chosen to perform selective and nonselective procedures are different. A selective fetal reduction, independent of the medical or nonmedical indication, will take place in the second trimester and can be achieved in different ways. When there are two separate placentas (dichorionic pregnancy), authors agree selective FR should be performed before 16 weeks of amenorrhea. This time frame is considered the most adequate because the procedures performed earlier correspond to a lower rate of prematurity for the healthy twin and a longer gestation, allowing for a higher weight at birth, FR is performed in this instance by an injection of potassium chloride (KCl) in the heart or in the umbilical cord of the fetus. When the twins share a single placental mass (monochorionic pregnancy), it seems that the healthy twin usually has a better chance of survival if selective termination is carried out after 18 weeks. If the disease is detected later, during the second trimester for example, medical teams prefer to wait for the fetuses to become viable, therefore allowing the healthy twin to survive in the event of a premature delivery. In case of monochorionic pregnancy, the usual approach is to use a bipolar forceps to burn the umbilical cord. Concerning nonselective FR, it is usually performed in the first trimester of pregnancy under ultrasound guidance, using a transabdominal injection of KCl into the region of the fetal thorax.

Today, to reduce the occurrence of high-risk multiple pregnancies around the world, several recommendations have been made with the purpose of implementing a policy of single-embryo transfer and monitoring ovulation induction (Legendre et al., 2013). Hence, high-risk multiple pregnancies will become increasingly rare, thus reducing the usage of fetal reduction. However, Evans and Britt (2010), highlight that with the growing availability of egg donors, the number of older women, many of whom already have children from previous relationships, using ART and seeking FR to a singleton will steadily increase. According to the same authors, the consequences of economic concerns resulting from high costs for every cycle and the pressure to achieve a very high pregnancy rate with each cycle should also be considered.

These issues place women/couples in an unprecedented position in relation to reproductive technologies, opening up an arena where medicine, social values, and culturally determined meaning are closely intertwined and should be under increasing ethical scrutiny. In fact, what began as an extreme medical procedure, to sacrifice some fetuses in order to save the lives of the others, appears to have led to a wide range of new indications that are not limited to medical reasons. As we have already stated before, FR can be used due to economic concerns and in order to avoid emotional and social burdens that could strain the entire family, thus taking on many of the contentious features of abortion. Indeed, selective reduction has turned into a clandestine practice that may be even more stigmatized than abortion. Many physicians openly refuse to perform "elective" reductions from a twin-pregnancy to a singleton, and even some people who proclaim themselves to be "prochoice" pass judgment upon those who knowingly choose to use assisted reproductive technologies (ART) and then opt for selective reduction (Rao, 2015). These issues will be analyzed in the next part of this text.

2. THE MORAL STATUS OF EMBRYO AND FETUS

Many factors influence how we feel about reproductive issues. Decisions in this area are very personal and go to the core of our beliefs and values. Because FR involves the destruction of a human fetus, the central ethical question relating to the use of FR is the moral status of the fetus and the potential eugenic dimension of these techniques. For some critics, FR is seen as "eugenic," because it may enable "frivolous choices," that is, facilitate the selection of fetuses with certain desirable physical characteristics, like sex, or the number of fetuses that will be further developed with only a social purpose. The moral status of the embryo and the fetus is a debate that remains controversial. No agreement has been reached as to what would be the best approach to determine it. Different assumptions about the moral status of an embryo have led to different conclusions about the appropriate respect for the embryo. As Callahan points out, "Science may eventually be able to empirically explain everything to be known about embryos, their genesis, and their development. But it is beyond the capacity of science to tell us how we ought to treat embryos or evaluate their moral status (2012, p. 206)." Thus, we have to look for answers to the ethical question regarding the moral status of the embryo in moral and empirical analysis. It might seem too wide a room to find moral principles and fundaments, but in the end we can reach some common ground just by observing that there is uneasiness concerning the uncertainty about the moral status of the embryo. This uneasiness stems from the potentiality criteria (before reaching our current stage of life, we were all embryos) and from the fact that, in spite of the difference of value we might give to the beginning of life, *that beginning is hard to ignore* (Callahan, 2012, p. 207). Three main moral positions have been identified and described: full moral respect, gradualistic respect (without any division), and no moral respect (Martinho da Silva et al., 2015).

Two leading theories of moral status appear to lead to diametrically opposed positions concerning the respect for the human embryo. If the threshold condition for full moral status is the range property of being a person, then assuming that the embryo is neither rational nor self-aware, it would seem that the embryo would fall well below the threshold of moral respect. On the other hand, if the threshold condition for full moral status is being human, then assuming that the embryo is a human being, it would seem to be entitled to the same basic rights as any other human being.

For those who consider that human life, as represented by the embryo or fetus, always takes precedence over the freedom of the individual (reproductive autonomy) and that the human life deserves respect since fertilization, FR is ethically unacceptable. This position too, which assumes that an embryo possesses human dignity and the right to life from the beginning, could include the view that some FR uses, specially the medical ones, may be ethically acceptable in certain specific conditions, in the sense of a lesser evil. A graduation of the protection of unborn life can be based on a number of different initial positions. From this perspective, the use of FR may be considered as acceptable depending on the developmental stages of the fetus. If no moral respect is attributed to the embryo, nor would it be regarded as having a right to life, the previous presented uses, medical and nonmedical, would be ethically acceptable.

3.   FETAL REDUCTION: "A SOFT COVER FOR HARD CHOICES"

As previously described, FR is the practice of terminating one or more fetuses to reduce a multiple pregnancy, e.g., from quadruplets to twins or a singleton. Therefore the possible connections between FR and abortion should be analyzed. For some, fetal reduction appears to be "a soft cover for hard choices," while for others the name is not sufficiently neutral because it conveys undesirable connotations.

Fetal reduction and abortion both involve the destruction of fetal life, but they are classified by different designations to underscore that they are regarded as fundamentally different medical procedures: both are performed using distinct techniques by different types of physicians upon women in very different circumstances: selective reduction often entails the deliberate termination of lives that were purposefully created at great effort and expense, whereas abortion is usually the response to an unplanned or accidental event. Focusing upon the difference in intent, Judith Daar states, "A woman undergoing a "traditional" abortion intends that her entire pregnancy will be terminated, i.e., following successful completion of the procedure, she will no longer be pregnant. In contrast, a woman undergoing fetal reduction intends that her pregnancy will not be terminated, but rather will be enhanced by creating a better environment for her remaining fetus(es) to develop" (1991). Using the lifeboat metaphor portrayed by Judith Daar (1991), we may consider fetal reduction as a desperate act of self-defense (and defense of others) that requires some fetuses to be jettisoned in order to safeguard the lives of the remaining fetuses and the mother. However, sinking the lifeboat is the consequence of a voyage that the ship's captain undertook after having knowingly permitted too many passengers to come on board, aware of the risk of a dangerous lifeboat situation. Here lies the foundation to include in this debate the principle of Responsibility as well as the Aristotelian virtue of Prudence.

A further issue that could also be included within this metaphoric language is that what began as an extreme "medical" procedure to sacrifice some fetuses in order to save the lives of others appears to have evolved into an option sought for a wide variety of reasons that are not limited to preserving the life or the health of the remaining fetuses and the mother but is also considered for social and economic reasons of the remaining passengers. Therefore, as FR crossed the boundary between a procedure that was justified solely for medical reasons and another one increasingly performed for social considerations, it appears to have taken on many of the contentious features of abortion. One could argue that these two techniques represent a distinction without a difference. After all, in both instances one or more fetuses are killed. While fetal death is a fact in both instances, the technical differences and the intended outcomes do distinguish the procedures. Therefore the opposite outcomes of each the procedures should also be considered; a woman is no longer pregnant following an abortion; a woman remains pregnant following FR. Hence, even if there are reasons to consider FR and abortion different medical procedures we must questioning if the two procedures appear to call for a distinct ethical evaluation. Fetal reduction seemed distinct from abortion when the decision to reduce the number of fetuses was dictated by purely medical criteria (although some abortions are also based upon medical criteria), but this boundary began to blur when the choice expanded to encompass economic, emotional, and social burdens. Therefore the labeling of FR as a medical procedure cannot constitute an excuse for the physicians and health systems not be accountable for the implantation of an excessive number of embryos, the numbers of FR performed, the time of pregnancy at which the FR has been performed, the motives and consequences of this practice.

4. PARENTAL AUTONOMY AND PARENTAL RESPONSIBILITY

Although everyone recognizes the strong desire of human beings to beget children and the important part this plays in the fulfillment of most people, the existence of an actual right to procreate, including the right to obtain reproductive assistance for the purpose of having a child, is still debated (Magalhães and Carvalho, 2010). The principle of respect for autonomy is a central tenet of modern medical ethics and it is often considered one of the most important principles in this area. Freedom cannot be equated with the unlimited, arbitrary exercise of the will, but always involves an element of responsibility. At present the key question concerns the extent of reproductive autonomy or, in other words, the limits to reproductive freedom. While one would strive to encourage autonomous decision-making, one recognizes that there may be problems with unfettered and unregulated choices. Autonomy is often used to justify the translation of the wish to have a child into the right to have one. But this is fallacious, because autonomy can only be ethically acceptable when it is applied to oneself or to the respect for the other, never as a claim for the other, in this case, a claim to have a specific child. In order to enact this respect for autonomy, one should not take this principle as an isolated one, ruling over the other bioethical principles, as if it were autonomous from beneficence, nonmaleficence, integrity, solidarity, and vulnerability. Perhaps the best formulation is that, while autonomous decision-making should be supported and encouraged, it is legitimate to limit this autonomy in certain situations. For example, where it exercises an unreasonable impact on the autonomy of others, or threatens others with significant harm. Respect for autonomous decision-making does not need to be incompatible with the recognition some hold that protection of the embryo/fetus or future child should be enforced even if it interferes with the parents' freedom. In ethical terms, a "couple's reproductive autonomy" is in fact indissolubly linked with its willingness to accept parental responsibility, which implies responding to and for the child before and after birth.

Elements considered as a necessary condition for autonomous decision-making include the presence of valuable options, clear information about the alternatives, and the absence of actual constraints. This consideration entails that external as well as internal constraints should be absent, such as emotions, doubts, or scruples that may limit the capacity for reflecting critically upon the available options. These constraints should also not impede to act in accordance with the individuals' underlying values and preferences. Often however, the decisions taken by couples to reduce the number of fetuses can be seen as lacking true personal autonomy because of parental desperation, medical coercion, and a lack of informed consent (Ring-Cassidy and Gentles, 2003). Due respect for parental autonomy requires that parents have access to all the information concerning the decision they are making and that they are not subject to coercion. A concern has been expressed that the mere availability of fetal reduction (or other selective technologies) might result in a form of coercion for the parents, who may come under a form of social pressure to use it.

As previously observed by Magalhães and Carvalho (2010) in relation to PGD the boundary between choice and coercion is not a clear cut one for health practitioners, with an amount of slippage occurring between these concepts. When giving the information following ART, slippage can occur due to the use of languages of risk, probability, and social implications that may be inferred in such classifications as "affected," "unaffected," "normal," and "best/good" embryos.

As a result, several authors (Legendre et al., 2013) affirm that, before every assisted reproduction treatment, patients should be duly informed of the risks they face and of the potentiality of having to resort to fetal reduction following a multiple pregnancy. As a matter of fact, the official authorities of several countries responsible for monitoring or regulating assisted reproduction treatment indeed caution and inform via their websites about the risks associated with multiple pregnancy induced by such procedures. The concept of preparedness refers to the state of being prepared or ready for specific or unpredictable events. Therefore, before beginning a fertility treatment, all couples should be informed that they may have to decide whether or not to resort to fetal reduction and be asked about the number of embryos that will be implanted (Legendre et al., 2013). However, despite some precautions, several studies reveal that couples requesting assisted reproduction treatment assimilate the information on the risks of multiple pregnancy poorly and hence also the possibility of having to decide about the recourse to fetal reduction; couples are too absorbed in their desire to have children and tend to consider the possibility of multiple pregnancy as unlikely and some women consider themselves as "too infertile" to develop such pregnancies.

5. CONCLUSION

Because FR affects the integrity of more than one human being, presenting a high degree of uncertainty, it imposes a highly complex decision-making process on the pregnant woman, her partner and the health practitioners during the prenatal period (Legendre et al., 2013).

Even with the impossibility of reaching consensus and the difficulty of compromised view about both moral status and moral respect, we have an obligation to continue to engage in the debate, not only in order to give those with whom we disagree a fair hearing but just as importantly in order to respect the other attributions of moral status, even if their fundaments cannot be recognized. If a compromise cannot be reached regarding the moral status and moral respect in terms of the individual position as such, it is appropriate to try to work in order to respect the attributions of moral status by other moral agents (Martinho da Silva et al., 2015). Respect for human life is the common ground upon which the ethical debate must stem from. Bioethical principles such as autonomy, responsibility, vulnerability, and respect should always protect human rights, which are the only consensual fundament to be considered in the pluralistic society we live in. Restoring Aristotelian Prudence within debates on FR issues means to consider four main obligations (ACOG Committee Opinion No. 369: Multifetal Pregnancy Reduction, 2017) to tackle the real causes of infertility, placing this debate within public health domain; to improve the informed consent process, by assuring that it is dialogical, diachronic, and integrative, i.e., it integrates into the present deliberation procedure the possible consequences that loom in the future; to assume from the beginning that terminology matters, and the words we use will surely influence the decisions we make; to rethink parenthood as a web of responsibilities, vulnerabilities, duties, and rights, which are highly visible when it comes to deciding about using medically assisted reproduction techniques to procreate. This web also includes the vulnerability and responsibility of health professionals and scientists, who have to be aware that the obligations of each of them will amplify all of them in the end. Society cannot use the word respect as a way of *domesticating the uneasiness* caused by FR in many cases. When it comes to medical procedures, one should not forget that the mission of physicians and other health professionals is to alleviate suffering. Therefore ART can only be medical if they aim at alleviating suffering, rather than causing more harm than good. Every reproduction technology deals with anthropological and ethical issues, because it has an impact on the way we understand ourselves ethically. Only by facing the ethical issues posed by ART, can infertile couples achieve all the benefits that science provides them with.

REFERENCES

ACOG Committee Opinion No. 369: Multifetal Pregnancy Reduction, 2017. Obstet. Gynecol. 130 (3), e158–e163.

Callahan, D., 2012. The Roots of Bioethics. New York: Oxford University Press.

Council of Europe, (1997). [online] Available at: http://Convention for the Protection of Human Rights and Dignity of the Human Being with regard to the Application of Biology and Medicine [Accessed 6 Dec. 2017].

Daar, J.F., 1991. Selective reduction of multiple pregnancy: lifeboat ethics in the womb. U.C Davis Law Rev. 25, 773.

Evans, M., Andriole, S., Britt, D., 2014. Fetal reduction: 25 Years' experience. Fetal Diagn. Ther. 35 (2), 69–82.

Evans, M., Britt, D., 2010. Multifetal pregnancy reduction: evolution of the ethical arguments. Semin. Reprod. Med. 28 (04), 295–302.

Legendre, C., Moutel, G., Drouin, R., Favre, R., Bouffard, C., 2013. Differences between selective termination of pregnancy and fetal reduction in multiple pregnancy: a narrative review. Reprod. BioMed. Online 26 (6), 542–554.

Magalhães, S., Carvalho, A., 2010. Searching for otherness: the view of a novel. Hum. Reprod. Genet. Ethics 16 (2), 139–164.

Martinho da Silva, P., Silvestre, M., Carvalho, A.S., Araújo, J., 2015. In: Ten Have, H. (Ed.), Embryo. Encyclopedia of Global Bioethics. Springer International Publishing, Switzerland, pp. 1–12.

Rao, R., 2015. Selective reduction: "a soft cover for hard choices" or another name for abortion? J. Law Med. Ethics 43 (2), 196–205.

Razaz, N., Avitan, T., Ting, J., Pressey, T., Joseph, K., 2017. Perinatal outcomes in multifetal pregnancy following fetal reduction. Can. Med. Assoc. J. 189 (18), E652–E658.

Ring-Cassidy, E., Gentles, I., 2003. Women's health after abortion: the medical and psychological evidence. Natl. Catholic Bioethics Quart. Spring 3 (1), 175–188.

Silvestre, M., 2012. Sex selection and assisted reproduction technologies. Lex Med. 9 (17), 157–163.

Tse, W., Law, L., Sahota, D., Leung, T., Cheng, Y., 2017. Triplet pregnancy with fetal reduction: experience in Hong Kong. Hong Kong Med. J. 23 (4), 326–332.

Zipori, Y., Haas, J., Berger, H., Barzilay, E., 2017. Multifetal pregnancy reduction of triplets to twins compared with non-reduced triplets: a meta-analysis. Reprod. BioMed. Online 35 (3), 296–304.

Chapter 10

Stem Cell Therapies for Neurodegenerative Disorders: An Ethical Analysis

Sorin Hostiuc*, Ionut Negoi†, Mugurel Constantin Rusu‡ and Mihaela Hostiuc§

*Department of Legal Medicine and Bioethics, Faculty of Dental Medicine, "Carol Davila" University of Medicine and Pharmacy, Bucharest, Romania, †Department of Surgery, Faculty of Medicine, "Carol Davila" University of Medicine and Pharmacy, Bucharest, Romania, ‡Department of Anatomy, Faculty of Dental Medicine, "Carol Davila" University of Medicine and Pharmacy, Bucharest, Romania, §Department of Internal Medicine and Gastroenterology, Faculty of Medicine, "Carol Davila" University of Medicine and Pharmacy, Bucharest, Romania

1. INTRODUCTION

Neurodegenerative disorders are characterized by progressive morphological or functional neuronal alterations, most of them without any curative treatments, and being often associated with a negative outcome. The most well-known neurodegenerative disorders are Alzheimer's disease (AD), Parkinson's disease (PD), Huntington's disease (HD), Friedreich's ataxia, amyotrophic lateral sclerosis, spinal muscular atrophy, and Lewy body disease (Barker and de Beaufort, 2013; Paulsen et al., 2013). These disorders affect distinct types of neural cells, and alter numerous physiopathological mechanisms, including (but not being limited to): neuronomuscular junction degradation, synaptic defects, proteasome inhibition, axonal transport defects, mitochondrial disorders, neurofilament accumulation, microglial/astrocyte activation/toxicity, glutamate-mediated toxicity, nuclear protein aggregation, cytoplasmic protein aggregation, and endoplasmic reticulum disorders (Blasco et al., 2014; Brown and Neher, 2010; Haston and Finkbeiner, 2016; Hetz and Mollereau, 2014; Mason et al., 2014; Van Damme et al., 2005).

Even though, at least currently, there are limited treatment options for these disorders, as they have "caught" the public eye due to their debilitating nature and rise in prevalence associated with increased life expectancy in recent years, numerous clinical trials were initiated to potentially find cures (Robillard et al., 2011), or at least to deepen the understanding of their physiopathology.

Clinical Ethics at the Crossroads of Genetic and Reproductive Technologies.
https://doi.org/10.1016/B978-0-12-813764-2.00010-6

Many trials went beyond classical, pharmacological, and surgical therapies by using stem cells (Hostiuc et al., 2016c), deep brain stimulation (Fukushi, 2012), and genetic engineering technologies such as crispr/cas9 (Yang et al., 2016).

In this chapter, we will focus our attention toward three main ethical issues surrounding stem cell therapies for neurodegenerative disorders. First, we will talk about the ethical acceptability of sham surgery, as clinical trials aimed toward testing them need a control group, and sham surgery was seen as the most plausible candidate. Next, we will focus our attention on the risk-benefit analysis of using stem cell therapies in neurodegenerative disorders. In the last part of the chapter, we will analyze the ethical issues surrounding the use of induced pluripotent stem cells for neurodegenerative disorders, with an emphasis on their moral status.

2. AN OVERVIEW ON STEM CELL THERAPIES FOR NEURODEGENERATIVE DISORDERS

2.1 Parkinson's Disease

Parkinson's disease is a neurodegenerative disorder of the central nervous system that mostly affects the motor system, characterized by shaking, rigidity, and difficulty walking followed, usually in later stages, by thought and behavioral disorders, depression, anxiety, and dementia (Hostiuc et al., 2016a). Pathologically, it is characterized by cell death in the basal ganglia (mostly affecting dopaminergic neurons) and the accumulation of alpha-synuclein proteins (Lewy bodies) in many of the remaining neurons, associated with a decreased density of astrocytes and an increased density of microglia in locus niger (Jankovic and Tolosa, 2007, pp. 271–274).

Stem cell therapies for PD have been tested since the 1970s, following the seminal works of Brundin et al. (1988) and Björklund et al. (1983). In various studies, they showed that dopaminergic neurons, obtained from the embryonic ventral mesencephalon, could be grafted in animal models of this disease (Brundin et al., 1988). Moreover, they were able release dopamine, and link to other neurons through synapses, leading to positive clinical results (Barker et al., 2016; Brundin et al., 2010). Soon after, the first clinical trials were developed using tissue from 12- to 14-week-old (Madrazo et al., 1988), or 8- to 10-week-old (Lindvall et al., 1989) embryos. Even though the initial studies showed a minimal clinical improvement, follow-up studies fine-tuned the methodologies, leading to important positive outcomes, with the subjects being included in three main categories: some showed significant improvement of the disease, even allowing a discontinuation of the pharmacological therapy; others showed a slight improvement, but still needed pharmacological therapy; in the last group were included subjects that failed to show any improvements related to the motor aspects of their disease (Brundin et al., 2010). As the results were extremely variable, in the late 1990s two studies were developed that aimed to test the usefulness of stem cell transplantation in PD, using a double-blind approach. Freed et al. decided to run a double-blind placebo-controlled trial to determine if the benefit of stem cell transplantation was greater compared to placebo. For this purpose, they selected 40 patients, from 34 to 75 years old, with severe PD, for the study. To preserve the impression that the surgical intervention was performed for the placebo group, the investigators used sham surgery. They drilled holes in the skulls of the subjects without penetrating the dura. The patients were followed for a year after surgery. The physicians who did the follow-up did not know if their patients had the sham or the actual surgery. The consent form detailed the risks and potential benefits, and it was approved by the IRBs from the University of Colorado, Columbia University, and North Shore University Hospital. It is not clear if the patients were aware that they could receive sham surgery. The investigators obtained fetal tissue after getting a written consent from the women who requested the abortion procedure. In patients who had the surgery and were younger than 60, there was a significant improvement as measured by standardized Parkinson's tests. However, five patients with transplant showed late dystonia and dyskinesia, suggesting the procedure still needed refinement (Freed et al., 2001; Hostiuc et al., 2016c). In 2003, Olanov et al. published another study that was conducted at Mount Sinai (New York) and Rush (Chicago) and included 34 patients with advanced PD, from 30 to 75 years old. In the study were included subjects with at least two cardinal signs of PD (tremor, rigidity, bradykinesia), a good response to levo-dopamine, stable doses of antiparkinsonian drugs, and motor complications that were not managed properly using pharmacological therapy. Patients with conditions affecting decisional capacity were excluded. All patients signed the informed consent, after which they were separated into a cases cohort, which received the stem cell therapy, and controls, which received a sham surgery. The primary outcome was represented in a change in the motor component of the Unified Parkinson's Disease Rating Scale (UPDRS) in the practically defined "off" state, between the T0 and T0+24 months. The study failed to show significantly improved responses in the subjects from the active arm, even though there was a trend toward a positive response (Olanow et al., 2003). These inconclusive results, associated with the development of deep brain stimulation techniques, and some inherent issues associated with these stem-cell-based regimens (see Table 1) have limited the development of follow-up studies until recently, when new projects, such as the EU-funded Transeuro project (TRANSEURO | Home, n.d.), and the potential usage of induced pluripotent stem cells (iPSCs) rebooted the interests in this technique.

2.2 Stem Cell Treatments for Other Neurological Disorders

Huntington's disease is an autosomal dominant disease caused by a mutant gene for huntingtin, causing disseminated neuronal dysfunction leading to complex neurological abnormalities (including cognitive, motor, psychiatric,

TABLE 1 Major Issues With Stem Cell Therapy in PD (Barker et al., 2016; Barker and de Beaufort, 2013; Brundin et al., 2010)

The need to obtain stem cells from human embryos (therefore generated by the termination of a pregnancy)
The need to obtain ventral mesencephalon tissue from multiple embryos to obtain enough biological material for transplant
A limited number of embryos were useful for transplantation: those too little (below 14 mm) were difficult to dissect to obtain the needed cells, those too big (above 28 mm) led to a poor survival of the neurons in transplanted hosts.
Patient particularities, which rendered basically impossible the standardization of the technique, which would have been required to obtain reproducible results
The development of apparently better technologies (such as deep brain stimulation)
Many PD-related symptoms could not be alleviated by this treatment, as the underlying substrate was not represented by dopaminergic-motor neurons
The appearance of an important complication, namely graft-induced dyskinesia (GID) in many subjects
A potential risk of tumorigenesis

metabolic, and sleep related), and ultimately, death. HD usually appears in middle aged persons, and has a progressive negative course for around 20 years (Wijeyekoon and Barker, 2011). The earliest, and most severe alterations, appear in the striatum, which made it the main target for potentially curable treatments (Wijeyekoon and Barker, 2011). In 1992, after some initial positive trials on animal subjects, Sramka et al. published a report regarding fetal neural transplantation in the basal ganglia of the subjects with HD, followed by immunosuppression with cyclosporine, showing a slight symptomatic relief of the hyperkinesia (Sramka et al., 1992). In the following years many studies were published regarding stem cell transplantation in Huntington's disease, most of them with a very low number of subjects (usually less than 10), using variable methodologies, and with variable success (for a review, see Wijeyekoon and Barker (2011)).

Amyotrophic lateral sclerosis is characterized by dysfunction and death of motor neurons located in multiple locations (spinal cord, brainstem, cerebral cortex), causing progressive muscle weakness and death within a few years. Recently, embryonic stem cells were shown to be useful to differentiate motor neurons (Wichterle et al., 2002), which were shown to establish functional synapses with muscle fibers (Harper et al., 2004). However, there are no significant clinical trials on human subject aimed to test stem cell-based therapies for the ALS.

Alzheimer's disease (AD) is characterized by neuronal and synaptic loss throughout the brain (both cortex and some subcortical regions), caused by protein misfoldings due to abnormal accumulation of amyloid beta protein and tau protein (Arias and Karlawish, 2014; Becker, 2005; Muratore et al., 2014). Clinically, AD leads to progressive memory loss, impairment in semantic memory, apathy, apraxia, agnosia, paraphasia, illusionary misidentification, sundowning, irritability, and anosognosia (Förstl and Kurz, 1999). As, in AD, the neuronal loss is widespread, a focal replacement of neurons, as done for PD is of limited use (Lindvall and Kokaia, 2006). However, there have been some studies suggesting that transplantation of fibroblasts in the forebrain could lead to the production of nerve growth factor, which could counteract neuronal death, stimulate cell function, and improve memory, at least in animal models (Fan et al., 2014; Tuszynski et al., 2005).

3. USE OF PLACEBO CONTROLS IN SHAM SURGERY[1]

Placebo is a form of simulated medical treatment intended to deceive the patient/subject who believes that he/she received an active, beneficial, medical intervention. In clinical medicine, the use of placebo is allowed in particular circumstances, and for particular pathologies, such as depression, anxiety, or surgical-related pain. The use of placebo is limited in clinical medicine because it can generate distrust within the physician-patient relationship. Siegler argued that placebo could be used in clinical practice if four conditions are simultaneously met: (1) the condition is known to respond well to placebo, (2) the alternative to placebo is either continued illness or the use of a drug with known adverse risks or addiction, (3) the patient wishes to be treated, and (4) the patient insists on obtaining a prescription from the physician (2010).

In medical research, placebo is more often used, in the control arm, if there is not a best-alternative-treatment (BAT) to which the active intervention could be compared. Investigators use placebo to differentiate the real effect, caused by an active intervention, from subjective effects, caused by the belief of the subject that he/she has received a beneficial medical intervention. A major advantage of placebo-controlled trials compared to BAT-controlled trials is the difference in effect sizes between the cases and control groups. Compared to BAT, in placebo-controlled trials the effect-size is higher, and the number of subjects needed to reach a certain statistical power is lower. Other reasons in favor of placebo-controlled trials include: (1) a new therapy might not be better concerning the primary outcome compared with the best available therapy, but it might be advantageous in other ways (safety, compliance, tolerability, cost), (2) the best available therapy control might show an inconsistency in effects caused by methodological differences (e.g., the inclusion criteria, that might be different compared to the ones used to prove its usefulness), and (3) the presence of

---

1. An initial version of this subchapter was published in Hostiuc et al. (2016c).

methodological limitations in using the best available therapy option (Castro, 2007). For these reasons, researchers tend to prefer placebo instead of best alternative randomized controlled trials, and often try to "bend the rules" and develop research protocols using placebo when an alternate plan could be designed.

The use of placebo in surgery trials is even more controversial. Clark, for example, considered that, unlike placebo-controlled trials in pharmacological research, sham surgeries "fail the test of beneficence" (Clark, 2002). Weijer gave some persuasive arguments for this, including the absence of a therapeutic purpose, important scientific disadvantages, and a noteworthy risk increase (2002). Some authors developed specific ethical frameworks for the use of sham procedures in trials. For example, Horns and Miller provided a six-step ethical framework for assessing the acceptability of sham surgeries, which included: *"1) there is a valuable, clinically relevant question to be answered by the research, 2) the placebo control is methodologically necessary to test the study hypothesis, 3) the risk of the placebo control itself has been minimized, 4) the risk of a placebo control does not exceed a threshold of acceptable research risk, 5) the risk of the placebo control is justified by valuable knowledge to be gained, and 6) the misleading involved in the administration of a placebo control is adequately disclosed and authorized during the informed consent process"* (2003).

3.1   Is Sham Surgery Accepted by Potential Subjects?

Even though the potential subjects' acceptability of a clinical trial is not a de facto mandatory condition for its initiation, a low acceptability score could render the process of finding/selecting subjects cumbersome. Therefore, one of the first questions that should be asked by any investigator who aims to perform sham surgery should be related to the subjects' acceptability of the procedure. There have been a few studies dealing specifically with this issue. Frank et al. performed a study on subjects with and without PD to assess their willingness to participate in neurosurgical trials for this disease. The investigators selected three groups of patients: with PD, without PD but with other neurological diseases (dementia excluded), and patients from primary care. They then gave each the option to be either included in an unblinded trial, a blinded trial, or not to participate at all. Most subjects from each group preferred to be included in the unblinded trial. The highest number of subjects selecting nonparticipation was in the PD group (34%, while in the other two the maximum nonparticipation portion accounted for 10.4%). Also, the PD group of subjects was the least willing one to be involved in a blinded study (24.5%, while the other groups favored this option in a percent from 35% to 40%) (Frank et al., 2008). The authors concluded: *"patients with PD, when compared with patients with non-PD neurology or primary care, may have adapted to their chronic illness and may not be so desperate that they would be more eager to participate in risky research. In fact, they appear more cautious"* (Frank et al., 2008). The conclusion is subjective, and not based on the actual study. Maybe, for example, the PD patients did not want to be included in the blinded trial because they were directly affected by the procedure, and felt the risks were too high. Whenever we would like to analyze the opinions of patients regarding a certain medical procedure, we must take into account all the possible reasons for a certain response, and whether their replies are in line with their actual beliefs. Such patient surveys tend to be performed more frequently and used as objective proofs, suggesting that patients agree with more controversial issues, not taking into account the validity of the questionnaires that are used or the mere fact that, for a certain procedure to be agreed upon, needs to be obtained an individual consent and not a population agreement regarding its usefulness. Moore et al., in the TransEuro project, found that subjects enrolled in PD clinical trials tend to be more educated, younger, with a higher cognitive score, and better motor function compared to patients who were eligible, but not included in the trial, and argued that this could raise problems regarding a parity of access to clinical trials (Moore et al., 2014). However, this could also be a method of protecting vulnerable subjects. Swift, in a qualitative study about the perspective of the patients and their relatives about sham surgery in PD, showed that participation was acceptable for a small majority of interviewers, but the main reasons for accepting it seemed to be the severity of the disease and the lack of good treatment options. Moreover, the surveyed persons preferred real to sham surgery; this comes to support the idea that subjects see themselves primarily as patients, that the acceptance for participating in clinical trials is not based on altruistic reasons (Swift, 2012), and that therapeutic misconception might be significant.

3.2 Sham Surgery as a Form of Mitigated Trolley

Albin believes that sham surgery could be seen as a form of mitigated trolley (2005). The trolley dilemma is often used to test the moral intuition for circumstances in which a few people are put at risk to save more. In this problem, a runaway trolley goes down a track toward five men who will be killed if it is not stopped or diverted. The trolley cannot be stopped, but it can be diverted. However, on the secondary line, there is another person who will be killed by the trolley. Therefore should one save five with the price of one life through action, or save one and have the other five die by not acting? Albin considers that sham surgery can be partially assimilated with a trolley in which the conductor diverts the line but puts a padding on the front, to cushion the impact of the trolley. The decision similar to the one made by the trolley conductor is to perform a clinical trial, and subsequently to put some people at a mitigated risk (the surgery is partially simulated, so cushioned), to aid many (Albin, 2005). What is wrong with this approach? Even if it apparently leads to a maximization of the benefits, and is often used by the supporters of utilitarian ethics in health care, we believe that it contradicts the utilitarian moral theory.

Jeremy Bentham, in his book Deontology of the Science of Morality, said that an action is correct or incorrect, deserving or not, receiving approval or disapproval, reported to the tendency in which it causes the increase or decrease in the quantity of public happiness (Bentham, 1834). This means that, when we analyze whether an act is moral or immoral from a utilitarian perspective, we should examine not only the good done to the ones directly affected by our actions but also the good generated by them in the general population. If a physician saves a few lives at the expense of sacrificing one, apparently, he does more good directly; however, his actions might cause a decrease in the trust in physicians in general—Why should I, as a patient, go to a doctor if he might sacrifice me for the good of others? This decrease in trust would cause a decreased addressability of patients toward the health-care system, and a decreased therapeutic compliance, therefore causing more harm overall. Similarly, this could decrease the trust in clinical trials, and subsequently the number of potential subjects willing to participate.

3.3   Risk to Benefit Analysis

One of the major reasons for accepting sham surgeries is represented by the fact that minor risks for a few patients are considered to be less important than the overall potential benefit the results of the study might lead to, for both the subject and the population potentially benefiting from it. For example, Olanow et al. stated, as a conclusion of their study, that *"This study did not confirm the clinical benefits reported in open-label trials. Furthermore, unanticipated and potentially disabling off-medication dyskinesias developed in greater than 50% of patients. We cannot therefore recommend fetal nigral transplantation as a therapy for PD at this time. It is possible, however, that enhanced benefits can be obtained in patients with milder disease, with transplantation of higher numbers of cells, and with more prolonged immunosuppression"* (Olanow et al., 2003). Therefore even if more than 50% of all subjects developed "unanticipated and potentially disabling off-medication dyskinesias," the authors concluded that more studies should be performed to search for additional potentially beneficial effects, and not to specifically search for methods of minimizing the risks for subjects, and possibly, patients.

The potential benefit for patients with PD is significant if the therapy will actually have a positive clinical effect. Another question we must ask is how would the subjects from the control group benefit from it? Normally, if a medical intervention is shown to be beneficial during a clinical trial, the subjects from the placebo group would receive the same procedure after the end of the trial; therefore, they would benefit from all the positive results of the trial without risking any unforeseen complication generated by the implantation of fetal stem cells in their brains. However, for this purpose, they would suffer two surgical interventions (one for the trial and one for the therapy), which might be associated with significant risks, especially taking into account the fact that most patients are old, with a severe pathology, and subsequently have a higher surgical risk. Performing sham surgery on patients with PD, puts them at a more than minimal risk, especially regarding anesthesia-related, which should warrant a careful analysis of the benefit differential compared to less invasive measures. Weijer considered that sham surgical procedures should be analyzed as nontherapeutic interventions, and argued that a nonsurgery control is likely a better option. According to him, therapeutic procedures should pass a test of clinical equipoise, and a harm-to-benefit analysis should be performed. Nontherapeutic procedures do not offer the prospect of benefits to the individual subject, and therefore, the harm-to-benefit analysis is not appropriate. Instead, this analysis should be replaced with the minimization of the risks consistent with a sound design and the reasonability of the risks in relation to the knowledge to be gained (Weijer, 2002). By using these principles regarding risk analysis, the use of sham surgery should be forbidden. Minimization of risks, in nontherapeutic interventions, cannot be correlated with the magnitude of the benefit; therefore, any risks that are more than minimal are in contradiction with the principle of nonmaleficence, and should not be allowed in clinical trials using sham controls.

3.4 What Role Should Collateral Benefits Have in the Decision to Allow Sham Surgery?

Beside a direct benefit for the subjects, derived from the actual therapeutic intervention, the possibility for clinical trial subjects to receive collateral benefits generated by the inclusion in the trials is sometimes discussed. In this case, all subjects would benefit from the best possible treatment (Emanuel and Emanuel, 1992; Hostiuc, 2014a,b; King, 2000). Nancy King argued that collateral benefits should not be used as types of benefits included in the risk analysis of clinical trials for two main reasons: (1) by providing a potentially higher standard of care for subjects compared to patients, we might potentially discourage the improvement of the standard treatment, and (2) as collateral benefits are under the control of the investigators/sponsors, they might become a means of manipulating or even coercing vulnerable patients to enter the clinical trial (King, 2000). We agree with her opinion, that these collateral benefits should not be taken into account as an element of the risk to benefit analysis. However, if the study generates such a benefit, the participants should receive it, as a reward for their altruistic participation in the study.

3.5 Are Placebo-Controlled Trials Actually Needed in Surgery Studies?

We presented above a few reasons for which investigators prefer placebo to BAT in randomized clinical trials. These reasons (and others) might cause them to try a slight bend of the strict ethical rules governing the use of placebo in clinical trials. For example, Dekkers and Boer argued that, for Parkinson's trials, an alternate design could consist of a core assessment procedure, in which measurement protocols are applied to the subjects before and after the surgical intervention (2001). Avins argues that unbalanced randomization might be less morally problematic in some instances, with the risk, however, of losing statistical power (1998). Macklin suggested that "Cellular-based surgical therapies have much in common with pharmacologic treatments and lend themselves to evaluation in randomized, double-blind, placebo-controlled trial" (1999). Even if this approach is viable from a scientific point of view, this is not necessarily the case from a moral viewpoint. The risks associated with anesthesia and a surgical procedure are inherently higher than in most pharmacological clinical trials, especially for older individuals with severe associated pathologies. If they are to be accepted by the potential subjects, this is because of an inherent wish to get better and to receive an experimental treatment with significant benefits from a health-care point of view (Swift, 2012; Weijer, 2002). Between these two types of therapies are also substantial differences regarding the way concepts like autonomy, therapeutic misconception, and trust are perceived by the patients. Therefore we should not try to minimize, but rather emphasize, the differences between surgical and clinical trials in order to reveal the particularities of the latter. Only by doing this the ethical issues raised in practice by sham surgery are maximized and the therapeutic misconception can be minimized.

3.6 Autonomy Versus Therapeutic Misconception

According to Lidz and Applebaum, therapeutic misconception occurs "when a research subject fails to appreciate the distinction between the imperatives of clinical research and of ordinary treatment, and therefore inaccurately attributes therapeutic intent to research procedure" (Lidz and Appelbaum, 2002). The therapeutic misconception may be caused by the patient's expectations that the investigation will act in his/hers best interest even during a clinical trial, by the lack of understanding regarding the concept of randomization, by treatment constraints associated with clinical trials, or by a wish that the study will be beneficent to them (Byrne and Thompson, 2006). There are two main responses to therapeutic misconception: to accept it as an inevitable consequence of clinical trials or to implement measures whose purpose is to reduce it, including the use of the "neutral discloser," rewriting the informed consent forms, changes in the information algorithm used by physicians when trying to enroll a patient in a clinical trial, changes in monetary rewards, and research advertisements (Emanuel, 2008, pp. 633–644). Therapeutic misconception is especially high in fields in which the patients are highly vulnerable like oncology or psychiatry (Hostiuc, 2015). We believe that surgery could also fit in this category as a patient scheduled for a surgical procedure most likely expects a direct benefit resulting from the intervention. Moreover, sometimes even the proper information might not change his/her preconception regarding the clinical utility of the surgery. Therefore, to minimize therapeutic misconception, subjects must be explicitly informed about the sham surgery and the understanding by the subjects of this issue should be tested explicitly before they sign the informed consent.

The above-presented list is by no means exhaustive; it shows, however, the complexity of the problem and the difficulties of its ethical analysis. A series of guidelines have been developed regarding the possibility of using placebo (sham surgeries) in research, of which one of the easiest and most useful for surgical investigators is the one by Tenery et al., who based their ethical analysis on the following elements: (1) Placebo-controlled trials should only be used in surgery if there are no other designs that could lead to the necessary information. (2) Special care should be given to the obtaining of the informed consent. The potential subjects should clearly know the risk and the particularities of each arm of the study, with an emphasis on the interventions that would/ would not be performed. It is recommended for a third party (not the investigator) to obtain the informed consent. (3) Placebo controls should not be used when the investigators investigate the usefulness of a slightly modified surgical procedure. (4) Placebo controls should be allowed when a surgical procedure is developed for an affliction for which there is no surgical treatment, or if the efficiency of the standard surgical procedure is questionable, and if it is known that the affliction is potentially influenced by placebo, or if the risks of the placebo intervention are small. (5) If the surgical treatment has high risks, and the standard, nonmedical treatment is efficient and acceptable to the patients, it should be offered in all the arms of the study (2002). Additionally, we believe that a first step should consist of a proper analysis of the acceptability of the sham surgery by potential subjects. Moreover, specific measures should be taken to minimize issues like coercion generated by potential collateral benefits, or therapeutic misconception.

4.   RISK TO BENEFIT ANALYSIS FOR STEM CELL THERAPIES IN NEURODEGENERATIVE DISORDERS

One of the first elements that has to be clarified when initiating any new therapy is represented by the ethical acceptability of the study, which in turn is essentially influenced by the risk-benefit ratio. Both risk and benefits are defined probabilistically: risks—as the probability of a physical, psychological, economic, or legal harm that can affect either the subjects/patients or the society at large, and benefits—something of a positive value potentially generated by the study, for either the subjects/patients or society (Master et al., 2007).

In general, risks are seen as a combination between the probability of appearance and the severity of the unfavorable events potentially generated by an intervention (see Fig. 1).

A method to assess the ethical acceptability of the risks that has been intensely used in recent years, is the component analysis, or the double track

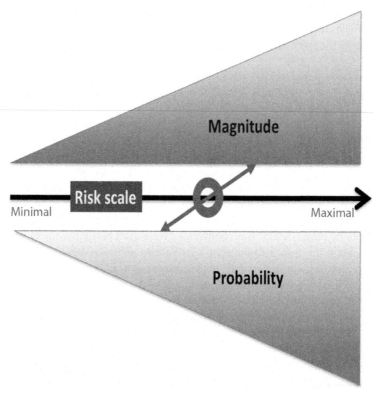

FIG. 1   Risk analysis. Risk is mainly composed of two elements: magnitude and probability. To assess the ethical acceptability of a risk, we should see where these two elements intersect a line (risk scale). Cutoffs to determine their probability, magnitude, and their intersection of the two can be used. *(From Hostiuc, 2014a. Ethics of Medical Research [Etica cercetarii medicale]. Casa Cartii de Stiinta, Cluj-Napoca, with permission.)*

method (Weijer, 2000; Weijer and Miller, 2004, 2007). The first stage of this method is to assess the presence of a therapeutic benefit. If the therapy generates a therapeutic benefit, the next stage is the analysis of equipoise. If the intervention satisfies equipoise (at the start of the study there is a state of genuine uncertainty regarding the preferred treatment within the community of expert practitioners), the study is acceptable. Otherwise, it is not. If there is no therapeutic benefit, the first step is to minimize risks (by taking advantage of therapeutic procedures whenever possible, to use procedures consistent with a correct scientific methodology), followed by an assessment regarding their reasonability if analyzed conjointly with the knowledge to be gained. If the ratio is favorable, the study is acceptable. Otherwise, it is not (Weijer, 2000). A decisional algorithm, based on the double track methodology, which also takes into account some other fundamental principles of clinical research ethics and clinical ethics (including the principles of the Helsinki declaration), is presented in Fig. 2.

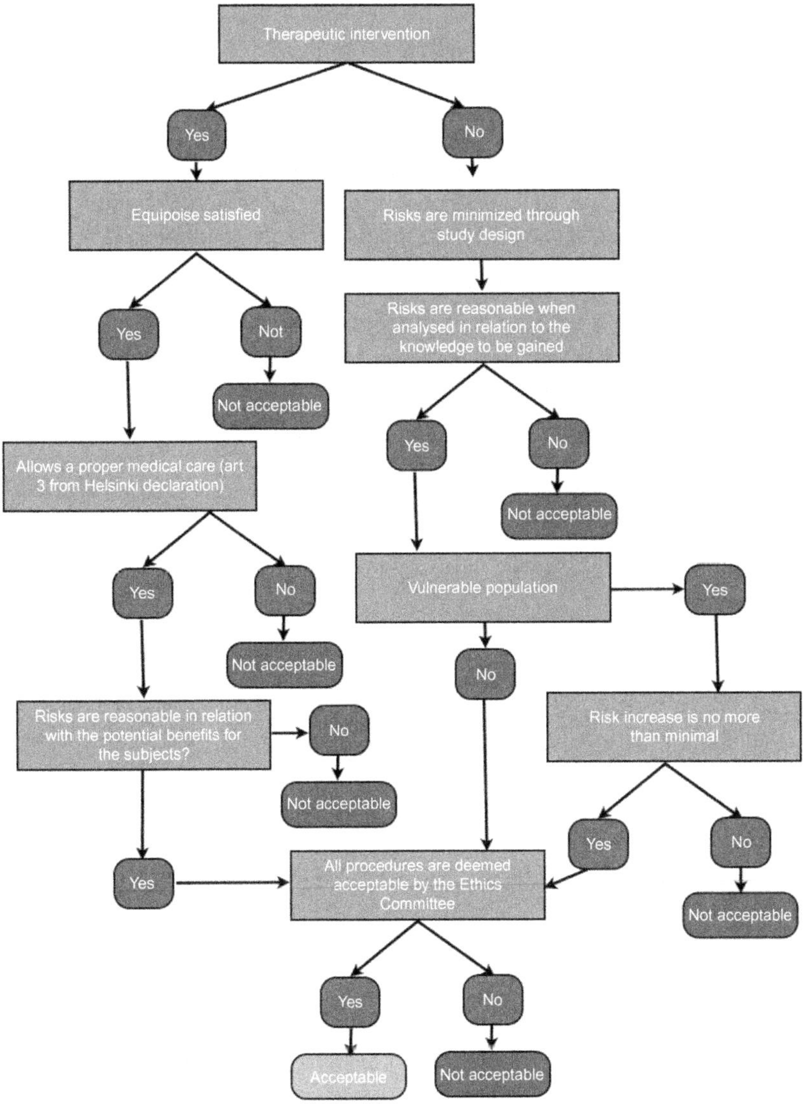

FIG. 2   Double track, risk assessment algorithm. *(Based on Hostiuc, 2014a. Ethics of Medical Research [Etica cercetarii medicale]. Casa Cartii de Stiinta, Cluj-Napoca; Wendler, D., Miller, F.G., 2007. Assessing research risks systematically: the net risks test. J. Med. Ethics 33, 481–486. doi:10.1136/jme.2005.014043, with permission.)*

4.1   What Are the Potential Benefits of Stem Cell Therapy in Neurodegenerative Disorder

Neurodegenerative disorders are usually characterized by an invariable, negative outcome, with progressive degradation, which cannot be reversed using currently available therapies. Therefore, any therapy that can lead to a slowdown of the degenerative process, or which could reverse it, has a tremendous importance in neurology. Initial studies, using stem cell therapies, performed mostly in patients with severe PD, have shown a variable response. However, it appears that the main issues with these therapies are related to fine-tuning and improving the methodology, as positive results have been obtained in subgroups of patients, and as pathology and molecular biology studies have shown positive biological effects of cell stem transplantation. Therefore these study, which are therapeutic, seem to respect equipoise, and should be ethically permissible from this point of view.

However, these therapies have numerous potential complications, which have to be properly addressed and, ideally, this should be made on animal or cell culture models. For example, in the initial human trials, together with dopaminergic neurons, other cells initially believed to have support functions were transplanted. Some of them, such as serotoninergic neurons, have been associated with graft-induced dyskinesia (Barker et al., 2016), which affected more than 50% of all subjects in some studies (Olanow et al., 2003). Many protocols included immunosuppression, which caused a plethora of side effects. For example, in a study on two patients who received stem cell treatment for HD, Reuter et al. showed the presence of renal impairment and gait disturbances (2008). Rossner showed the presence of immunosuppression therapy related complications (renal impairment, anemia, and lymphopeny) in all four subjects with HD (2002). An additional, unfavorable secondary effect of stem cell therapy for PD, that was identified in average more than a decade after the transplantation, is represented by the presence of ubiquinated Lewy bodies within the graft (Kordower et al., 2008), suggesting the possibility for the disease to propagate from host to graft cells, and the possibility that stem cell therapy might have a limited duration of efficacy (Li et al., 2008). However, this complication seems to have limited clinical consequences, as the number of affected neurons was small, and the Lowy bodies were identified more than a decade after transplantation, a period in which the therapy had positive effects. Bjorklund, in an animal study trying to show the transformation of embryonic stem cells in dopaminergic neurons, found that, out of the 25 rats that received the treatment, five died due to teratoma-like tumors (Björklund et al., 2002). Therefore at least on animal models, there seems to be a substantial risk of tumorigenesis after undifferentiated stem-cell transplantation. This risk was not shown for neural stem cells, but they are also associated with a limited capacity to spontaneously transform into dopaminergic neurons in vivo (Yang et al., 2004), a fact which raised the need to develop more complex procedures, such as the predifferentiation of undifferentiated stem cells (Master et al., 2007). Another potential complication is represented by the possible migration of the stem cells in a distant location from the transplanted area, which could lead to significant complications. Cicchetti et al. showed, through MRI, the rostral migration of transplanted superparamagnetic iron oxide-labeled neural stem cells (Cicchetti et al., 2007). The side effects should be carefully scrutinized before developing practical applications of these therapies. For example, even if immunosuppression-related complications are only causing minor inconveniences in clinical trials, they could be associated with important morbidity and even mortality in certain subgroups of patients. The development of Levy bodies suggest that the transplanted neurons could be affected by the disease, an effect that could develop over time, and therefore patients should be made aware of the possibility of reappearance/worsening of the symptoms after more than a decade, even if immediately after the treatment the effect would seem beneficial. Even if, in case of an apparently beneficial results of future clinical trials in this area, it would be a strong urge to recommend clinical use of this technique by skipping a few steps (Hostiuc et al., 2016b), we would recommend a proactive approach to risk management.

5. INDUCED PLURIPOTENT STEM CELL THERAPIES FOR NEURODEGENERATIVE DISORDERS

Induced pluripotent stem cells are somatic (adult, differentiated) cells that can be reprogrammed into either pluripotent stem cells or directly into mature cells (Zhou et al., 2008). See Fig. 3 for details. They were initially developed by Takahashi and Yamanaka (2006), who showed that a small number of transcription factors, if expressed ectopically on a somatic cell, can generate its transition to a pluripotent state (2006).

For neurodegenerative disorders, iPSCs have been used with two main purposes in mind: to develop new therapeutic strategies and to model diseases, as they allowed the study of the effects of both familial mono-allelic and multifactorial disorders (Hockemeyer and Jaenisch, 2016). Recent studies showed that iPSCs can be used for maturing midbrain dopaminergic neurons (Arenas et al., 2015), and numerous studies, currently in different stages, aim to test the usefulness of this new therapy in Parkinson's disease (Barker et al., 2017).

Even though PD is most likely the first candidate for a successful iPSC treatment, other neurodegenerative diseases could also benefit significantly. For example, numerous studies have been published showing the usefulness of iPSCs for Alzheimer's disease modeling. Muratore et al. developed four iPSCs lines from two patients with familial Alzheimer's disease with the APP V717I mutation, and differentiated them in forebrain neurons (2014). Virumbrales et al. recently presented an optimized in vitro protocol for generating human basal forebrain cholinergic neurons from iPSCs derived from presenilin-2 mutation carriers, followed by a CRISPR/Cas9 correction of the presenilin-2 mutation, which led to the normalization of the electrophysiological phenotype of those cells (2017). Xu et al. showed that in Huntington's disease, human-induced pluripotent stem cells can be corrected using CRISPR/Cas9 and the piggyback approach (2017).

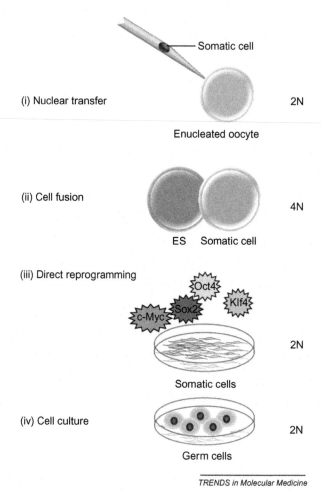

FIG. 3   Strategies to induce nuclear reprogramming. Somatic cells have been converted to a pluripotent state using all of the four displayed mechanisms. (i) Somatic cells can be reprogrammed through injection into a previously enucleated oocyte (nuclear transfer) (reviewed in Hochedlinger and Jaenisch, (2006). (ii) When a pluripotent cell such as an ES cell is fused with a somatic cell, it will generate a tetraploid cell that has acquired a pluripotent state (cell fusion) (Tada et al., 2001; Cowan et al., 2005). (iii) Ectopic expression of the transcription factors Oct4, Sox2, Klf4, and c-Myc via viral (retro-, lenti-, or adenovirus) or plasmid-based vectors is sufficient to reprogram somatic cells (direct reprogramming) (Takahashi and Yamanaka, 2006; Wernig et al., 2007; Maherali et al., 2007). (iv) Under specific culture conditions at a low frequency, it has been shown that germ cells can reprogram into pluripotent cells (culture-induced reprogramming) (Jaenisch and Young, 2008; Kanatsu-Shinohara et al., 2004). Abbreviations: *2N*, diploid cell; *4N*, tetraploid cell; *ES*, embryonic stem. *(From Amabile, G., Meissner, A., 2009. Induced pluripotent stem cells: current progress and potential for regenerative medicine. Trends Mol. Med. 15, 59–68. doi:10.1016/j.molmed.2008.12.003, with permission.)*

5.1 Moral Status of IPSCs[2]

According to Warren, "to have a moral status is to be morally considerable, or to have moral standing. It is to be an entity towards which moral agents have, or can have, moral obligations. If an entity has moral status, then we may not treat it in just any way we please" (1997). Moral status is a central issue in many bioethical debates, including reproductive ethics, pediatric ethics, and stem cell research. In embryonic stem cell research, it is essential to establish the moment in which the embryo/fetus acquires moral status, as from that moment on, the types of interventions that are allowed on it are limited. Regarding the moment when an embryo acquires a moral status, there are more than a few theories. According to Kant, the moral status is a fundamental attribute of any moral agent, "respect is always directed towards persons and never toward things" (Kant and Braileanu, 1930). By persons he understood moral agents, or beings, capable of having moral duties and obligations (Kant, 2008). Moral agents, according to the Kantian theory, are represented by normal adults who are capable of self-governing in moral issues and are autonomous (Schneewind, 1992). The Kantian theory grants all adults full autonomy and a full moral status. However, as it links it to rationality, it considers all persons not capable of moral reasoning to be outside this concept, included here being infants, embryos, animals, or persons with mental/intellectual deficits. Therefore, according to this theory, embryos do not have a moral status, causing us not to have direct moral obligations toward them (we do have indirect moral obligations, however, but not causally linked with the presence of their moral status) (Hostiuc, 2014a,b). At the other end of the spectrum there are authors, like Taylor, arguing that all beings, including bacteria, try to maintain their existence and reproduce, and therefore have a final purpose of their life; as this is an attribute of moral agents, all organisms should have a moral status (Taylor, 2011). Even more, some authors have classified the moral status depending on its intensity, further complicating the issue. For example, Baertschi argued that there are four types of moral status: (1) complete—giving moral agents all the rights and duties associated with it, (2) incomplete—giving moral agents all or some of the rights associated with it, and some or no obligations, (3) intrinsic—giving moral status based on their intrinsic value, or potentiality, and (4) conferred—in which beings are awarded a moral status based on certain characteristics, or properties (Baertschi, 2008). As stem cells (at least totipotent stem cells) can lead to complete human beings, we can either consider them to have intrinsic moral status or conferred moral status. The most important intrinsic value causing beings to be considered moral agents is dignity. Aquinas for example said that "dignity means the goodness a thing possesses because of itself; utility, because of another thing" (Thomas de Aquino, Scriptum super Sententiis, pr., n.d.). Kant considered that there are two kinds of entities, each having a proper value. However, only entities that are ends in themselves (human beings)

---

2. Parts of this section were previously published in Hostiuc, (2014b).

have intrinsic value (Kant and Braileanu, 1930), which can equate to dignity (Baertschi, 2008). Another important intrinsic value is the potentiality to become a human being; if an entity has this potential, it may be considered intrinsically a moral agent (Charo, 1994; Freeman, 1997; Streiffer, 2005). Based on this approach, embryonic stem cells can be considered as having inferred moral status, as they have the potential of becoming human beings. Another possibility for embryonic stem cells to have moral status is by it being conferred to them, based either on the potentiality of generating a human being, on their genetic structure, or on the general aspects of morality from the society we live in (Hostiuc, 2014a,b). Based on this argument of potentiality, some authors have argued that iPSCs also could have an inferred moral status. An important argument supporting it is a recent decision of the Court of Justice of the European Union, in a case regarding the patenting of neural precursor stem cells produced from human embryonic stem cells. The ruling stated, "The use of human embryos for scientific research purposes is not patentable. A 'human embryo' within the meaning of Union law is any human ovum after fertilisation or any human ovum not fertilised but which, through the effect of the technique used to obtain it, is capable of commencing the process of development of a human being" (Court of Justice of the European Union, 2011; Spranger, 2012). Moreover, the ruling extended its applicability on any type of cell that is capable of beginning the process of developing a human being, and linked it to human dignity (therefore it inferred moral status on these cells, including iPSCs), "the use of biological material originating from humans must be consistent with regard for fundamental rights and, in particular, the dignity of the person. It is from this perspective that the patentability of the human body is prohibited. As a result of this, in its opinion, the concept of human embryo, within the meaning of Article 6(2)(c) of the Directive, must be interpreted broadly as covering any cell capable of commencing the process of development of a human being. This includes, on the one hand, the human ovum as soon as fertilised and, on the other hand the non-fertilised human ovum into which the nucleus from a mature human cell has been transplanted and a non-fertilised human ovum whose division and further development have been stimulated by parthenogenesis" (Court of Justice of the European Union, 2011). Some authors argued that this approach is too excessive, as it could limit funding and development of iPSCs-based therapies, limiting the therapeutic options for the patients. Moreover, a limit that is imposed regionally does not mean that these therapies cannot be developed in other parts of the world (Barker and de Beaufort, 2013), generating two main effects: a redistribution of the funds for developing neurodegenerative disorders using iPSCs toward more "compliant" regions (Barker and de Beaufort, 2013) and a geographical redistribution of the clinical-trial related risks, with a significant potential for exploitation of the subjects or host communities (Emanuel et al., 2004).

Our opinion is that the inferred moral status of the iPSCs is an overreach, at least if we take into account the use of somatic cells. We must respect their potentiality, but not by assigning them moral status, which would significantly limit our possibilities to use them in research, and later on in clinical practice. A possible solution would be to apply the distinction between moral status and moral value. According to Steinbock, only sentient beings can have moral status. Other entities, including stem cells, can have a moral value, namely have the prospect of bringing forth something special, which would generate hesitation from persons to consider them as mere things (Steinbock, 2007, pp. 416–441). Moral value is attributed to two main things: symbols, such as flags, the Bible, the Qur'an; and material things, such as human remains and human embryos. "Dead bodies are owed respect both because of what they are–the remains of the once-living human organism–and because of what they symbolize the human person who is no more. Human embryos deserve respect for similar reasons: they are a developing form of human life, and also a symbol of human existence" (Steinbock, 2007, p. 436). Based on this distinction, we can use things with moral value for morally significant purposes (Steinbock, 2007, p. 438), but we should not use them for morally insignificant purposes. As developing a new therapy for disorders without efficient treatments is obviously a morally significant purpose, the use of iPSCs should be allowed, and even more, encouraged.

6. CONCLUSIONS

As many other technologies discussed in this book, stem cell therapies for neurodegenerative disorders are not yet routinely used in clinical medicine. However, due to the recent advances of medicine in the last years, it is expected that they could be mainstream in less than a decade, making debates about their morality highly relevant. In this chapter we have only analyzed a few of the major issues surrounding the topic that we thought should be taken into account by any investigator in the field.

REFERENCES

Albin, R.L., 2005. Sham surgery controls are mitigated trolleys. J. Med. Ethics 31, 149–152. https://doi.org/10.1136/jme.2003.006155.

Arenas, E., Denham, M., Villaescusa, J.C., 2015. How to make a midbrain dopaminergic neuron. Development 142, 1918–1936. https://doi.org/10.1242/dev.097394.

Arias, J.J., Karlawish, J., 2014. Confidentiality in preclinical Alzheimer disease studies: when research and medical records meet. Neurology 82, 725–729. https://doi.org/10.1212/WNL.0000000000000153.

Avins, A.L., 1998. Can unequal be more fair? Ethics, subject allocation, and randomised clinical trials. J. Med. Ethics 24, 401–408.

Baertschi, B., 2008. In: The question of the embryo's moral status.Presented at the Bioethica Forum, pp. 76–80.

Barker, R.A., de Beaufort, I., 2013. Scientific and ethical issues related to stem cell research and interventions in neurodegenerative disorders of the brain. Prog. Neurobiol. 110, 63–73. https://doi.org/10.1016/j.pneurobio.2013.04.003.

Barker, R.A., Parmar, M., Kirkeby, A., Björklund, A., Thompson, L., Brundin, P., 2016. Are stem cell-based therapies for Parkinson's disease ready for the clinic in 2016? J Parkinsons Dis 6, 57–63. https://doi.org/10.3233/JPD-160798.

Barker, R.A., Parmar, M., Studer, L., Takahashi, J., 2017. Human trials of stem cell-derived dopamine neurons for Parkinson's disease: dawn of a new era. Cell Stem Cell 21, 569–573. https://doi.org/10.1016/j.stem.2017.09.014.

Becker, R.E., 2005. Lessons from Darwin: evolutionary biology's implications for Alzheimer's disease research and patient care. Curr. Alzheimer Res. 2, 319–326.

Bentham, J.B.J., 1834. Deontology or The Science Of Morality: In Which the Harmony and Co-incidence of Duty and Self-Interest, Virtue and Felicity, Prudence and Benevolence, Are Explained and Exemplified. Longman, Rees, Orme, Browne, Green, and Longman, London.

Björklund, A., Stenevi, U., Schmidt, R.H., Dunnett, S.B., Gage, F.H., 1983. Intracerebral grafting of neuronal cell suspensions. I. Introduction and general methods of preparation. Acta Physiol. Scand. Suppl. 522, 1–7.

Björklund, L.M., Sánchez-Pernaute, R., Chung, S., Andersson, T., Chen, I.Y.C., McNaught, K.S.P., Brownell, A.-L., Jenkins, B.G., Wahlestedt, C., Kim, K.-S., Isacson, O., 2002. Embryonic stem cells develop into functional dopaminergic neurons after transplantation in a Parkinson rat model. PNAS 99, 2344–2349. https://doi.org/10.1073/pnas.022438099.

Blasco, H., Mavel, S., Corcia, P., Gordon, P.H., 2014. The glutamate hypothesis in ALS: pathophysiology and drug development. Curr. Med. Chem. 21, 3551–3575.

Brown, G.C., Neher, J.J., 2010. Inflammatory neurodegeneration and mechanisms of microglial killing of neurons. Mol. Neurobiol. 41, 242–247. https://doi.org/10.1007/s12035-010-8105-9.

Brundin, P., Strecker, R.E., Clarke, D.J., Widner, H., Nilsson, O.G., Astedt, B., Lindvall, O., Björklund, A., 1988. Can human fetal dopamine neuron grafts provide a therapy for Parkinson's disease? Prog. Brain Res. 78, 441–448.

Brundin, P., Barker, R.A., Parmar, M., 2010. Neural grafting in Parkinson's disease problems and possibilities. Prog. Brain Res. 184, 265–294. https://doi.org/10.1016/S0079-6123(10)84014-2.

Byrne, M.M., Thompson, P., 2006. Collective equipoise, disappointment, and the therapeutic misconception: on the consequences of selection for clinical research. Med. Decis. Making 26, 467–479. https://doi.org/10.1177/0272989x06290499.

Castro, M., 2007. Placebo versus best-available-therapy control group in clinical trials for pharmacologic therapies: which is better? Proc. Am. Thorac. Soc. 4 (7), 570–573.

Charo, R., 1994. Hunting of the snark: the moral status of embryos, right-to-lifers, and third world women. Stanford Law Pol. Rev. 6, 11.

Cicchetti, F., Gross, R.E., Bulte, J.W.M., Owen, M., Chen, I., Saint-Pierre, M., Wang, X., Yu, M., Brownell, A.-L., 2007. Dual-modality in vivo monitoring of subventricular zone stem cell migration and metabolism. Contrast Media Mol. Imaging 2, 130–138. https://doi.org/10.1002/cmmi.138.

Clark, P.A., 2002. Placebo surgery for Parkinson's disease: do the benefits outweigh the risks? J. Law Med. Ethics 30 (1), 58–68.

Court of Justice of the European Union, 2011. C-34/10 Oliver Brüstle v Greenpeace eV, judgment of 18 October 2011 [WWW Document]. http://ec.europa.eu/dgs/legal_service/arrets/10c034_en.pdf. Accessed 12 October 2017.

Cowan, C.A., et al., 2005. Nuclear reprogramming of somatic cells after fusion with human embryonic stem cells. Science 309, 1369–1373.

Dekkers, W., Boer, G., 2001. Sham neurosurgery in patients with Parkinson's disease: is it morally acceptable? J. Med. Ethics 27, 151–156.

Emanuel, E.J., 2008. The Oxford Textbook of Clinical Research Ethics [WWW Document]. http://public.eblib.com/EBLPublic/PublicView.do?ptiID=665420.

Emanuel, E.J., Emanuel, L.L., 1992. Four models of the physician-patient relationship. JAMA 267, 2221–2226.

Emanuel, E.J., Wendler, D., Killen, J., Grady, C., 2004. What makes clinical research in developing countries ethical? The benchmarks of ethical research. J Infect Dis 189, 930–937.

Fan, X., Sun, D., Tang, X., Cai, Y., Yin, Z.Q., Xu, H., 2014. Stem-cell challenges in the treatment of Alzheimer's disease: a long way from bench to bedside. Med. Res. Rev. 34, 957–978. https://doi.org/10.1002/med.21309.

Förstl, H., Kurz, A., 1999. Clinical features of Alzheimer's disease. Eur. Arch. Psychiatry Clin. Neurosci. 249, 288–290. https://doi.org/10.1007/s004060050101.

Frank, S.A., Wilson, R., Holloway, R.G., Zimmerman, C., Peterson, D.R., Kieburtz, K., Kim, S.Y., 2008. Ethics of sham surgery: perspective of patients. Mov. Disord. 23, 63–68.

Freed, C.R., Greene, P.E., Breeze, R.E., Tsai, W.-Y., DuMouchel, W., Kao, R., Dillon, S., Winfield, H., Culver, S., Trojanowski, J.Q., Eidelberg, D., Fahn, S., 2001. Transplantation of embryonic dopamine neurons for severe Parkinson's disease. N. Engl. J. Med. 344, 710–719. https://doi.org/10.1056/NEJM200103083441002.

Freeman, M.D., 1997. The Moral Status of Children: Essays on the Rights of the Children. Martinus Nijhoff Publishers, The Hague.

Fukushi, T., 2012. Ethical practice in the era of advanced neuromodulation. Asian Bioeth. Rev. 4, 320–329.

Harper, J.M., Krishnan, C., Darman, J.S., Deshpande, D.M., Peck, S., Shats, I., Backovic, S., Rothstein, J.D., Kerr, D.A., 2004. Axonal growth of embryonic stem cell-derived motoneurons in vitro and in motoneuron-injured adult rats. Proc. Natl. Acad. Sci. U. S. A. 101, 7123–7128.

Haston, K.M., Finkbeiner, S., 2016. Clinical trials in a dish: the potential of pluripotent stem cells to develop therapies for neurodegenerative diseases. Annu. Rev. Pharmacol. Toxicol. 56, 489–510. https://doi.org/10.1146/annurev-pharmtox-010715-103548.

Hetz, C., Mollereau, B., 2014. Disturbance of endoplasmic reticulum proteostasis in neurodegenerative diseases. Nat. Rev. Neurosci. 15, 233–249. https://doi.org/10.1038/nrn3689.

Hochedlinger, K., Jaenisch, R., 2006. Nuclear reprogramming and pluripotency. Nature 441, 1061–1067.

Hockemeyer, D., Jaenisch, R., 2016. Induced pluripotent stem cells meet genome editing. Cell Stem Cell 18, 573–586. https://doi.org/10.1016/j.stem.2016.04.013.

Horng, S., Miller, F.G., 2003. Ethical framework for the use of sham procedures in clinical trials. Crit. Care Med. 31, S126–130. https://doi.org/10.1097/01.CCM.0000054906.49187.67.

Hostiuc, S., 2014a. Ethics of Medical Research Etica cercetarii medicale Casa Cartii de Stiinta, Cluj-Napoca.

Hostiuc, S., 2014b. Moral status of the embryo. Clinical and legal consequences. Gineco.eu 10, 102–104.

Hostiuc, S., 2015. Does the additional protocol concerning biomedical research of the Oviedo convention protect psychiatry subjects from therapeutic misconception? Eur. Psychiatry 30, 662.

Hostiuc, S., Drima, E., Buda, O., 2016a. Shake the disease. Georges Marinesco, Paul Blocq and the pathogenesis of parkinsonism, 1893. Front. Neuroanat. 10.

Hostiuc, S., Moldoveanu, A., Dascălu, M.-I., Unnthorsson, R., Jóhannesson, Ó.I., Marcus, I., 2016b. Translational research—the need of a new bioethics approach. J. Transl. Med. 14, 1–10. https://doi.org/10.1186/s12967-016-0773-4.

Hostiuc, S., Rentea, I., Drima, E., Negoi, I., 2016c. Placebo in surgical research: a case-based ethical analysis and practical consequences [WWW document]. Biomed. Res. Int. https://doi.org/10.1155/2016/2627181.

Jaenisch, R., Young, R., 2008. Stem cells, the molecular circuitry of pluripotency and nuclear reprogramming. Cell 132, 567–582.

Jankovic, J., Tolosa, E., 2007. Parkinson's Disease and Movement Disorders. Lippincott Williams & Wilkins, Philadelphia, PA.

Kanatsu-Shinohara, M., et al., 2004. Generation of pluripotent stem cells from neonatal mouse testis. Cell 119, 1001–1012.

Kant, I., 2008. Intemeierea metafizicii moravurilor. Editura Humanitas, Bucharest, pp. 1–48.

Kant, I., Braileanu, T., 1930. Critica ratiunii pure. Editura Casei Şcoalelor, Bucharest.

King, N.M., 2000. Defining and describing benefit appropriately in clinical trials. JL Med. Ethics 28, 332.

Kordower, J.H., Chu, Y., Hauser, R.A., Olanow, C.W., Freeman, T.B., 2008. Transplanted dopaminergic neurons develop PD pathologic changes: a second case report. Mov. Disord. 23, 2303–2306. https://doi.org/10.1002/mds.22369.

Li, J.-Y., Englund, E., Holton, J.L., Soulet, D., Hagell, P., Lees, A.J., Lashley, T., Quinn, N.P., Rehncrona, S., Björklund, A., Widner, H., Revesz, T., Lindvall, O., Brundin, P., 2008. Lewy bodies in grafted neurons in subjects with Parkinson's disease suggest host-to-graft disease propagation. Nat. Med. 14, 501–503. https://doi.org/10.1038/nm1746.

Lidz, C.W., Appelbaum, P.S., 2002. The therapeutic misconception—problems and solutions. Med. Care 40, 55–63. https://doi.org/10.1097/01.Mlr.0000023956.25813.18.

Lindvall, O., Kokaia, Z., 2006. Stem cells for the treatment of neurological disorders. Nature 441, 1094–1096. https://doi.org/10.1038/nature04960.

Lindvall, O., Rehncrona, S., Brundin, P., Gustavii, B., Astedt, B., Widner, H., Lindholm, T., Björklund, A., Leenders, K.L., Rothwell, J.C., Frackowiak, R., Marsden, D., Johnels, B., Steg, G., Freedman, R., Hoffer, B.J., Seiger, A., Bygdeman, M., Strömberg, I., Olson, L., 1989. Human fetal dopamine neurons grafted into the striatum in two patients with severe Parkinson's disease. A detailed account of methodology and a 6-month follow-up. Arch. Neurol. 46, 615–631.

Macklin, R., 1999. The ethical problems with sham surgery in clinical research. N. Engl. J. Med. 341, 992.

Madrazo, I., León, V., Torres, C., Aguilera, M.C., Varela, G., Alvarez, F., Fraga, A., Drucker-Colín, R., Ostrosky, F., Skurovich, M., 1988. Transplantation of fetal substantia nigra and adrenal medulla to the caudate nucleus in two patients with Parkinson's disease. N. Engl. J. Med. 318, 51. https://doi.org/10.1056/NEJM198801073180115.

Maherali, N., et al., 2007. Directly reprogrammed fibroblasts show global epigenetic remodeling and widespread tissue contribution. Cell Stem Cell 1, 55–70.

Mason, A.R., Ziemann, A., Finkbeiner, S., 2014. Targeting the low-hanging fruit of neurodegeneration. Neurology 83, 1470–1473. https://doi.org/10.1212/WNL.0000000000000894.

Master, Z., McLeod, M., Mendez, I., 2007. Benefits, risks and ethical considerations in translation of stem cell research to clinical applications in Parkinson's disease. J. Med. Ethics 33, 169–173. https://doi.org/10.1136/jme.2005.013169.

Moore, S.F., Guzman, N.V., Mason, S.L., Williams-Gray, C.H., Barker, R.A., 2014. Which patients with Parkinson's disease participate in clinical trials? One centre's experiences with a new cell based therapy trial (TRANSEURO). J Parkinsons Dis 4, 671–676. https://doi.org/10.3233/JPD-140432.

Muratore, C.R., Rice, H.C., Srikanth, P., Callahan, D.G., Shin, T., Benjamin, L.N.P., Walsh, D.M., Selkoe, D.J., Young-Pearse, T.L., 2014. The familial Alzheimer's disease APPV717I mutation alters APP processing and tau expression in iPSC-derived neurons. Hum. Mol. Genet. 23, 3523–3536. https://doi.org/10.1093/hmg/ddu064.

Olanow, C.W., Goetz, C.G., Kordower, J.H., Stoessl, A.J., Sossi, V., Brin, M.F., Shannon, K.M., Nauert, G.M., Perl, D.P., Godbold, J., Freeman, T.B., 2003. A double-blind controlled trial of bilateral fetal nigral transplantation in Parkinson's disease. Ann. Neurol. 54, 403–414. https://doi.org/10.1002/ana.10720.

Ortiz-Virumbrales, M., Moreno, C.L., Kruglikov, I., Marazuela, P., Sproul, A., Jacob, S., Zimmer, M., Paull, D., Zhang, B., Schadt, E.E., Ehrlich, M.E., Tanzi, R.E., Arancio, O., Noggle, S., Gandy, S., 2017. CRISPR/Cas9-correctable mutation-related molecular and physiological phenotypes in iPSC-derived Alzheimer's PSEN2 N141I neurons. Acta Neuropathol. Commun. 5. https://doi.org/10.1186/s40478-017-0475-z.

Paulsen, J.S., Nance, M., Kim, J.-I., Carlozzi, N.E., Panegyres, P.K., Erwin, C., Goh, A., McCusker, E., Williams, J.K., 2013. A review of quality of life after predictive testing for and earlier identification of neurodegenerative diseases. Prog. Neurobiol. 110, 2–28. https://doi.org/10.1016/j.pneurobio.2013.08.003.

Reuter, I., Tai, Y.F., Pavese, N., Chaudhuri, K.R., Mason, S., Polkey, C.E., Clough, C., Brooks, D.J., Barker, R.A., Piccini, P., 2008. Long-term clinical and positron emission tomography outcome of fetal striatal transplantation in Huntington's disease. J. Neurol. Neurosurg. Psychiatry 79, 948–951. https://doi.org/10.1136/jnnp.2007.142380.

Robillard, J.M., Federico, C.A., Tairyan, K., Ivinson, A.J., Illes, J., 2011. Untapped ethical resources for neurodegeneration research. BMC Med. Ethics 12, 9. https://doi.org/10.1186/1472-6939-12-9.

Rosser, A.E., Barker, R.A., Harrower, T., Watts, C., Farrington, M., Ho, A.K., Burnstein, R.M., Menon, D.K., Gillard, J.H., Pickard, J., Dunnett, S.B., 2002. Unilateral transplantation of human primary fetal tissue in four patients with Huntington's disease: NEST-UK safety report ISRCTN no 36485475. J. Neurol. Neurosurg. Psychiatry 73, 678–685. https://doi.org/10.1136/jnnp.73.6.678.

Schneewind, J., 1992. 10 autonomy, obligation, and virtue: an overview of Kant's moral philosophy. Cambridge Comp. Kant 3, 309.

Siegler, M., 2010. Clinical Ethics: A Practical Approach to Ethical Decisions in Clinical Medicine, 7th ed. McGraw-Hill.

Spranger, T.M., 2012Case C-34/10, Oliver Brustle v. Greenpeace e.V., Judgment of the Court (Grand Chamber) of 18 October 2011, nyr. Common Market Law Rev. 49, 1197.

Sramka, M., Rattaj, M., Molina, H., Vojtassák, J., Belan, V., Ruzický, E., 1992. Stereotactic technique and pathophysiological mechanisms of neurotransplantation in Huntington's chorea. Stereotact. Funct. Neurosurg. 58, 79–83.

Steinbock, B. (Ed.), 2007. The oxford handbook of bioethics. In: Oxford Handbooks. Oxford University Press, Oxford, NY.

Streiffer, R., 2005. At the edge of humanity: Human stem cells, chimeras, and moral status. Kennedy Inst. Ethics J. 15, 347–370.

Swift, T.L., 2012. Sham surgery trial controls: perspectives of patients and their relatives. J. Empir. Res. Hum. Res. Ethics 7, 15–28. https://doi.org/10.1525/jer.2012.7.3.15.

Tada, M., et al., 2001. Nuclear reprogramming of somatic cells by in vitro hybridization with ES cells. Curr. Biol. 11, 1553–1558.

Takahashi, K., Yamanaka, S., 2006. Induction of pluripotent stem cells from mouse embryonic and adult fibroblast cultures by defined factors. Cell 126, 663–676. https://doi.org/10.1016/j.cell.2006.07.024.

Taylor, P.W., 2011. Respect for Nature: A Theory of Environmental Ethics. Princeton University Press, Princeton.

Tenery, R., Rakatansky, H., Riddick Jr., F.A., Goldrich, M.S., Morse, L.J., O'Bannon III, J.M., Ray, P., Smalley, S., Weiss, M., Kao, A., 2002. Surgical "placebo" controls. Ann. Surg. 235, 303.

Thomas de Aquino, Scriptum super Sententiis, pr n.d. [WWW Document] http://www.corpusthomisticum.org/snp0000.html (Accessed 12 October 2017).

TRANSEURO | Home n.d. [WWW Document], http://www.transeuro.org.uk/ (Accessed 30 November 2017).

Tuszynski, M.H., Thal, L., Pay, M., Salmon, D.P., HS, U., Bakay, R., Patel, P., Blesch, A., Vahlsing, H.L., Ho, G., Tong, G., Potkin, S.G., Fallon, J., Hansen, L., Mufson, E.J., Kordower, J.H., Gall, C., Conner, J., 2005. A phase 1 clinical trial of nerve growth factor gene therapy for Alzheimer disease. Nat. Med. 11, 551. https://doi.org/10.1038/nm1239.

Van Damme, P., Dewil, M., Robberecht, W., Van Den Bosch, L., 2005. Excitotoxicity and amyotrophic lateral sclerosis. Neurodegener Dis 2, 147–159. https://doi.org/10.1159/000089620.

Warren, M.A., 1997. Moral Status: Obligations to Persons and Other Living Things: Obligations to Persons and Other Living Things. Oxford University Press, Oxford.

Weijer, C., 2000. The ethical analysis of risk. J. Law Med. Ethics 28, 344–361. https://doi.org/10.1111/j.1748-720X.2000.tb00686.x.

Weijer, C., 2002. I need a placebo like i need a hole in the head. J. Law Med. Ethics 30, 69–72.

Weijer, C., Miller, P.B., 2004. When are research risks reasonable in relation to anticipated benefits? Nat. Med. 10, 570–573.

Weijer, C., Miller, P.B., 2007. Refuting the net risks test: a response to Wendler and Miller's "assessing research risks systematically" J. Med. Ethics 33, 487–490. https://doi.org/10.1136/jme.2006.016444.

Wernig, M., et al., 2007. In vitro reprogramming of fibroblasts into a pluripotent ES-cell-like state. Nature 448, 318–324.

Wichterle, H., Lieberam, I., Porter, J.A., Jessell, T.M., 2002. Directed differentiation of embryonic stem cells into motor neurons. Cell 110, 385–397.

Wijeyekoon, R., Barker, R.A., 2011. The current status of neural grafting in the treatment of Huntington's disease. A review. Front. Integr. Neurosci. 5. https://doi.org/10.3389/fnint.2011.00078.

Xu, X., Tay, Y., Sim, B., Yoon, S.-I., Huang, Y., Ooi, J., Utami, K.H., Ziaei, A., Ng, B., Radulescu, C., Low, D., Ng, A.Y.J., Loh, M., Venkatesh, B., Ginhoux, F., Augustine, G.J., Pouladi, M.A., 2017. Reversal of phenotypic abnormalities by CRISPR/Cas9-mediated gene correction in Huntington disease patient-derived induced pluripotent stem cells. Stem Cell Rep. 8, 619–633. https://doi.org/10.1016/j.stemcr.2017.01.022.

Yang, M., Donaldson, A.E., Marshall, C.E., Shen, J., Iacovitti, L., 2004. Studies on the differentiation of dopaminergic traits in human neural progenitor cells in vitro and in vivo. Cell Transplant. 13, 535–547.

Yang, W., Tu, Z., Sun, Q., Li, X.-J., 2016. CRISPR/Cas9: implications for modeling and therapy of neurodegenerative diseases. Front. Mol. Neurosci. 9. https://doi.org/10.3389/fnmol.2016.00030.

Zhou, Q., Brown, J., Kanarek, A., Rajagopal, J., Melton, D.A., 2008. In vivo reprogramming of adult pancreatic exocrine cells to beta-cells. Nature 455, 627–632. https://doi.org/10.1038/nature07314.

FURTHER READING

Amabile, G., Meissner, A., 2009. Induced pluripotent stem cells: current progress and potential for regenerative medicine. Trends Mol. Med. 15, 59–68. https://doi.org/10.1016/j.molmed.2008.12.003.

Wendler, D., Miller, F.G., 2007. Assessing research risks systematically: the net risks test. J. Med. Ethics 33, 481–486. https://doi.org/10.1136/jme.2005.014043.

Chapter 11

Predictive Genetic Testing in Multifactorial Disorders

Sorin Hostiuc

Department of Legal Medicine and Bioethics, Faculty of Dental Medicine, "Carol Davila" University of Medicine and Pharmacy, Bucharest, Romania

1. INTRODUCTION

Human diseases are caused by three main categories of factors: genetic, infectious, and environmental. Genetic diseases are determined by different types of alterations of the normal human genome. Environmental diseases are determined by environmental factors that can be related to the personal lifestyle (smoking, alcohol/substance abuse, abnormal eating patterns), physical factors from the environment (UV radiation, cold, heat, air pressure, electricity), or exposure the irritant or toxic chemicals from the environment (heavy metals, halogens, organic compounds, or noxious gases). Various agents such as prions, viruses, and bacteria cause infectious diseases. Very often however, a certain disorder is not caused by a single type of risk factor, in which case the disease is called multifactorial.

Multifactorial diseases, such as arterial hypertension, coronary heart disease, cerebrovascular diseases, low-back syndrome, diabetes, cancer, are caused by complex and variable interactions between multiple genetic, environmental (Kirch, 2008), and infectious (Coulehan, 1979) factors. Genetic risk factors may alter the natural history of a disease (prognosis, complications, and severity), affect the risk of actually developing a certain disease, and the therapeutic response (adverse effects, response to medication, and efficacy) (Campbell and Rudan, 2007; Becker et al., 2011). Recent advances in human genetic began to elucidate the genetic "profile" of many multifactorial disorders, but significant gaps still exist in fully understanding their etiology.

Predictive genetic testing is used to detect gene mutations associated with disorders in patients not presenting signs/symptoms at the time of the testing. The classical example in this regard is BRCA1 testing for breast cancer. In the breast, BRCA1 is normally expressed in epithelial cells, and is needed for repairing damaged DNA (if possible), or to destroy affected cells

Clinical Ethics at the Crossroads of Genetic and Reproductive Technologies.
https://doi.org/10.1016/B978-0-12-813764-2.00011-8

229

(if DNA corruption is beyond repair). BRCA1 mutations deactivate this gene, increasing the susceptibility for breast cancer. Women with mutations in the BRCA1 gene have a risk of up to 65% to develop breast cancer until the age of 70 (Antoniou et al., 2003).

Predictive tests are of three main types: presymptomatic, susceptibility, and carrier testing.

In *presymptomatic* testing are analyzed genetic mutations with a high penetration. Presymptomatic testing can be done for either diseases with a mendelian inheritance [such as HTT mutations for Huntington disease (Sathasivam et al., 2013) APP, PSEN1 and PSEN2 mutations for early-onset Alzheimer's disease (Bird, 2012)], or multifactorial inheritance. The tests can identify healthy persons with a very high risk of developing the disease (close to 100%) during a normal lifetime.

Susceptibility testing analyzes genetic mutations that increase the risk of developing a certain disease, such as BRCA 1 and 2 for breast cancer. The presence of a particular mutation can augment the risk of developing the disease, without causing it with certainty. Therefore by screening for that particular mutation, we can intervene before the development of the disease, potentially preventing the clinical effects (Goel, 2001).

Carrier testing can be also considered a form of predictive testing, as it tries to identify persons who are at risk of transmitting a particular disease to their offspring, without actually having the disease. This is currently done for some diseases such as cystic fibrosis (Watson et al., 2004), Tay-Sachs, familial dysautonomia for Ashkenazi Jewish population (Gross et al., 2008), spinal muscular atrophy (Prior, 2008), and so on. Ethical issues surrounding carrier testing will be analyzed in other chapters of the book.

Predictive testing can be performed for both monogenic and multifactorial diseases. Monogenic diseases are caused by mutations in a single gene, and predictive testing in this case is extremely useful as it detects a significant risk factor for a particular disorder. For example, a positive detection of the HTT mutation indicates, with a probability of almost 100%, the fact that the subject will have Huntington disease, in the course of a normal lifespan. The detection of MLH1 gene mutations incurs a 48% risk of developing Lynch syndrome (hereditary nonpolyposis colorectal cancer syndrome), the detection of MSH2 gene mutations—a 48% risk increase, and for MSH6—a 12% risk increase for the same disease (Bonadona et al., 2011). As there is a very high difference in the chance of acquiring the disease between carriers and noncarriers, genetic testing can identify persons at high risk, and can allow the development of useful preventive strategies (Janssens and Khoury, 2006).

In multifactorial diseases, the predictive value of a single gene mutation is limited. Most mutations only incur a slight risk increase compared to noncarriers. For example, Sladek et al. analyzed some mutations associated with diabetes, and found modest increases for the studied genes, both in the heterozygous and homozygous status: TCF7L2 (OR = 1.65 and 2.77, respectively), SLC30AB (OR = 1.18 and 1.53 respectively), HHEX (OR = 1.19 and 1.44 respectively), or EXT2 (1.25 and 1.50 respectively) (Sladek et al., 2007). Stacey et al. showed that mutations in 2q35 and TNRC9 are associated with small risk increases in breast cancer: for 2q35 the OR was 1.11 for heterozygous mutation and 1.44 for homozygous mutation, while for TNRC9 was 1.27 for heterozygous and 1.64 for homozygous mutations (Stacey et al., 2007). As each disease may have multiple risk factors, it is not enough to test for a single (Janssens and Khoury, 2006), but rather for a panel of multiple genes. A person may have one genetic risk factor for developing atherosclerosis, while another may have more than 10, each leading to a unique risk increase pattern. A person may have five genetic risk factors for developing lung cancer leading to a 5% risk increase, while another may have only two mutations, causing a 20% risk increase.

Many mutations have incomplete penetrance and/or variable expressivity, an issue usually studied in the context of Mendelian disorders. *Penetrance* is defined as the percent of persons with a particular mutation or genotype present-ing clinically/morphologically/physiologically the phenotype generated by that particular genotype (Shawky, 2014). Penetrance may be complete in which case all persons with a particular mutation will have the corresponding pheno-type, partial/reduced in which the allele may cause only sometimes the appear-ance of the corresponding phenotype, or pseudo-incomplete in which the nonappearance of the phenotype is caused by the fact that the phenotype is not clinically evident or has not yet been expressed (Shawky, 2014). Unlike penetrance, which assesses the presence of a certain phenotype, *expressivity* assesses the degree in which a phenotype is expressed—some individuals may have more extensive forms of the disease compared to others who have the same genetic profile (Lobo, 2008). For example, in 1993, Collins et al. stud-ied the distribution of holoprosencephaly in a large family with multiple cases, finding that, while for some members it was lethal, for others it caused only minor facial deformities such as a single upper central incisor, without any neu-rological complications (Collins et al., 1993).

Many known multifactorial diseases have some subgroups that are caused by mutations in single genes (Janssens and Khoury, 2006). For example, atheroscle-rosis, even if it is a known multifactorial disease, has a subset—familial hyper-cholesterolemia due to low-density lipoprotein receptor mutations that is caused by mutations in the low-density lipoprotein receptor (Nordestgaard et al., 2013; Goldstein and Brown, 1977). Lung emphysema is usually caused by a mixture of genetic and environment factors. However, the presence of alpha-1 antitrypsin can cause early onset emphysema with family inheritance (Goldstein and Brown, 1977; Fregonese and Stolk, 2008). Many susceptibility genes have pleio-tropic effects that are involved in the pathogenesis of more than one disease. For example, ApoE variants are associated with both changes in the lipid profile and Alzheimer's disease (the e4 allele). Therefore testing for CVD risk may yield information about the risk of developing Alzheimer's disease (Beeri et al., 2006). Sometimes, there are associations between susceptibility to various dis-eases and physical traits. For example, a height increase of 10 cm is associated with a 12% risk increase of developing colorectal cancer, and a 17% odds increase in developing breast cancer (Khankari et al., 2016). Some alleles can be risk factors for a disease and protective factors for others (Janssens and Khoury, 2006). For example, deletions occurring in the angiotensin converting enzyme gene may decrease the risk of Alzheimer's disease (Lehmann et al., 2005) and increase the risk of ischemic stroke (Casas et al., 2004). Some diseases require the presence of a particular genetic mutation, but its presence is not indicative for the development of the disease. For example, the presence of HLA-B27 is often associated with the development of ankylosing spondylitis; however, not all persons with HLA-B27 will develop the disease (Brewerton et al., 1973). Homozygous mutations in the phenylalanine hydroxylase gene may cause phenylketonuria, but only if the patients have a diet rich in phenylalanine (Burgard et al., 2017). All these technical issues increase significantly the difficulty of conceptualizing a bioethical, analytical framework for multifactorial diseases.

There have been various criteria used to assess the usefulness of a predictive test. For example, in 1968, Wilson and Jungner developed a set of criteria aimed toward screening a population for a certain disease, which included four main elements: knowledge of the disease (it must be an important problem, recognizable from pre- or early symptomatic stage, and a natural course that should be properly understood); knowledge of the test (suitable for examination, accepted by the population, and case finding should be a continuous process); treatment of the disease (there is an accepted treatment for that disease, there are facilities for diagnosis and treatment available and clear guidelines regarding who should be treated); and cost-related issues (the cost should be economically balanced in relation to the whole medical expenditure generated by the screening, diagnosis and treatment procedures) (1968). In later years, various authors proposed a series of new criteria that should be taken into account before proposing new screening procedures, including: a recognized need the screening program should address, the objectives of the screening that should be defined at the outset, the target population that should be clearly defined, there is scientific proof suggesting the effectiveness of the screening procedure. The program should include education, testing, clinical, and management procedures. There should be quality assurance procedures, including mechanisms of **minimizing potential risks**, the program should respect fundamental bioethical principles including **informed consent, confidentiality, justice (by promoting equitable access for the entire target population), and nonmaleficence (the overall benefits should outweigh the harms)** (Andermann et al., 2008). Khoury et al. developed a series of principles for screening the genetic susceptibility of a population for a particular disease, which had to include three main areas: (1) public health assessment (the screened disease should generate an important public health burden, the prevalence of the genetic trait in the target population and its attributable burden should be known, and the natural course of the disease should be properly understood); (2) the evolution of tests and interventions (there should be available data regarding the positive and negative predictive values for the test in the target population, and the safety and effectiveness of the test and associated interventions should be known); (3) policy development and screening implementation (screening procedures are deemed acceptable by the target population, there are adequate facilities for the prevention, treatment, education, counseling, social support, and surveillance procedures, there should be a consensus regarding the appropriateness of screening and intervention procedures. In addition, based on proper scientific proofs, the screening should be a continuous process, the cost effectiveness ratio should be established, screening and interventions procedures should be available for the target population, and there should be safeguards regarding respecting basic bioethics principles such as **privacy, respect for autonomy, or justice)** (Khoury et al., 2003). As seen from these criteria, each predictive test should respect some basic bioethical principles, including autonomy, trust, justice, beneficence, and nonmaleficence.

2. INFORMATION AND CONSENT IN PREDICTIVE TESTING FOR MULTIFACTORIAL DISORDERS

2.1 Respect for Autonomy

A person is autonomous if she/he can initiate actions based on an intrinsic power to initiate them (and therefore the initiation of those actions is not dependent on the power or desire of others). The initiation of the action is generated by the person's willingness to act and is based on his/her free will to act, on her judgment and possibly taking into account the possible outcomes or consequences of that action (Buss, 2008). When assessing the capacity of a patient to act autonomously, we must check whether he/she received proper/relevant information, understood it, took a decision based on internal convictions and was not subjected to significant external influences (Faden et al., 1986; Hostiuc, 2014).

Receiving proper information to make a properly autonomous choice is extremely hard in medical genetics in general, and especially in predictive testing for multifactorial diseases, mainly due to the fact that it is a specialized field, with progresses made at a very fast pace, and due to inherent uncertainties of the importance of the assessment of various genetic risk factors. An additional factor generating difficulties is represented by the fact that many genetic tests for multifactorial diseases bypass physicians altogether, within the context of direct to consumer (DTC) genetic testing.

2.2 Delivering Relevant Information

In recent years, the pattern of communicating information about genetic tests tended to shift from clinicians to genetic counselors to over-the-phone counseling, Internet-based counseling and self-access (Haga et al., 2014). The primary interaction of patients with physicians, for most multifactorial diseases, is still done through the general practitioner (GP) or a clinical specialist. For example, diabetes care is primarily handled by nutritionists, diabetologists, and GPs; hypertension care is handled by cardiologists, internal medicine physicians, and GPs; ankylopoietic spondylitis care is handled by rheumatologists and internal medicine physicians.

For nongeneticists to recommend predictive genetic testing, they have to properly understand the genetic information, and "translate" it to the patients, which is sometimes quite difficult. Often, the treating physician lacks enough knowledge to encompass all the subtleties of informing patients about the results of predictive testing. However, this limited scientific knowledge can be overcome through specific training in medical genetics, diagnostic testing, and genetic counseling (Plunkett and Simpson, 2002). If these issues cannot be surpassed, treating physicians should include in the information process, genetic counselors. Wilkes et al. showed that, primary care physicians from the United States have an inadequate understanding of the evidence-based screening, the proper communication able to encourage shared decision-making, the ethical, legal, and social consequences of screening for genetic mutations, and even the basics of clinical genetics (2017). A study performed by Zack et al. regarding the understanding of information needs of general practitioners managing Osteogenesis Imperfecta showed that 69% of all physicians lacked enough knowledge about the disease and its genetics characteristics (2006). A study by Menasha et al. showed that 71% of the physicians who responded to a questionnaire rated their knowledge regarding genetics and genetic testing as fair or poor (2000). Patients usually prefer to obtain their medical information (at least initially) from a general practitioner or a medical specialist other than a geneticist. A survey conducted by the American Medical Association showed that 71% of patients would initially consult a general practitioner for advice regarding a genetic pathology, and around 80% trusted the ability of their general practitioner in the field of clinical genetics (Goldberg, 1998). Teixeira et al., in a study testing the importance of the general practitioner as an information source for patients with hereditary hemochromatosis, found that, even if two-thirds of the patients obtained relevant data from a specialist, they would have preferred to obtain it initially from a general practitioner (2014).

Some types of genetic screening or genetic testing already entered in the domain of activity of nonclinical geneticists. For example, pediatricians are involved in the genetic screening for phenylketonuria, clinical oncologists— for BRCA1,2 and HER2, and so on (Becker et al., 2011). Knowledge about these tests, and how to conduct the medical counseling to allow a proper under-standing of relevant information should not be a significant issue. This is not the case for genetic testing in multifactorial disorders, where knowledge is much more "unstable." For example, Mihaescu et al. performed a study aimed at eval-uating the benefits of updating risk information in commercial genome-wide scans, by comparing type 2 diabetes risk predictions based on TCF7L2 alone, 18 polymorphisms alone, and 18 polymorphisms plus age, sex, and body mass index. By adding to the TCF7L2 gene another 17 polymorphism, the researchers found that 34% of the participants had to be reclassified in different risk groups. The inclusion of information about age, gender, and BMI generated a reclassification in 29% (2009).

Commercial companies promote the idea that genome-wide scans may aid the patients in assessing their risk of developing a certain disease. However, as seen from this study, the assessment, at least at the time, is extremely superfluous, especially for multifactorial diseases, the risk profile being highly dependent upon the panel or risk factors evaluated (Mihaescu et al., 2009). Genetic tests could, for many multifactorial disorders, reveal not only risk factors, but also show what are the best preventive measures to minimize the risks associated with a certain genetic profile, or even to allow personalized treatments, depending of the specific genetic profile of the risk factors (personalized pharmacogenomics) (Horgan et al., 2014; Mordini, 2004).

Medical genetics is a highly specialized medical field, using a medical jargon that is difficult to interpret by a nonspecialist. For example, in the "Standards and guidelines for the interpretation of sequence variants: a joint consensus recommendation of the American College of Medical Genetics and Genomics and the Association for Molecular Pathology," is said, in the chapter about PP2/BP1 variant spectrum (PP2—Missense variants in a gene that has a low rate of benign missense variation and in which missense variants are a common mechanism of disease; the BP1-Missense variant in a gene for which primarily truncating variants are known to cause disease) that "Many genes have a defined spectrum of pathogenic and benign variation. For genes in which missense variation is a common cause of disease and there is very little benign variation in the gene, a novel missense variant can be considered supporting evidence for pathogenicity (PP2). By contrast, for genes in which truncating variants are the only known mechanism of variant pathogenicity, missense variants can be considered supporting evidence for a benign impact (BP1). For example, truncating variants in ASPM are the primary type of pathogenic variant in this gene, which causes autosomal recessive primary microcephaly, and the gene has a high rate of missense polymorphic variants. Therefore missense variants in ASPM can be considered to have this line of supporting evidence for a benign impact." (Richards et al., 2015). This jargon determines many clinicians to look information about genetic testing elsewhere. Zack et al., for example, showed that 60% of general practitioners looked for information about osteogenesis imperfecta on websites, 60% in textbooks, and only 14% in specialized journals (Zack et al., 2006). The primary sources of information have some important limitations—website articles are often written by nonspecialists, and therefore are more prone to errors, while textbooks rarely have the latest information, which can be an important issue, especially in a highly dynamic fields such as medical genetics (see also the Mihaescu study presented above, 2009).

The minimal information that has to be delivered by physicians (and understood by patients) is presented in Table 1.

TABLE 1 Relevant Information in Multifactorial Predictive Genetic Testing

Information	Details
General information	<ul><li>The purpose</li><li>Method of investigation (including type of tissue, invasiveness, accuracy)</li><li>Information about the interpretation of the results—risk pattern increase/decrease, uncertainty of the results, penetrance, expressivity, age of onset, severity of symptoms)</li><li>Current status of knowledge about that particular battery of tests (especially regarding its actual predictive value)</li><li>Risks (related to the procedure, and if possible health-related consequences of the tested diseases)</li><li>Benefits (related with potential lifestyle changes, personalized disease management)</li><li>Alternatives (other types of tests, clinical follow-ups, not taking the test, other providers of genetic tests, if available)</li></ul>
Information regarding potential horizontal and vertical effects	<ul><li>What will the results mean for family members, recommendations regarding disclosure/nondisclosure</li><li>Contact information for family members, if both parties agree to disclose relevant information to family members and the patient does not wish to inform them him/herself</li><li>Potential effects on reproduction (recommendations for prenatal diagnosis, preimplantation diagnosis, gamete donation)</li><li>The fact that these tests might bring information about nonmaternity or nonpaternity</li></ul>
Management of biological samples and genetic data	<ul><li>How will the biological samples be managed—storage, secondary uses, use in research, duration of storage, what to do after the duration of storage expires</li><li>Who will have access to the data, how is the data kept confidential, in which condition can data be disclosed</li></ul>

Modified from Human Genetic Society Australasia, 2014. Guideline. Pre-symptomatic and Predictive Testing for Genetic Disorders. Human Genetic Society Australasia, Sydney, NSW; Hostiuc, S., 2015. Forensic autopsy. The particularities of consent for research. Online J. Health Ethics, 11, 8.

From these data, we can infer that: (1) patients trust more the physicians who are closest to them (usually general practitioners), and they should be the ones managing the information process regarding predictive genetic testing; (2) clinical geneticists should work, as much as possible, as a team with the clinicians when transmitting genetics-related information; (3) highly specialized, but relevant information should be transmitted from clinical geneticists to the clinicians working daily with the patients, using a toned down medical genetics jargon.

2.3 Understanding Relevant Information

The second fundamental issue that has to be fulfilled for respecting the autonomy of the patient is represented by him/her properly understanding the information that is needed to make an autonomous choice. There are three main elements of understanding that have to be reached: understanding how, that, and what. Understanding "how" implies that the person has the power to act according to the obtained knowledge (and is sometimes called competence as a status) (Huibers and van 't Spijker, 1998). Understanding "that" means that the patient knows that he/she has to do something. Understanding "what" implies the understanding of what is communicated by the physician (Hostiuc, 2014)—"the typical pattern of understanding in informed consent settings is for patients or subjects to come to understand that they must consent to or refuse a particular proposal by understanding what is communicated in an informational exchange with a professional" (Faden et al., 1986). Understanding "how" and "what" are particularly challenged in predictive genetic testing. Understanding "how" is disputed by the fact that often "how" to act based on received information that has many unknowns—we often do not know either that predictive genetic testing is indeed useful, due to frequent shift in recommendations regarding the panel of markers that are useful for a particular disorder, or if, in case of a positive test result, we can actually intervene in a significant manner to reduce the estimated risk associated with a particular constellation of markers suggesting an increased risk. For example, let us take a phrase like: "you have five genetic risk factors for developing arterial hypertension, namely A, B, C, D, E. Overall, the risk increase generated by all these factors is 12% compared to a person with a similar age and BMI without these five factors. A and B can be prevented through diet, C through exercise. D is also a risk factor for early onset Alzheimer's disease, while E is a protective factor for type II diabetes, decreasing the risk by 1.2%. New research may change the data about these risk factors, and we should discuss them again at a later date." The information is clear, as much as possible, in an easy to understand language, and has embedded practical recommendations. However, due to a very high number of variables, the patient may not be able to grasp the actual meaning encompassed in the given information, and discern what is really important and what should he do. Therefore he does not really obtain the ability to understand "how."

Understanding "what" is challenged by both increased difficulty of conceptualizing genetic information by the patient, and by the difficulties of clinicians to actually grasp the intricacies of these tests, and to be able to present them in a coherent and easy to understand fashion.

The relevancy of properly understanding the information about genetic predictive testing for multifactorial disorders is extremely high for preventive medicine. Knowing specific risks for major multifactorial disorders may generate lifestyle changes (eating habits, work-related changes), the usage of drugs that are personalized for a specific genetic profile, and so on. The patients must understand not only the medical information, but also the long-term consequences it may generate (Cameron et al., 2012). Some authors found, for example, that patients are prone to believe that genes are immutable, so the genetic risk factors are not modifiable through lifestyle changes (Marteau and Lerman, 2001). Even if this may be true for some genetic disorders, such as Huntington's disease, in numerous multifactorial disorders there are many risk factors that may be counterbalanced with proper lifestyle changes. For example, the enzyme CYP2D6 is known to catabolyze polycyclic hydrocarbons and arylamines into mutagenic substances (Marcy et al., 2002). Around 90%–95% of all persons have alleles causing either fast or intermediate metabolism of these substances, and 5%–10% are slow metabolizers (Amos et al., 1992). A fast metabolism of these substances is associated with an increased risk for lung cancer; for examples, the fast metabolizing type is associated with an OR increase of 2.28 [1.60–3.40] for lung cancer (Amos et al., 1992). Fast metabolizers are therefore more prone to developing lung cancer, but the risk may be significantly decreased by not smoking or not living in a highly polluted area. A person with a fast metabolizer phenotype for CYP2D6 must therefore understand not only that he has a risk factor, but also that important lifestyle changes can significantly decrease his chances of developing this disease. Or, if he/she does not want to quit smoking, more frequent screenings could reduce the stage in which such a cancer will be identified.

2.4   Capacity to Act Voluntarily

For a person to act voluntarily, his/her actions must be generated by their own intentions/wishes, not by those of various third parties. The capacity to act voluntarily depends on two main factors: controls and resistance. Controls are those external elements that are able to pressure persons to act against their intrinsic wishes; resistance—the intrinsic capacity of a person to resist external controls (Hostiuc, 2014; Faden et al., 1986). If internal resistance is basically identical for nongenetic and genetic testing (as it mainly depends on the psyche of the individual), controls are significantly different for these two groups of tests. For example, DTC companies promote various testing through often "blurry" messages, recommending patients to obtain a risk profile for various disorders, without clearly stating the limits of these tests. A very convincing advertisement might alter importantly the capacity of a person to objectively analyze the usefulness of a genetic test. Huibers et al. argued that physicians could be significant controls in predictive genetic testing, as physicians are often inclined to "please" them, making it hard to differentiate a willingness to act from a willingness to please (1998). Families might also act as significant controls, either by pressuring a person to obtain a genetic risk profile (e.g., for a disease that is common in a particular family), or by pressuring them not to get tested (usually for more sensitive information, which could generate discrimination). Physicians should be very attentive for signs suggesting that families act as significant controls. For example, phrases like "I would feel guilty if …", or "my wife/husband thinks I should …" (Huibers and van 't Spijker, 1998). Employers might try to force a person to be tested for some work-related pathologies. Even if these cases are not yet common, new laws might soon make, at least in some states, these tests mandatory for the employees (see, e.g., the debates around the HR1313/2017 from the United States). Moreover, there are already published studies presenting algorithms on how to use genetic testing in work environments. For example, MacDonald and Williams-Jones argued for six main conditions that have to be met for genetic testing to be offered for employees, including: (1) the genetic test must be available, highly sensitive and specific, with a low incidence of false positives, and sufficiently penetrant for the test results to have health implications; (2) testing should be carried by an independent lab and the results should be directly given to the employees; the results should be kept confidential from the employer, who can only access the test results at the request of the employee; (3) pre- and post-genetic counseling should be made available, irrespective of the outcome of the test; (4) the genes being tested should not be associated with identifiable or historically disadvantaged groups; (5) if relevant, the employer must guarantee access to group insurance; and (6) if the employee chooses to reveal the results of the test, suitable policies should be put in place to ensure a reasonable degree of job security (MacDonald and Williams-Jones, 2002). Even if such guidelines seem benign, the potential of external controls is significant. For example, even if the employee is the only one who could divulge genetic information toward the employer, at his specific request, the employer can easily "induce" this behavior through apparently unrelated actions. This would generate a perceived lack of alternativity, automatically breaching the principle of respect for autonomy (Huibers and van 't Spijker, 1998).

3. NONDIRECTIVENESS IN PREDICTIVE TESTING FOR MULTIFACTORIAL DISORDERS

One of the cornerstones of bioethics in genetic testing has been, at least until recently, the principle of nondirectiveness, stating that genetic counselors should provide relevant information about the genetic risk, to explain the choices regarding the testing and management, but to refrain in providing any kind of recommendations about the best course of action. The controls generated by the physicians should be therefore minimized, and the decision should lay solely in the hands of the patient. This principle was derived from the idea that genetic testing often is a social rather than a medical issue. Initially, the main focus of clinical genetic testing was the reproductive risk (Raes et al., 2016) (the risk of having a baby with a particular genetic disorder). The families with this risk often faces difficult choices such as whether or not to undertake a pregnancy, to terminate the pregnancy, or to give the future child to adoption (Burke et al., 2001). These choices should be analyzed within the family core, not within the physician-patient relationship. This principle was afterwards extended for genetic testing for monogenic disorders such as Huntington's disease (Mundluru et al., 2016), in which test results may generate life-altering changes, and later on became one of the main bioethical principles associated with genetic testing.

As an alternative, shared decision-making (SDM) involves at least two parties (the physician and the patient); both take steps to participate in the decision-making process, that is mainly based on information sharing, and the decision is taken conjointly (Elwyn et al., 2000). Elwyn et al. suggested that there are eight steps/competencies in involving patients in SDM: (1) starting the SCM process, either implicitly or explicitly, (2) identification of ideas, fears, and expectations regarding the issue at hand, and possible therapeutic alternatives; (3) identification of the main options; (4) identification of the preferred option, followed by targeted information based on the preferred option; (5) reviewing the process (the understanding of relevant information, verifying the reactions to the preferred alternative); (6) acceptance of the process; (7) taking the actual decision, followed by discussion about it; and (8) arrangement of follow-ups (2000).

Janssens and Khoury argued that, due to a "lower predictive value of the results, the pleiotropic effects of the susceptibility genes, and the low inheritance of the genomic profiles, genetic testing for multifactorial disorders is closer to nongenetic tests than monogenic diseases, in areas such as associated psychological burden, anxiety caused by the presence of a risk factor or disclosure to the families, insurance related issues (Janssens and Khoury, 2006). In the Safer case, the panel of judges concluded that "there is no essential difference between the type of genetic threat at issue here and the menace of infection, contagion, or a threat of physical harm" (New Jersey Superior Court, Appellate Division, 1996). From here, we may infer that the main function of predictive testing for multifactorial disorders is medical rather than social, making shared decision-making the best choice in many instances (see below), depending on the employed type of physician-patient relationship (Emanuel and Emanuel, 1992).

4. DIRECT TO CONSUMER GENETIC TESTING

In recent years, it has become possible to obtain genetic tests for some disorders from commercial distributors rather than physicians (often from pharmacies or the internet). Tests range from single gene tests, such as BRCA1 and 2, to genes associated with multifactorial disorders, or even genome-wide scans for some diseases (often oncological or cardiovascular) (Becker et al., 2011; Mihaescu et al., 2009).

Sometimes, commercial companies are marketing a specific genetic test immediately after a potential correlation with a certain disorder is suggested, without expecting the publication of highly relevant studies to properly assess the correlation. We must take into account that, at least for multifactorial disorders, there is a low predictive value for many genetic tests (Becker et al., 2011).

Genetic testing require biological samples taken from the subject; however, as sampling is often done by the consumers, the ones actually doing the testing may be induced in error as related to the identity of the one from whom the biological samples were taken. MDC companies often induce the idea that their tests have medical validity. An example in this regard was presented by Howard and Bory regarding the now defunct Navigenics website: Under the "what we offer tab" on the home page, a consumer is welcomed with this message: "Navigenics gives you a view into your DNA, revealing your genetic predisposition for important health conditions. This level of personalization may help you take action to detect health conditions early, reduce their effects or prevent them entirely" (Navigenics, 2009). Meanwhile the disclaimer "Navigenics does not provide medical advice, diagnosis, or treatment" (Navigenics, 2009) is present, in much smaller and paler print, at the bottom of each page of the company's website" (Howard and Borry, 2009).

Direct to consumer genetic testing has a series of potential advantages that could, in theory at least, decrease the morbidity associated with various disorders, and even the mortality (see Table 2). If patients go directly to these companies, physicians are bypassed, which could be considered, at least in more traditional medical systems, as a severe blow to one of the main functions of physicians—namely to develop prophylactic strategies for their patients. Besides the obvious benefits of predictive genetic testing, this "bypass" can generate nonautonomous behaviors from the patients (who could accept predictive testing without fully knowing the consequences of the medical information that will be obtained), or even maleficence (an incorrect assessment of the identified risk factors could generate active harms to the patients). If proper medical counseling is done before or even after the testing, these harms can be minimized; however, this would make DTC a de facto part of a medical intervention, endorsed by the treating physician. Medical recommendations should be based strictly on objective, medical criteria. Therefore, if a certain panel of markers are proved to be useful to assess certain risks for multifactorial diseases, they

TABLE 2 Advantages and Disadvantages of Direct to Consumer Genetic Testing

Advantages	Disadvantages
Increases the addressability to genetic testing due to smaller costs compared to genetic testing within the physician-patient relationship	Inadequate counseling from physicians
Increases the addressability to physicians if the results are in doubt	Improper understanding of the results
Increases the chances of identifying risk factors, especially those modifiable through lifestyle changes	Changing the importance of various risk factors
	Decreases the prestige of the medical profession
	Risk of unknowingly performing tests on third parties

could be endorsed by the physicians. If, however, the medical data supporting a certain association is weak, or inconclusively proven, physicians should refrain in recommending the testing, based on the principle of nonmaleficence. Some more advantages and disadvantages of direct to consumer genetic testing are presented in Table 2.

5.   WHEN TO RECOMMEND PREDICTIVE GENETIC TESTING?

As seen in the introductory part of this chapter, there are numerous factors that have to be taken into account before recommending predictive genetic testing for multifactorial disorders. Some authors have developed algorithms to assess the ethical acceptability of predictive testing based on various criteria.

For example, Bunnick et al. argued that there are four main disease characteristics that are especially relevant for the ethical evaluation of predictive testing (2012): severity, actionability, the age of onset for the disorder, and the distinction between somatic and psychiatric disorders. Severity quantifies the mortality and morbidity associated with a particular disease (including the emotional, psychological, and social impact of the results of the genetic tests) (Bunnik et al., 2012). Actionability quantifies the usefulness of a potential medical procedure, and is mainly based on the risk/benefit assessment. The age of onset—the average age at which a disorder may become clinically manifest. See also Table 3.

Burke et al. presented four main categories of genetic tests, based on two main criteria: clinical validity and the presence of an effective treatment (Burke et al., 2001) (see Table 4). Each group has distinct associated ethical issues, which are presented in Table 4.

TABLE 3 Main Disease Characteristics Relevant for the Ethical Evaluation of Predictive Testing

Characteristic	Underlying Ethical Principles	Modifiers
Severity	Beneficence—duty to provide genetic counseling and psychological support increases with the severity increase Autonomy/nondirectiveness—the strictness of enforcing nondirectiveness is directly correlated with the severity of the disease and/or potential social/psychological impact	Different magnitude of severity depending on additional factors, penetrance)
Actionability	Nonmaleficence—increases when are present associated health and nonhealth related risks (social risks such as financial costs, psychological risks) Beneficence—increases when associated health and nonhealth related risks are decreased by the testing Justice—decreases if the tests can cause discrimination (e.g., gender based, or racial based markers), unequal employment opportunities, unequal access to health care	Generation of therapeutic alternatives (including preventive) increases actionability Personal utility (e.g., relief from uncertainty, solace, value of knowing)
Age of onset	Nonmaleficence—a positive test may act as a stressor, with negative consequences for the wellbeing of a person, that can be directly correlated with the difference between the expected age of onset and the moment of testing Autonomy—the right not to know, especially in children	Difficult assessment of early versus late-onset disease, especially in multifactorial disorders, where preclinical changes can be identified well before a clear clinical syndrome can be identified
Psychiatric/psychosomatic distinction	Maleficence generated by (1) knowledge regarding psychiatric disorders is higher compared to the one about nonpsychiatric pathologies (car alter the perception about the world, behaviors, desires, relationships, etc., and (2) self-fulfilling prophecies, with genetic tests acting as trigger events for psychiatric disorders, and (3) stigma associated with psychiatric disorders is higher, potentially causing various familial, psychological or social issues.	

Based on Bunnik, E.M., Schermer, M.H.N., Jw Janssens, A.C., 2012. The role of disease characteristics in the ethical debate on personal genome testing. BMC Med. Genet. 5, 4, modified.

TABLE 4 Main Categories of Genetic Tests

Group	Underlying Ethical Principles
High clinical validity, lack of an effective treatment	Autonomy—absolute, full enforcement of the nondirectiveness principle
	Beneficence—minimal, mainly nonmedical (life choices)
	Maleficence—significant, due to decreased actionability, increased potential to alter personal interactions
	Justice—moderate risk of stigma and discrimination for neurological disorders (such as Huntington's disease) (Tibben et al., 1993), high risk of stigma for psychiatric disorders
High clinical validity, effective treatment	Autonomy—variable, depending on the disorder and the type of treatment. For example, for diabetes, a shared decision-making should be preferable, while for BRCA1 positive breast cancer—nondirectiveness is the best option (Surbone, 2001)
	Beneficence—significant; the test should be recommended whenever the nondirectiveness principle must be enforced
	Maleficence—minimal, mainly nonmedical
Low clinical validity, lack of an effective treatment	Autonomy—strong. Physicians should emphasize the lack of an effective treatment and the low clinical validity, and should not recommend these tests
Low clinical validity, effective treatment	Autonomy—a shared decision-making approach should be preferable, as the issues are difficult to analyze by a nonspecialist. A positive recommendations should be given to test for low emotional weight disorders such as hypertension or diabetes, and if the treatment is safe/associated with low risks
	Beneficence—potentially high
	Maleficence—generated by increased anxiety associated with a positive test result, which may prove to be false positive, or by a lack of implementing preventive measures in case of a false negative test

Based on Burke, W., Pinsky, L.E., Press, N.A., 2001. Categorizing genetic tests to identify their ethical, legal, and social implications. Am. J. Med. Genet. A, 106, 233–240, modified.

6. NONMALEFICENCE AND RISK ASSESSMENT

Nonmaleficence, or do-no-harm, is one of the fundamental principles of medical ethics, identifiable since the Hippocratic era of medicine. According to this principle, physicians have an absolute obligation not to harm their patients. Normally, this principle should prevail when in conflict with other fundamental ethical principles. In clinical practice, nonmaleficence is usually assessed through risk analysis and cost-benefit analysis. Risk assessment is done by analyzing the probabilities of negative outcomes for all potential courses of preventive, diagnosis or therapeutic management. To perform risk assessment, one should identify the risks, estimate, and then evaluate them (Beauchamp and Childress, 2005).

In clinical genetics, the way the information about the risk profile is taken into account by patients varies extensively, with the mention that most studies were performed on monogenic or major gene disorders (Mihaescu et al., 2009). Pjil et al. showed that informing the familial risk for diabetes failed to increase fatalism or overall well-being, and led to the adoption of a healthier habits, at least up to 3 months after the consultation, a possible cause being the fact that the subjects considered the information as highly relevant, being tailored to their particular risk profiles (2009). Some other examples of consequences of testing positive for various genetic markers, taken from two systematic review performed by Heshka et al. (2008) and McBride et al. (2010) are presented in Table 5.

When analyzing whether to recommend a genetic test for a multifactorial disorder, the clinician should take into account all these risks and to recommend it (or not) based on a more holistic view of the risks, without necessarily emphasizing the medical ones (unless they are highly significant, of course).

7. CONFIDENTIALITY AND FAMILY SHARING OF THE GENETIC RESULTS

Confidentiality of the medical act is, even today, considered one of the most important ethical principles that must be respected in clinical medicine. Accordingly, any type of data that is obtained within a physician-patient relationship is secret, and its disclosure can be allowed only in very particular circumstances. As Kottow said "Clinicians' work depends on sincere and complete disclosures from their patients; they honor this candidness by confidentially safeguarding the information received … Limitations or exceptions put on confidentiality would destroy it, for the confider would become suspicious and uncooperative, the confidant would become untrustworthy and the whole climate of the clinical encounter would suffer irreversible erosion" (1986). In the past, confidentiality was seen by many medical ethicians/deontologists as the cornerstone of medical morality (Hostiuc et al., 2014; Brouardel, 1887; Herzlich, 1982; Larkin et al., 1994), and its breach was nonimaginable (see, e.g., the discussion about

TABLE 5 Potential Consequences of Predictive Genetic Testing for Multifactorial Disorders

Consequences	Details
General distress	Overall, there is no increased general distress in carriers versus noncarriers, even though some studies suggested a small increase in distress for carriers in the short term (1 week-4 months) (Watson et al., 2004; Lodder et al., 2001)
Anxiety	Most published study revealed no significant differences between carriers and noncarriers for hereditary breast and ovarian cancer. For HNPCC—there was a short term increase in anxiety followed by a return to baseline levels in carriers and a decrease in short term, followed by a return to normal levels in noncarriers. For Alzheimer's disease—there were no differences nor in the short or the long term regarding anxiety levels
Depression	Most studies showed no effect
Screening behavior	Mammography screening increased for BRCA1/2 carriers at 12 months posttest, the overall rate being higher compared to noncarriers. Intravaginal ultrasound screening was much higher posttest for CA-125 positive women (risk factor for ovarian cancer). Colonoscopy increased in carriers for HNPCC genes at 12 months.
Lifestyle changes	Smoking cessation seemed to increase (at least in some subgroups), in patients with a positive testing for L-myc gene (Ito et al., 2006), or GSTM1-missing (Sanderson et al., 2008), both genes increase the risk of lung cancer. Risk factors for conditions such as hypercholesterolemia or obesity did not yield lifestyle changes (Hicken and Tucker, 2002; Marteau et al., 2004)

confidentiality in French in the 19th century from Hostiuc and Buda, 2015). Open debates about the limits of confidentiality stemmed in post war Western medicine with the Tarasoff case (Gutheil, 2001; Furrow, 1997). The treating physician is the one mainly responsible for respecting the confidentiality. For example, the Declaration of Geneva, currently published by the World Medical Association, states that "I will respect the secrets that are confided in me, even after the patient has died" (World Medical Association, 1948). However, recent advances in medicine made many authors to consider that confidentiality, as was classically understood, is not feasible in modern medicine. For example, Mark Siegler presented the case of a patient with mild chronic obstructive pulmonary disease, who became concerned about the confidentiality of his medical records. When analyzing the persons who had access to the medical chart of the patient, Siegler found that at least 75 health professionals and hospital personnel had access to it (Siegler, 1982). Without access, the other health-care professionals involved in the therapeutic management of that patient would not be able to perform their tasks. Confidential information was shared nonetheless outside the physician-patient binomial structure, making it an obvious breach, both in the eyes of the patient, and, retrospectively, of the physician. Since the 1980s, when the article written by Mark Siegler was published, until nowadays we witnessed an unprecedented explosion of technologies that are able to use our private information, including in the medical field—implantable/wearable devices, sensors, and trackers. All these makes basically impossible for physicians to actually enforce a strict concept of confidentiality.

In the last decades, the society embraced a culture of openness, of disclosure of private information even to strangers, the most obvious case and point being the explosive development of social media tools, such as Facebook, YouTube, or Instagram (Bertot et al., 2010; Eysenbach, 2008). Often patients themselves openly share, either knowingly or unknowingly, sensible medical information. For example, in the last years has emerged the concept of patient-led research (PLR), in which patients organize themselves to perform medical studies. One such driven study that became famous, was the lithium study for amyotrophic lateral sclerosis, organized by patients on the PatientsLikeMe platform (Armon, 2010; Wicks et al., 2014). In such studies, sensitive medical information are openly shared between patients, but also third parties. Often, not even patients seem to wish to keep the medical record private. Genetic information associates a potentially significant risk of harming the patient in the future—may generate employment difficulties, insurance-related difficulties, discrimination, and so on. To limit these consequences, which are rarely obvious to patients, physicians should have a duty of recommending nondisclosure of these information, based on the ethical principle of nonmaleficence even if patients seem open about disclosing medical data.

A major topic of debate, especially regarding predictive testing, is whether the families should be informed about the results. Normally, the physician-patient relationship must be agreed upon by both parties, with some notable exceptions (emergency medicine, patients lacking decisional capacity, expert medicine). The physician enters in a contractual relationship with the patient in order to recommend (and perform) predictive genetic testing. To inform the family of the patient two basic conditions must be simultaneously met: (1) the patient must agree to the disclosure (or alternatively the physician has a duty to warn), and (2) family members agree to enter in a professional relationship with the physician. Moreover, the first condition must be fulfilled before taking steps to fulfill the second one. Namely, the physician must have a valid medical reason for entering in a professional relationship with the family, reason that normally can be detailed to it only after the patient accepted disclosure. Not acting in this manner would breach the principle of autonomy, which can only be superseded in particular circumstances.

One such circumstance is the duty to warn, according to which physicians may disclose confidential information to third parties (in this case family members) to prevent a foreseeable harm (Godard et al., 2006). This duty can be performed with or without the consent of the patient.

If the **physician has the consent of the patient for disclosure**, his course of action is mainly dependent on the analysis of the risk/benefit ratio, with the mention that we should include in the analysis not only medical, but also social, psychological, or economical factors. The risk of multifactorial inheritance in first-degree relatives equals with the square root of the incidence of the disease in the general population (Turnpenny and Ellard, 2016). The risk for second degree relatives equals $p^{3/4}$, and for third degree is $p^{7/8}$(Turnpenny and Ellard, 2016). Therefore if a certain disease/risk factor is present in the patient, the closest relatives have the highest risk increase, and this increase becomes negligible in more distant relatives. The duty to warn should therefore take into account this pattern or risk increase in relatives, and disclosure should be done mainly to those at the highest risk, taking of course into account the other factors that were detailed above. For example, schizophrenia has an overall incidence of 10/1000. Therefore the risk of a first degree relative to have a risk factor/disease is 10%. Second degree relatives have a risk of 3%, while third degree relatives have a risk of 1.7%. We believe that it would be justifiable to disclose this risk to first degree relatives; to second degree relatives, we would refrain in recommending disclosure as, even if the risk increases threefold compared to the general population, we have to also take into account potential issues that may be generated for the patient, who would be considered, by more distant relatives, as the bearer of a social stigma (Bunnik et al., 2012). For a disease such as lung cancer, for the same numeric values, we would recommend disclosure to second, or even third degree relatives, as proactive lifestyle changes could significantly decrease the risk, and there is no associated stigma (Bunnik et al., 2012).

Family members might not want to know about the results of the genetic testing, as it can lead to a need of life-altering changes to prevent further harm or to decrease potential harm. A person might not get married, have children, accept a certain job, eat whatever he/she wants, and so on, based on that particular information. Both informing and not informing could generate maleficence—informing could generate maleficence through telos altering choices (altering the right to an open future) that are forced upon that person, while not informing could generate maleficence through a potential increase in morbidity or early mortality (Fulda and Lykens, 2006).

Such an example was presented by Rita Kielstein and Hans-Martin Saas, regarding a family in which a 55-year-old woman developed renal failure caused by autosomal dominant polycystic kidney disease (ADPKD). After she began dialysis, all her four sons underwent ultrasonography screening for the disease, and were found to be presymptomatic carriers of ADPKD. Her father and two brothers died due to end-stage renal disease, most likely caused by ADPKD. After screening, the sons criticized their parents of having four kids in full knowledge of the fact that a severe disease was present in the family.

Three years after the ultrasound screening, the oldest brother committed suicide after finding out he had symptoms of the disease. The second son sold his half-completed new house when he learned about the diagnosis. The third son broke up the engagement as he did not want to burden his fiancée with the disease, nor to have children who might have it. The fourth son quit the university when finding out the diagnosis, took a good paying job to make money and enjoy life as long as it lasts (Kielstein and Hans-Martin, 2002). This right not to know is also based on the principle of autonomy, and in practice it has to be analyzed conjointly with beneficence and nonmaleficence. Informing the family about the results of a genetic test might generate a harmful outcome (social, psychological, financial, or even medical). As Godard et al. stated: "There are positive aspects to ignorance, particularly in a disease where there is no prospect of prevention or treatment" (Godard et al., 2006). If there are no efficient therapeutic alternatives, as both principles of autonomy and nonmaleficence dictate nondisclosure, this should be the norm. If however an efficient treatment is available for the disease of interest, beneficence would dictate disclosure, a course of action that is contrary to the one generated by respecting the autonomy and nonmaleficence principles. Physicians should, in these instances, try to shift from nondirectiveness to SDM, and recommend disclosure, eventually in joint discussions with patients and their families. A refusal from the relatives should be respected, and the initial discussions should only tackle the issue of disclosing material medical information, and not the actual medical information that should, according to the physician, be disclosed.

Sometimes, even if the patient agrees to share genetic information with the relatives, this cannot be easily done. Basttistuzzi et al. found that often genetic services from Italy needed to use letters to inform relatives about a genetic risk, with the acceptance of the patient, due to various reasons such as geographical distance, fear of sharing bad news, and difficult family relationships (Battistuzzi et al., 2012).

If the **physician does not have the agreement of the patient to disclose genetic information,** the duty to warn can be implemented only if the risk is significant. What exactly is a significant risk is impossible to say. The Declaration on the Human Genome Project of the WMA states that "even if family members of the patient may be at risk, medical secrecy has to be kept unless there is a serious harm and this harm could be avoided by disclosing the information; the confidentiality can be breached only as a last resort when all trials to convince the patient to pass on the information by himself, have failed" (World Medical Association, 1992). Following the Tarasoff case, have been developed a series of general guidelines, the most well known analyzing four main issues: (1) the magnitude, imminence and type of harm; (2) the capacity of an individual to carry out the harmful behavior; (3) the likelihood that a person would act harmfully on a third party, and (4) the identification of the likely recipients of the potential harm (Rosner, 2003). In a similar fashion, for genetic disorders, the Institute of Medicine recommended that genetic information should be disclosed if the following conditions are simultaneously met: (1) the harm done to the relative is irreversible and highly likely; (2) there have been failed attempts to elicit a voluntary disclosure; (3) the disclosure will prevent the harm; (4) the harm generated by the disclosure is less than the harm generated by nondisclosure; and (5) there are no other ways to remove that particular harm. If the harm is not irreversible/fatal, disclosure without consent should be forbidden, and the duty to warn should be replaced with the recommendation to encourage patients to discuss the identified risk factors/diseases with their relatives (Motulsky et al., 1994).

Other authors suggest that genetic information should be shared by default with the relatives (Callier and Simpson, 2012), at least for some risk factors/disorders, as the genetic information is, fundamentally, shared within the family (the "genetic information is familial" thesis, or GIFT) (Liao, 2009). Liao et al., for example, argues that sharing genetic information with the relatives should be dependent upon two main factors: P (the probability that a person has a particular constellation of genes if another person has them) and Q (the probability of a person to develop a disease if she/he has a certain constellation of genes). Based on these two factors, they built four types of GIFTs: (1) stronger deterministic (high P, high Q), stronger nondeterministic (high P, medium-to-low Q), weaker deterministic (medium-to-low P, high Q), and weaker nondeterministic (medium-to-low P, medium-to-low Q). For stronger deterministic GIFTs the pressure to disclose is the highest, and for weaker nondeterministic—the lowest (2009). We believe that, even if this algorithm has certain advantages, such as quantifying the QxP values and giving cut-off values for disclosure, it is not feasible in practice. Let us take the example of Huntington's disease. According to this model, it has a strong deterministic GIFT and therefore there is a high pressure to divulge the information to the potentially affected relatives. However, as there is no viable treatment, disclosure does not have any kind of objective, medical benefit; moreover, it could generate maleficence through psychological harm.

8. PREDICTIVE GENETIC TESTING IN CHILDREN

Predictive testing in children has a series of important differences compared to adults, due to many reasons, among which we can include: (1) the interpretation of the genetic tests and their results is highly dependent on the cognitive and emotional development (Fanos, 1997); (2) often children who are tested come from families in which the disease is already present, and parents may be unable to cope emotionally and help their children (Fanos, 1997), which is essential to surpass a life-altering diagnosis; (3) a tendency to ignore the moral status of the children, (4) children can easily breach the confidentiality, which can lead to negative social consequences (Borry et al., 2006); and (5) a tendency to ignore the possibility that children can make autonomous choices (Suter, 1993).

The right to an open future. According to Joel Feiberg, children have four kind of rights: (1) rights that are common to the adults (such as the right to life); (2) dependency rights, or rights appertaining only to children, such as the right to food, shelter, protection, education; (3) rights that can only be exercised by adults (or, sometimes by children that are close to adulthood), such as the right to freely exercise the religion of choice, and (4) right-in-trust, which are saved for the children until they became adults, such as the reproductive rights: "look like adult autonomy rights … except that the child cannot very well exercise his free choice until later when he is more fully formed and capable … rights that are to be saved for the child until he is an adult, but which can be violated 'in advance,' so to speak, before the child is even in a position to exercise them. … His right while he is still a child is to have these future options kept open until he is a fully formed self-determining adult capable of deciding among them" (Curren, 2012). These right-in-trust are called by Feinberg the "child's right to an open future" (Curren, 2012). One such example is represented by a Jehovah's Witness child that needs blood transfusion that is refused by the parents (Davis, 1997). Based on the right to an open future, the physicians are obliged to provide blood transfusion, even if this intervention is not accepted by the religion of the parents (Curren, 2012).

Limited horizons. According to Wertz et al. parents usually want their children to be genetically tested in the hope of a negative result, and are often not well prepared for a positive one. If this happens, they argue that parents could divert resources from the children who tested positive, which would limit their future horizons—they would not be financed, for example, to access a college education as it would be seen as pointless from an economic point of view (Wertz et al., 1994). Malpas argued against this point of view, stating that parents want genetic testing not necessarily expecting for a negative result (even though hoping for it), but rather to be able to prepare, as best as they can, for a potential negative outcome. Knowledge about a positive diagnosis could lead to a better adjustment to the disorder as they grow old, to make important life choices based on that information, to undertake extracurricular trainings, for parents to be able to plan their economic, social, and even personal life through, for example, deciding whether to have more children, to move closer to the extended family, and to relocate in cities or even countries with health-care systems more suited to aid the child in need (Malpas, 2008).

Damage to the self-esteem of the child. Some authors argue that a positive diagnosis could alter the self-esteem of the child. For example, Bloch and Hayden argued that predictive testing could significantly alter the self-esteem and sense of worth of a developing child (1990). Similarly, Clarke and Flinter argued that predictive genetic testing could affect the deepest levels of self-esteem (1996). This hypothesis is not supported by studies, mainly done on terminally ill children, showing that open communication decreases anxiety, increases the levels of self-control, and prepares the children for future treatments and hospital admissions (Koocher, 1986; Gibbons, 2001; Malpas, 2008).

The acceptability of predictive testing in children was mainly analyzed based on the time of onset of the disease of interest—childhood onset versus delayed onset.

For *adult-onset disorders*, predictive testing in children is controversial. The right to an open future would dictate against testing for a particular disorder/risk factor unless there is a positive benefit to risk ratio. The main benefits generated by predictive testing include: the presence of beneficial therapies, the possibility to alter the course of the disease through behavior changes, targeted surveillance, refinement of prognosis, clarification of diagnosis, reduction of uncertainty and associated anxiety, the opportunity for psychological adjustment to the disorder, sharing information with family members, the possibility to develop realistic life-plans, avoiding to have children with genetic disease, and having time to cope with giving birth to a child with a genetic disorder (Ross et al., 2013). The most important negative effects include ambiguous results generated by insufficient knowledge, incomplete penetrance, multiple risk and protective factors, ineffective or even harmful preventive or therapeutic interventions (Ross et al., 2013), diseases associated with a significant social stigma, alteration of self-image, increased anxiety and guilt, familial stress, difficulties in obtaining life and disability insurance, change life-planning decisions based on social pressure, and distortion of the parental perception of the child (Ross et al., 2013). We have to emphasize that this risk/benefit ratio has to be analyzed on a per case basis, especially as many arguments, either pro or against predictive testing in childhood for adult onset disorders are based on empirical evidence. As an example, many authors argued in the past against testing for Huntington's disease in childhood, suggesting that it could generate significant psychological burden (Bloch and Hayden, 1990; Kessler, 1987). However, a landmark clinical study done by Wiggins et al. showed that predictive testing for Huntington's disease decreased the scores for distress and depression for both patients included in the increased risk and those in the decreased risk group for having the disease (Wiggins et al., 1992). Moreover, unlike in monogenic disorders, where the identification of the genetic risk factor is usually associated with a significant increase in the risk of developing them, in multifactorial disorders, early preventive strategies, or even personalized treatments may be implemented early on, limiting later morbidity and mortality. For example, current guidelines recommend testing children for LDL-R mutations if they are found in a first degree relative, or if the child has a father who died of coronary heart disease and he/she has even a moderate increase in cholesterol levels (Wiegman et al., 2015). Even if coronary heart disease will occur in adulthood, the identification of familial hypercholesterolemia may lead to recommendations regarding adherence to a certain lifestyle, treatment with a fat-modified, health diet, and statins, which should be introduced by age 8–10. For a person with nonfamilial hypercholesterolemia, the cumulative low-density lipoprotein cholesterol burden of a 55-year-old person is 160 mmol.

This level is reached, by a person with familial hypercholesterolemia by age 35; persons who are treated since they are 18 years of age reach the 160 mmol threshold at 48 years, and persons treated since age 10—at 53 years, a value very close to the one from nonfamilial hypercholesterolemia (Wiegman et al., 2015). Therefore we consider that predictive testing for multifactorial disorder should be recommended for minors for well established risk factors, and only in exceptional circumstances should they not be recommended by physicians.

For *child-onset diseases*, most relevant national guidelines suggest that testing should be done when the results are of immediate relevance for the health of the child, or may offer a timely medical benefit (Borry et al., 2006). Therefore, for these childhood-onset diseases, predictive testing has immediate, direct benefits for the patient, and morally it should be preferable to respecting the right to an open future. More debatable is the issue of testing children for disorders without a potentially curative treatment. Some national guidelines suggest that, in the absence of a medically objective benefit, predictive testing should not be performed, based on the principle of nonmaleficence. As there is no curative treatment, the medical benefit does not exist (even though there can be significant psychosocial benefits—see above); the procedure may incur potential harms to the patient (again mainly psychological, but with a potential to somatization). The recommendations of the European Society of Human Genetics are similar, "In the case of presymptomatic and predictive genetic testing for conditions which become manifest in childhood and which can be effectively treated or prevented, there are good reasons to comply or to actively bring up the possibility of a test. However, if the preventive and therapeutic measures will be deferred to a later time, the justification for immediate testing is less compelling and careful, supportive counseling will often be appropriate whether or not testing happens at that time" … "In the case of presymptomatic and predictive genetic testing for conditions which become manifest in childhood and which can not be effectively treated or prevented, there are both benefits and risks, and usually neither the benefits nor the risks completely outweigh each other. Genetic testing could be considered if this would be to the psychological or social benefit of the child and his family" (European Society of Human Genetics, 2009).

Predictive testing for adoption. There have been cited cases in which adoption agencies, or even prospective adoptive parents, have requested predictive testing for some disorders. Moreover, Bloch and Hayden cite a case in which a 25-week pregnant woman who had a 50% risk for Huntington's disease, which requested a prenatal testing. She decided that, if the test would yield a low risk, she would keep the baby; if the risk was higher, she would rather give the child for adoption (Bloch and Hayden, 1990). We believe that these kind of tests should not be allowed as they not only have the potential of significantly altering the right to an open future, but also can easily lead to eugenic choices.

9.   GENETIC EXCEPTIONALISM AND MULTIFACTORIAL DISORDERS

Genetic exceptionalism is the practice of treating genetic information distinct from any other type of medical information (Evans et al., 2010). The main reason for this distinction is generated by the fact that genetic data is directly associated not only with the identity of an individual, but also with the genetic information of the entire family. An information obtained from a relative can have positive (but also adverse effects for other members of the family) (Fulda and Lykens, 2006). This is not unique for genetic information, as medical conditions are often shared between family members (see, e.g., the transmission of the common flu within a family); what is particular to this information in the vertical transmission (from parents to children), and the fact that it may allow making relevant assumptions about the medical condition of an individual without him/her being tested (Green and Botkin, 2003). Other arguments favoring genetic exceptionalism include: potential psychological harm, potential usage for discrimination and stigmatization, and the possibility that it may aid in predicting the medical future of a person. Even if they all seem relevant, there are important counterarguments for each (for a detailed discussion see Green and Botkin, 2003); moreover, clinical studies often contradicted these, mainly empirical based claims (see above the example regarding the psychological risk associated with disclosure about Huntington's disease). Genetic exceptionalism was mainly analyzed within the context of monogenic diseases (Liao, 2009; Janssens and Khoury, 2006; Green and Botkin, 2003; Fulda and Lykens, 2006). Predictive tests for multifactorial disorders have the potential of being much more deeply embedded in routine medical practice, as they can be used for a wide array of common disorders, can lead to personalized treatments and lifestyle changes. These tend to generate a phenomenon entitled routinization of genetics (Foster et al., 2006) that, even it has moral and legal particularities, will generate a "normalization" of this exceptionalism. Therefore predictive testing for multifactorial disorders should be seen, and presented to the patients, as something normal, routinized, part of the everyday medical practice, and only in particular circumstances these tests should be presented as "exceptional."

10.   CONCLUSIONS

Predictive testing for multifactorial diseases will most likely enter in routine clinical practice in the near future. Most common diseases might benefit from an early identification of the risk pattern, and especially potential preventive strategies. Therefore physicians should learn how to include it in the current clinical practice, and ethical norms, previously associated with genetic testing should be "normalized," of course implementing safeguards against improper usage of the results.

REFERENCES

Amos, C.I., Caporaso, N.E., Weston, A., 1992. Host factors in lung cancer risk: a review of inter-disciplinary studies. Cancer Epidemiol. Prevent. Biomark. 1, 505–513.

Andermann, A., Blancquaert, I., Beauchamp, S., Déry, V., 2008. Revisiting Wilson and Jungner in the genomic age: a review of screening criteria over the past 40 years. Bull. World Health Organ. 86, 317–319.

Antoniou, A., Pharoah, P.D.P., Narod, S., Risch, H.A., Eyfjord, J.E., Hopper, J.L., Loman, N., Olsson, H., Johannsson, O., Borg, Å., Pasini, B., Radice, P., Manoukian, S., Eccles, D.M., Tang, N., Olah, E., Anton-Culver, H., Warner, E., Lubinski, J., Gronwald, J., Gorski, B., Tulinius, H., Thorlacius, S., Eerola, H., Nevanlinna, H., SyrjÄkoski, K., Kallioniemi, O.P., Thompson, D., Evans, C., Peto, J., Lalloo, F., Evans, D.G., Easton, D.F., 2003. Average risks of breast and ovarian cancer associated with Brca1 or Brca2 mutations detected in case series unselected for family history: a combined analysis of 22 studies. Am. J. Hum. Genet. 72, 1117–1130.

Armon, C., 2010. Is the lithium-for-ALS genie back in the bottle? Not quite. Neurology 75, 586–587.

Battistuzzi, L., Ciliberti, R., Forzano, F., De Stefano, F., 2012. Regulating the communication of genetic risk information: the Italian legal approach to questions of confidentiality and disclosure. Clin. Genet. 82, 205–209.

Beauchamp, T., Childress, J., 2005. Principles of Biomedical Ethics. Oxford University Press, Oxford.

Becker, F., Van El, C.G., Ibarreta, D., Zika, E., Hogarth, S., Borry, P., Cambon-Thomsen, A., Cassiman, J.J., Evers-Kiebooms, G., Hodgson, S., Janssens, A.C.J.W., Kaariainen, H., Krawczak, M., Kristoffersson, U., Lubinski, J., Patch, C., Penchaszadeh, V.B., Read, A., Rogowski, W., Sequeiros, J., Tranebjaerg, L., Van Langen, I.M., Wallace, H., Zimmern, R., Schmidtke, J., Cornel, M.C., 2011. Genetic testing and common disorders in a public health framework: how to assess relevance and possibilities. Eur. J. Hum. Genet. 19, S6–S44.

Beeri, M.S., Rapp, M., Silverman, J., Schmeidler, J., Grossman, H., Fallon, J., Purohit, D., Perl, D., Siddiqui, A., Lesser, G., 2006. Coronary artery disease is associated with Alzheimer disease neuropathology in Apoe4 carriers. Neurology 66, 1399–1404.

Bertot, J.C., Jaeger, P.T., Grimes, J.M., 2010. Using ICTs to create a culture of transparency: E-government and social media as openness and anti-corruption tools for societies. Gov. Inf. Q. 27, 264–271.

Bird, T.D., 2012. Early-Onset Familial Alzheimer Disease. https://www.ncbi.nlm.nih.gov/books/NBK1236/, last accessed 4 June 2018.

Bloch, M., Hayden, M., 1990. Opinion: predictive testing for Huntington disease in childhood: challenges and implications. Am. J. Hum. Genet. 46, 1.

Bonadona, V., Bonaïti, B., Olschwang, S., Grandjouan, S., Huiart, L., Longy, M., Guimbaud, R., Buecher, B., Bignon, Y.-J., Caron, O., 2011. Cancer risks associated with germline mutations in MLH1, MSH2, and MSH6 genes in lynch syndrome. JAMA 305, 2304–2310.

Borry, P., StultiËns, L., Nys, H., Cassiman, J.J., Dierickx, K., 2006. Presymptomatic and predictive genetic testing in minors: a systematic review of guidelines and position papers. Clin. Genet. 70, 374–381.

Brewerton, D.A., Hart, F.D., Nicholls, A., Caffrey, M., James, D.C.O., Sturrock, R.D., 1973. Ankylosing spondylitis AND HL-A 27. Lancet 301, 904–907.

Brouardel, P., 1887. Le secret médical. Baillière et fils, Paris.

Bunnik, E.M., Schermer, M.H.N., Jw Janssens, A.C., 2012. The role of disease characteristics in the ethical debate on personal genome testing. BMC Med. Genet. 5, 4.

Burgard, P., Ullrich, K., Ballhausen, D., Hennermann, J.B., Hollak, C.E., Langeveld, M., Karall, D., Konstantopoulou, V., Maier, E.M., Lang, F., 2017. Issues with European guidelines for phenyl-ketonuria. Lancet Diabetes Endocrinol. 5, 681–683.

Burke, W., Pinsky, L.E., Press, N.A., 2001. Categorizing genetic tests to identify their ethical, legal, and social implications. Am. J. Med. Genet. A 106, 233–240.

Buss, S., 2008. Personal autonomy. Stanf. Encycl. Philos. https://plato.stanford.edu/archives/fall2008/entries/personal-autonomy/.

Callier, S., Simpson, R., 2012. Genetic diseases and the duty to disclose. Virtual Mentor 14, 640.

Cameron, L.D., Marteau, T.M., Brown, P.M., Klein, W.M., Sherman, K.A., 2012. Communication strategies for enhancing understanding of the behavioral implications of genetic and biomarker tests for disease risk: the role of coherence. J. Behav. Med. 35, 286–298.

Campbell, H., Rudan, I., 2007. Study design in mapping complex disease traits. In: Wright, A., Hastie, N. (Eds.), Genes and Common Diseases: Genetics in Modern Medicine. Cambridge University Press, Cambridge, pp. 92–112.

Casas, J.P., Hingorani, A.D., Bautista, L.E., Sharma, P., 2004. Meta-analysis of genetic studies in ischemic stroke: thirty-two genes involving approximately 18 000 cases and 58 000 controls. Arch. Neurol. 61, 1652–1661.

Clarke, A., Flinter, F., 1996. The genetic testing of children: a clinical perspective. In: Marteau, T., Richards, M. (Eds.), The Troubled Helix: Social and Psychological Implications of the New Human Genetics. Cambridge University Press, Cambridge, England, pp. 164–176.

Collins, A.L., Lunt, P.W., Garrett, C., Dennis, N., 1993. Holoprosencephaly: a family showing dominant inheritance and variable expression. J. Med. Genet. 30, 36–40.

Coulehan, J.L., 1979. Multifactorial etiology of disease. JAMA 242, 416.

Curren, R.R., 2012. Philosophy of Education: An Anthology. Blackwell Publishing, Malden.

Davis, D.S., 1997. Genetic dilemmas and the child's right to an open future. Hastings Cent. Rep. 27, 7–15.

Elwyn, G., Gray, J., Clarke, A., 2000. Shared decision making and non-directiveness in genetic counseling. J. Med. Genet. 37, 135–138.

Emanuel, E.J., Emanuel, L.L., 1992. Four models of the physician-patient relationship. JAMA 267, 2221–2226.

European Society of Human Genetics, 2009. Genetic testing in asymptomatic minors: recommendations of the European Society of Human Genetics. Eur. J. Hum. Genet. 17, 720–721.

Evans, J.P., Burke, W., Khoury, M., 2010. The rules remain the same for genomic medicine: the case against [ldquo]reverse genetic exceptionalism[rdquo]. Genet. Med. 12, 342–343.

Eysenbach, G., 2008. Medicine 2.0: social networking, collaboration, participation, apomediation, and openness. J. Med. Internet Res. 10(3), e22.

Faden, R.R., Beauchamp, T.L., King, N.M., 1986. A History and Theory of Informed Consent. Oxford University Press, New York.

Fanos, J.H., 1997. Developmental tasks of childhood and adolescence: implications for genetic testing. Am. J. Med. Genet. A 71, 22–28.

Foster, M.W., Royal, C., Sharp, R.R., 2006. The routinisation of genomics and genetics: implications for ethical practices. J. Med. Ethics 32, 635–638.

Fregonese, L., Stolk, J., 2008. Hereditary alpha-1-antitrypsin deficiency and its clinical consequences. Orphanet J. Rare Dis. 3, 16.

Fulda, K.G., Lykens, K., 2006. Ethical issues in predictive genetic testing: a public health perspective. J. Med. Ethics 32, 143–147.

Furrow, B.R., 1997. Doctor's dirty little secrets: the dark side of medical privacy. Washburn Law J. 37, 283.

Gibbons, M., 2001. Psychosocial aspects of serious illness in childhood and adolescence. In: Armstrong-Dailey, A. (Ed.), Hospice Care for Children. vol. 2. Oxford University Press, Oxford, pp. 54–73.

Godard, B., Hurlimann, T., Letendre, M., Egalite, N., Brca, I., 2006. Guidelines for disclosing genetic information to family members: from development to use. Fam. Cancer 5, 103–116.

Goel, V., 2001. Appraising organised screening programmes for testing for genetic susceptibility to cancer. B. Med. J. 322, 1174.

Goldberg, S., 1998. American Medical Association: Genetic Testing: A Study of Consumer Attitudes. Survey Center, Chicago.

Goldstein, L., Brown, S., 1977. The low-density lipoprotein pathway and its relation to atherosclerosis. Annu. Rev. Biochem. 46, 897–930.

Green, M.J., Botkin, J.R., 2003. Genetic exceptionalism in medicine: clarifying the differences between genetic and nongenetic tests. Ann. Intern. Med. 138, 571–575.

Gross, S.J., Pletcher, B.A., Monaghan, K.G., For The Professional, P., Guidelines, C., 2008. Carrier screening in individuals of Ashkenazi Jewish descent. Genet. Med. 10, 54–56.

Gutheil, T.G., 2001. Moral justification for Tarasoff-type warnings and breach of confidentiality: a clinician's perspective. Behav. Sci. Law 19, 345–353.

Haga, S.B., Barry, W.T., Mills, R., Svetkey, L., Suchindran, S., Willard, H.F., Ginsburg, G.S., 2014. Impact of delivery models on understanding genomic risk for type 2 diabetes. Public Health Genomics 17, 95–104.

Herzlich, C., 1982. The evolution of relations between French physicians and the state from 1880 to 1980. Sociol. Health Illn. 4, 241–253.

Heshka, J.T., Palleschi, C., Howley, H., Wilson, B., Wells, P.S., 2008. A systematic review of perceived risks, psychological and behavioral impacts of genetic testing. Genet. Med. 10, 19–32.

Hicken, B., Tucker, D., 2002. Impact of genetic risk feedback: perceived risk and motivation for health protective behaviors. Psychol. Health Med. 7, 25–36.

Horgan, D., Jansen, M., Leyens, L., Lal, J.A., Sudbrak, R., Hackenitz, E., Busshof, U., Ballensiefen, W., Brand, A., 2014. An index of barriers for the implementation of personalized medicine and pharmacogenomics in Europe. Public Health Genomics 17, 287–298.

Hostiuc, S., 2014. Informed consent Consimtamantul informat Casa Cartii de Ştiinţă, Cluj-Napoca.

Hostiuc, S., Buda, O., 2015. Crystallization of the concept of the medical secret in 19th century France. Jahr-Eur. J. of Bioeth. 6, 329–339.

Hostiuc, S., Negoi, I., Buda, O., Hangan, T., 2014. Confidentiality in obstetrics in the XIXth century Romania. Gineco.eu, 20 (1), 20–23.

Howard, H.C., Borry, P., 2009. Personal genome testing: do you know what you are buying? Am. J. Bioeth. 9, 11–13.

Huibers, A.K., van 't Spijker, A., 1998. The autonomy paradox: predictive genetic testing and autonomy: three essential problems. Patient Educ. Couns. 35, 53–62.

Ito, H., Matsuo, K., Wakai, K., Saito, T., Kumimoto, H., Okuma, K., Tajima, K., Hamajima, N., 2006. An intervention study of smoking cessation with feedback on genetic cancer susceptibility in Japan. Prev. Med. 42, 102–108.

Janssens, A.C.J., Khoury, M.J., 2006. Predictive value of testing for multiple genetic variants in multifactorial diseases: implications for the discourse on ethical, legal and social issues. Ital. J. Publ. Health 4 (3), 35–41.

Kessler, S., 1987. Psychiatric implications of presymptomatic testing for Huntington's disease. Am. J. Orthopsychiatry 57, 212.

Khankari, N.K., Shu, X.-O., Wen, W., Kraft, P., Lindström, S., Peters, U., Schildkraut, J., Schumacher, F., Bofetta, P., Risch, A., Bickebӧller, H., Amos, C.I., Easton, D., Eeles, R.A.,

Gruber, S.B., Haiman, C.A., Hunter, D.J., Chanock, S.J., Pierce, B.L., Zheng, W., Colorectal Transdisciplinary Study (CORECT), Discovery, Biology, and Risk of Inherited Variants in Breast Cancer (DRIVE), Elucidating Loci Involved in Prostate Cancer Susceptibility (ELLIPSE), Transdisciplinary Research in Cancer of the Lung (TRICL), 2016. Association between adult height and risk of colorectal, lung, and prostate Cancer: results from meta-analyses of prospective studies and Mendelian randomization analyses. PLoS Med. 13, e1002118.

Khoury, M.J., McCabe, L.L., McCabe, E.R., 2003. Population screening in the age of genomic medicine. N. Engl. J. Med. 348, 50–58.

Kielstein, R., Hans-Martin, S., 2002. Nephrology ethics forum. Am. J. Kidney Dis. 39, 637–652.

Kirch, W. (Ed.), 2008. Multifactorial disease. In: Encyclopedia of Public Health. Springer Netherlands, Dordrecht.

Koocher, G.P., 1986. Psychosocial issues during the acute treatment of pediatric cancer. Cancer 58, 468–472.

Kottow, M.H., 1986. Medical confidentiality: an intransigent and absolute obligation. J. Med. Ethics 12, 117–122.

Larkin, G.L., Moskop, J., Sanders, A., Derse, A., 1994. The emergency physician and patient confidentiality: a review. Ann. Emerg. Med. 24, 1161–1167.

Lehmann, D.J., Cortina-Borja, M., Warden, D.R., Smith, A.D., Sleegers, K., Prince, J.A., Van Duijn, C.M., Kehoe, P.G., 2005. Large meta-analysis establishes the ACE insertion-deletion polymorphism as a marker of Alzheimer's disease. Am. J. Epidemiol. 162, 305–317.

LIAO, S.M., 2009. Is there a duty to share genetic information? J. Med. Ethics 35, 306–309.

Lobo, I., 2008. Same genetic mutation, different genetic disease phenotype. Nat. Educ. 1, 64.

Lodder, L., Frets, P.G., Trijsburg, R.W., Tibben, A., Meijers-Heijboer, E.J., Duivenvoorden, H.J., Wagner, A., Van Der Meer, C.A., Devilee, P., Cornelisse, C.J., 2001. Men at risk of being a mutation carrier for hereditary breast/ovarian cancer: an exploration of attitudes and psychological functioning during genetic testing. Eur. J. Human Genet. 9, 492.

Macdonald, C., Williams-Jones, B., 2002. Ethics and genetics: susceptibility testing in the workplace. J. Bus. Ethics 35, 235–241.

Malpas, P., 2008. Predictive genetic testing of children for adult-onset diseases and psychological harm. J. Med. Ethics 34, 275–278.

Marcy, T.W., Stefanek, M., Thompson, K.M., 2002. Genetic testing for lung Cancer risk: if physicians can do it, should they? J. Gen. Intern. Med. 17, 946–951.

Marteau, T.M., Lerman, C., 2001. Genetic risk and behavioural change. B. Med. J. 322, 1056.

Marteau, T., Senior, V., Humphries, S.E., Bobrow, M., Cranston, T., Crook, M.A., Day, L., Fernandez, M., Horne, R., Iversen, A., 2004. Psychological impact of genetic testing for familial hypercholesterolemia within a previously aware population: a randomized controlled trial. Am. J. Med. Genet. A 128, 285–293.

McBride, C.M., Koehly, L.M., Sanderson, S.C., Kaphingst, K.A., 2010. The behavioral response to personalized genetic information: will genetic risk profiles motivate individuals and families to choose more healthful behaviors? Annu. Rev. Public Health 31, 89–103.

Menasha, J.D., Schechter, C., Willner, J., 2000. Genetic testing: a physician's perspective. Mount Sinai J. Med. NY 67, 144–151.

Mihaescu, R., Van Hoek, M., Sijbrands, E.J., G Uitterlinden, A., Witteman, J.C., Hofman, A., Van Duijn, C.M., Janssens, A.C.J., 2009. Evaluation of risk prediction updates from commercial genome-wide scans. Genet. Med. 11, 588–594.

Mordini, E., 2004. Ethical considerations on pharmacogenomics. Pharmacol. Res. 49, 375–379.

Motulsky, A.G., Holtzman, N.A., Fullarton, J.E., Andrews, L.B., 1994. Assessing Genetic Risks: Implications for Health and Social Policy. National Academies Press, Washington, DC.

Mundluru, S.N., Therkelsen, K.E., Verscaj, C.P., Dasgupta, S., 2016. Conflicts between non-directive counseling and unbiased patient care: the influence of medical students' personal beliefs on proposed Huntington's disease genetic testing recommendations. Med. Sci. Educ. 26, 639–646.

Navigenics, 2009. Available at: http://www.navigenics.com/. Accessed 9 March 2009.

New Jersey. Superior Court, Appellate Division, 1996. Safer v. Estate of Pack. *A. 2d*. New Jersey. Superior Court, Appellate Division, Jersey City, New Jersey.

Nordestgaard, B.G., Chapman, M.J., Humphries, S.E., Ginsberg, H.N., Masana, L., Descamps, O.S., Wiklund, O., Hegele, R.A., Raal, F.J., Defesche, J.C., 2013. Familial hypercholesterolaemia is underdiagnosed and undertreated in the general population: guidance for clinicians to prevent coronary heart disease: consensus statement of the European atherosclerosis society. Eur. Heart J. 34, 3478–3490.

Pijl, M., Timmermans, D.R., Claassen, L., Janssens, A.C.J., Nijpels, G., Dekker, J.M., Marteau, T.M., Henneman, L., 2009. Impact of communicating familial risk of diabetes on illness perceptions and self-reported behavioral outcomes. Diabetes Care 32, 597–599.

Plunkett, K.S., Simpson, J.L., 2002. A general approach to genetic counseling. Obstet. Gynecol. Clin. North Am. 29, 265–276.

Prior, T.W., 2008. Carrier screening for spinal muscular atrophy. Genet. Med. 10, 840.

Raes, I., Ravelingien, A., Pennings, G., 2016. Donor conception disclosure: directive or non-directive counseling? J. Bioeth. Inq. 13, 369–379.

Richards, S., Aziz, N., Bale, S., Bick, D., Das, S., Gastier-Foster, J., Grody, W.W., Hegde, M., Lyon, E., Spector, E., Voelkerding, K., Rehm, H.L., 2015. Standards and guidelines for the interpretation of sequence variants: a joint consensus recommendation of the American College of Medical Genetics and Genomics and the Association for Molecular Pathology. Genet. Med. 17, 405–423.

Rosner, R., 2003. Principles and Practice of Forensic Psychiatry. CRC Press, London.

Ross, L.F., Saal, H.M., David, K.L., Anderson, R.R., Pediatrics, A.A.O., 2013. Technical report: ethical and policy issues in genetic testing and screening of children. Genet. Med. 15, 234–245.

Sanderson, S.C., Humphries, S.E., Hubbart, C., Hughes, E., Jarvis, M.J., Wardle, J., 2008. Psychological and behavioural impact of genetic testing smokers for lung cancer risk: a phase II exploratory trial. J. Health Psychol. 13, 481–494.

Sathasivam, K., Neueder, A., Gipson, T.A., Landles, C., Benjamin, A.C., Bondulich, M.K., Smith, D.L., Faull, R.L., Roos, R.A., Howland, D., 2013. Aberrant splicing of HTT generates the pathogenic exon 1 protein in Huntington disease. Proc. Natl. Acad. Sci. 110, 2366–2370.

Shawky, R.M., 2014. Reduced penetrance in human inherited disease. Egypt. J. Med. Hum. Genet. 15, 103–111.

Siegler, M., 1982. Confidentiality in medicine—a decrepit concept. N. Engl. J. Med. 307, 1518–1521.

Sladek, R., Rocheleau, G., Rung, J., Christian, D., Shen, L., Serre, D., Boutin, P., Vincent, D., Belisle, A., Hadjadj, S., 2007. A genome-wide association study identifies novel risk loci for type 2 diabetes. Nature 445, 881.

Stacey, S.N., Manolescu, A., Sulem, P., Rafnar, T., Gudmundsson, J., Gudjonsson, S.A., Masson, G., Jakobsdottir, M., Thorlacius, S., Helgason, A., 2007. Common variants on chromosomes 2q35 and 16q12 confer susceptibility to estrogen receptor-positive breast cancer. Nat. Genet. 39, 865.

Surbone, A., 2001. Ethical implications of genetic testing for breast cancer susceptibility. Crit. Rev. Oncol. Hematol. 40, 149–157.

Suter, S.M., 1993. Whose genes are these anyway?: familial conflicts over access to genetic information. Mich. Law Rev. 91, 1854–1908.

Teixeira, E., Borlido-Santos, J., Brissot, P., Butzeck, B., Courtois, F., Evans, R.W., Fernau, J., Nunes, J.A., Mullett, M., Paneque, M., 2014. The importance of the general practitioner as an information source for patients with hereditary haemochromatosis. Patient Educ. Couns. 96, 86–92.

Tibben, A., Frets, P.G., Van De Kamp, J.J., Niermeijer, M.F., Vegter-Van Der Vlis, M., Roos, R.A., Van Ommen, G.J.B., Duivenvoorden, H.J., Verhage, F., 1993. Presymptomatic DNA-testing for Huntington disease: pretest attitudes and expectations of applicants and their partners in the Dutch program. Am. J. Med. Genet. A 48, 10–16.

Turnpenny, P.D., Ellard, S., 2016. Emery's Elements of Medical Genetics E-Book. Elsevier Health Sciences, Philadelphia.

Watson, M.S., Cutting, G.R., Desnick, R.J., Driscoll, D.A., Klinger, K., Mennuti, M., Palomaki, G.E., Popovich, B.W., Pratt, V.M., Rohlfs, E.M., 2004. Cystic fibrosis population carrier screening: 2004 revision of American College of Medical Genetics mutation panel. Genet. Med. 6, 387.

Wertz, D.C., Fanos, J.H., Reilly, P.R., 1994. Genetic testing for children and adolescents: who decides? JAMA 272, 875–881.

Wicks, P., Vaughan, T., Heywood, J., 2014. Subjects no more: what happens when trial participants realize they hold the power? BMJ 348, g368.

Wiegman, A., Gidding, S.S., Watts, G.F., Chapman, M.J., Ginsberg, H.N., Cuchel, M., Ose, L., Averna, M., Boileau, C., Borén, J., Bruckert, E., Catapano, A.L., Defesche, J.C., Descamps, O.S., Hegele, R.A., Hovingh, G.K., Humphries, S.E., Kovanen, P.T., Kuivenhoven, J.A., Masana, L., Nordestgaard, B.G., Pajukanta, P., Parhofer, K.G., Raal, F.J., Ray, K.K., Santos, R.D., Stalenhoef, A.F.H., Steinhagen-Thiessen, E., Stroes, E.S., Taskinen, M.-R., Tybjærg-Hansen, A., Wiklund, O., Averna, M., Boileau, C., Borén, J., Bruckert, E., Catapano, A.L., Chapman, M.J., Cuchel, M., Defesche, J.C., Descamps, O.S., Gidding, S.S., Ginsberg, H.N., Hegele, R.A., Hovingh, G.K., Humphries, S.E., Kovanen, P.T., Kuivenhoven, J.A., Masana, L., Nordestgaard, B.G., Ose, L., Pajukanta, P., Parhofer, K.G., Raal, F.J., Ray, K.K., Santos, R.D., Stalenhoef, A.F.H., Steinhagen-Thiessen, E., Stroes, E.S., Taskinen, M.-R., Tybjærg-Hansen, A., Watts, G.F., Wiegman, A., Wiklund, O., Gidding, S.S., Watts, G.F., Chapman, M.J., Ginsberg, H.N., Cuchel, M., Ose, L., Chapman, M.J., Ginsberg, H.N., 2015. Familial hypercholesterolaemia in children and adolescents: gaining decades of life by optimizing detection and treatment. Eur. Heart J. 36, 2425–2437.

Wiggins, S., Whyte, P., Huggins, M., Adam, S., Theilmann, J., Bloch, M., Sheps, S.B., Schechter, M.T., Hayden, M.R., Canadian Collaborative Study of Predictive Testing, 1992. The psychological consequences of predictive testing for Huntingtons disease. N. Engl. J. Med. 327, 1401–1405.

Wilkes, M.S., Day, F.C., Fancher, T.L., McDermott, H., Lehman, E., Bell, R.A., Green, M.J., 2017. Increasing confidence and changing behaviors in primary care providers engaged in genetic counseling. BMC Med. Educ. 17, 163.

Wilson, J.M.G., Jungner, G., 1968. Principles and Practice of Screening for Disease. World Health Organization, Geneva.

World Medical Association, 1948. Declaration of Geneva. World Medical Association, Geneva, p. 1.

World Medical Association, 1992. Declaration on the Human Genome Project. World Medical Association, Geneva.

Zack, P., Devile, C., Clark, C., Surtees, R., 2006. Understanding the information needs of general practitioners managing a rare genetic disorder (osteogenesis imperfecta). Public Health Genomics 9, 260–267.

FURTHER READING

Hostiuc, S., 2015. Forensic autopsy. The particularities of consent for research. Online J. Health Ethics 11, 8.

Human Genetic Society Australasia, 2014. Guideline. Pre-symptomatic and Predictive Testing for Genetic Disorders. Human Genetic Society Australasia, Sydney, NSW.

Chapter 12

Whole-Genome Sequencing as a Method of Prenatal Genetic Diagnosis ☆

Fermín J. González-Melado
High Centre for Theological Studies, Badajoz, Spain

1. INTRODUCTION

In January 2017, Illumina, the largest maker of DNA sequencers, launched a new DNA sequencer with an architecture that can push the cost of decoding a human genome from $1,000 to $100 (Herper, 2017). At the same time, this enterprise announced that, by the end of the year, these new machines would run DNA six times faster than actual sequencers. In this context, whole-genome sequencing (WGS) was set to become the preferred method for prenatal diagnosis. The ethical issues raised by the use of this technology have received significant attention (Donley et al., 2012; Deans et al., 2015; Knoppers et al., 2015; Horn and Parker, 2017) but it is necessary to continue with an ethical reflection about the implementation of WGS on prenatal genetic diagnosis.

But what really is WGS? How reliable is it? What are the ethical implications of WGS? In this chapter, we will attempt to answer these questions. First, we will briefly discuss the latest research on prenatal genetic diagnosis, particularly in WGS, as well as the future applications of WGS. Second, we will explore the ethical implications of adopting WGS in the prenatal diagnosis and introduce pre-WGS test and post-WGS diagnostic genetic counseling as a necessary instrument to use WGS as the future preferred prenatal diagnosis test.

☆ "Over the course of the next few decades, the availability of cheap, efficient DNA sequencing technology will lead to a medical landscape in which each baby's genome is sequenced, and that information is used to shape a lifetime of personalized strategies for disease prevention, detection and treatment." (Collins, 2014)

Clinical Ethics at the Crossroads of Genetic and Reproductive Technologies.
https://doi.org/10.1016/B978-0-12-813764-2.00012-X

2. FROM THE STANDARD OF PRENATAL DIAGNOSIS TO WHOLE-GENOME SEQUENCING

Since the 1960s, identifying women at increased risk for pregnancies with Down syndrome has been the focus of prenatal screening programs. The current standard for noninvasive prenatal diagnosis combines maternal age, levels of specific markers in maternal serum, and ultrasound findings in the first or second trimester to generate a risk estimate for Down syndrome and, secondary, for trisomy 18 (Edwars syndrome) and trisomy 13 (Patau syndrome). Women at increased risk are offered a diagnostic amniocentesis or chorionic villus sampling. From 1970s to 1980s, the standard test on cultured cells from prenatally obtained amniotic-fluid or chorionic villus sampling has been a karyotype that can detect chromosomal aneuploidies and structural abnormalities larger than 5–10 megabases in size. Since the 1980s, the karyotype has been supplemented by fluorescence in situ hybridization (FISH) to rapidly test a few common aneuploidies if an expedited diagnosis is desired. FISH with locus-specific probes was also the method of choice to test for smaller structural chromosomal abnormalities but requires knowledge about which locus might be of interest, and only a few loci can be investigated in a single assay.

In the early 21st century a different genomic approach was introduced, namely comparative genomic hybridization (CGH), also called chromosomal microarrays analysis. In CGH, a fluorescently labeled DNA is hybridized to a slide that carries thousands of probes spread across the genome. CGH has a higher resolution than karyotyping, spanning from entire chromosomes to deletions and duplications of just several kilobases or even single exons. CGH is especially useful for the search of chromosomal abnormalities such as deletions, duplications, and translocations. It also does not require a cells culture thus results can be available faster. CGH is now the first-tier genetic diagnostic test for children and adults with multiple congenital anomalies, genetic syndromes, and intellectual and developmental disabilities, where its diagnostic yield is 15%–20% (Miller et al., 2010; Battaglia et al., 2013). The use of CGH is widespread, and it was demonstrated that it detects a clinically significant or potentially clinically significant copy number variations (CNVs) in pregnancies with a normal karyotype and no observable abnormalities. CGH also identifies CNVs that predispose to later-onset disorders in about 1% of cases. The American College of Obstetrics and Gynecology recommends that CGH be offered as the first-line test when fetal abnormalities are present and for stillbirth samples (The American College of Obstetrics and Gynecology, 2013). Although the higher detection rate of CGH tests is an important advance in prenatal genetic diagnosis, it still means that in the majority of cases a distinct genetic etiology for birth defects seen on prenatal ultrasound examination cannot be found (Hillman et al., 2015) (Table 1).

In recent decades, there has been intense interest in the development of risk-free-noninvasive alternatives to invasive methods of prenatal genetic diagnosis.

TABLE 1 Advances of Genomice Testing in Prenatal Medicine (Part 1) (Peters et al., 2015)

Technique	Description	Diseases Detected	Limitations
The existing standard of noninvasive prenatal diagnosis	Serum-based first and second-trimester screens studying differences on concentrations of specific protein markets associated with fetal malformations. Ultrasonography +age of mother	As such, these methods do not *directly* diagnose disease. For direct prenatal diagnosis, invasive tests are performed to collect fetal tissue that holds the cytogenetic and molecular information for prenatal genetic diagnosis	These methods do not provide desirable levels of accuracy. Their high false positive rate means that 5% of expectant mothers will unnecessarily undergo either amniocentesis or chorionic villus samples
1980s G-banding chromosome analysis	Traditional karyotyping using amniocentesis and chorionic villus sampling after cellular proliferation to obtain metaphase spreads	Aneuploidies, other numerical chromosome abnormalities, large genomic alterations such as balanced and unbalanced chromosomal rearrangements of at least 10–20 Mb in size and mosaicism	The need for cellular proliferation and the relative low-resolutions of G-banding
1990s Fluorescence in situ hybridization (FISH) analysis	Clinically significant chromosomal aberrations can be detected in metaphase or nondividing interphase cells. The use of specific DNA probes allows a rapid detection of a whole chromosome aneuploidy, large and submicroscopic	Most laboratories use FISH probes to identify genomic alterations in specific, targeted chromosomal regions, such as FISH panels for detection of trisomy 13, 18, 21, and monosomy X or to test for deletions in the DiGeorge critical region on chromosome 22q11.2	FISH-based tests do not provide genome-wide analyses, but are limited to the targeted genomic regions of interest

Continued

TABLE 1 Advances of Genomice Testing in Prenatal Medicine (Part 1)—cont'd

Technique	Description	Diseases Detected	Limitations
	rearrangements, including microdeletions and duplications within known disease-associated regions of the genome		
2000s Array comparative genomic hybridization (aCGH) and single-nucleotide polymorphism (SNP) microarrays	Microarray technology involves assessment of patient's DNA by thousands of DNA probes that have been previously selected from the human genome to generate a high-resolution karyotype	Genomic disorders caused by chromosome rearrangements that result in a gain or loss of dosage-sensitive gene, recurrent or not recurrent aberrations. Microdeletions and microduplications can also be detected. Whole-genome aCGH probes detect DNA copy number changes; whereas SNP microarrays detect single-nucleotide polymorphisms	The continued discovery of novel causal alleles and genes, as well as variable penetrance and expressivity of known mutations limits the clinical validity of this approach

Initially, these efforts were focused on the isolation of nucleate fetal cells from maternal blood (Boyer et al., 1976; Iverson et al., 1981). However, fetal cell isolation did not translate well into clinical practice due to the lack of robust methods for the recovery of these cells (Bianchi et al., 2002). In recent years, a number of reports with novel cell isolation approaches have been published (Huang et al., 2008; Zimmermann et al., 2013; Mouawia et al., 2012) but, at this moment, the current noninvasive prenatal genetic testing (NIPGT) for the detection of fetal chromosome anomalies relies on circulating cell-free fetal nucleic acids (cffDNA).

In 1997, Doctors Y.M. Lo and N. Corbetta showed the existence of free cffDNA in the maternal blood stream. Their study was based on the detection of Y chromosome sequences in the plasma and serum of pregnant women with a male fetus (Lo et al., 1997). This new discovery represented a breakthrough in the field of NIPGT (Lo et al., 1998). Free fetal DNA coexists with maternal DNA in the mother's blood. The presence of fetal DNA in maternal plasma becomes more noticeable as the pregnancy progresses, representing approximately 3% of the total DNA present in maternal plasma during the early stages of pregnancy and 6% at term (Lo et al., 1998). The coexistence in maternal blood of fetal and maternal DNA, and the fact that most aneuploidies have a maternal origin, has made it difficult to distinguish these two DNA populations. Fan et al. first demonstrated the proof of concept for the use of shotgun next-generation sequencing (NGS) of maternal plasma DNA for noninvasive aneuploidy detection in 2008 (Fan et al., 2008). This was followed rapidly by reports from Chiu et al. (2008) and Chu et al. (2009). These methods utilized NGS of plasma cell-free DNA. These initial proofs of concept were followed by a large-scale demonstration that the genome plasma sequencing (maternal genome + fetal genome) approach can be used routinely for the noninvasive detection of aneuploidies involving chromosomes 21, 18, and 13 and other disorders involving sex chromosome copy number abnormalities (Chiu et al., 2011; Palomaki et al., 2011, 2012; Bianchi et al., 2012; Sachs et al., 2015; Song et al., 2013; Mazloom et al., 2013; Stumm et al., 2014; Porreco et al., 2014; Bianchi et al., 2014). cffDNA-based tests have a detection rate and a false positive rate of 99.4% and 0.16%, respectively, for Down syndrome, 96.6% and 0.05% for trisomy 18, 86.4% and 0.09% for trisomy 13, and 89.05% and 0.20% for monosomy X (Benn et al., 2015). The development of NGS technologies (ACMG Board of Directors, 2012)[1] has revolutionized Mendelian disease gene identification and genetic diagnosis in pediatric and adult medicine because it has the ability to interrogate multiple genes at once, identifying deleterious variants that can be correlated with the clinical presentation to make a molecular diagnosis (Berg et al., 2011; Gilissen et al., 2011; Gonzaga-Jauregui et al., 2012).

None of these techniques can detect other types of mutations, such as point mutations and small insertion-deletion mutations that cause the now more than 4600 known single-gen disorders and others yet to be characterized (Xue et al., 2014). To find these kinds of mutations, targeted analysis of candidate genes was needed, requiring prior knowledge of clinical phenotypes caused by mutations in specific genes. This is a real challenge because, on the one hand, some phenotypes or genetic disorders can be caused by mutations in different genes,

---

1. This term encompasses a variety of technologies that permit rapid sequencing of large numbers of segments of DNA, up to and including entire genomes. Massively parallel sequencing, therefore, is not a test in itself or a specific sequencing technology. The term emphasizes a distinction from initial approaches that involved sequencing of one DNA strand at a time.

while the genetic causes of other phenotypes are not yet diagnosed. On the other hand, in prenatal diagnosis certain phenotypic features, such as intellectual disability or minor birth defects and dysmorphic features, cannot be ascertained in the fetus before birth because of the limitations of prenatal imaging and the developmental stage at which they become recognizable.

Different technologies, one based on massively multiplexed PCR and another based on selection and sequencing of specific tags from chromosome of interest, have also been developed and have similar performance (Zimmermann et al., 2012; Ashoor et al., 2012, 2013; Nicolaides et al., 2012, 2013; Pergament et al., 2014; Verweij et al., 2013). In 2010, the possibility of using NGS technologies toward whole-genome fetal recovery was described for the first time (Lo et al., 2010). More recent studies have supported the initial finding, providing new perspectives in the field of NIPT using WGS of fetal genome recovery from maternal blood (Fan et al., 2012; Kitzman et al., 2012). Chen et al. presented a novel approach that has some advantages toward noninvasive whole-genome fetal recovery that could be used for noninvasive prenatal WGS (Table 2).

TABLE 2 The Advances of Genomic Testing in Prenatal Medicine (Part 2) (Peters et al., 2015)

Technique	Description	Diseases Detected	Limitations
Noninvasive prenatal genetic testing from nucleated fetal cells from maternal blood	Isolation of nucleated fetal cells from maternal blood for karyotyping or whole-genome amplification and copy number analysis	Same diseases that can be detected by karyotyping or genome amplification and copy number analysis	Not translate well into clinical practice due to a lack of robust methods for fetal cell isolation
Noninvasive prenatal genetic testing using circulating cell-free fetal nucleic acids	Fetal DNA is detectable by PCR in maternal plasma and serum	Prediction of Rhesus D blood group status, diagnosis of paternal inherited thalassemia and achondroplasia	The difficulty to distinguish between maternal inherited fetal alleles and their endogenous maternal counterparts
Whole-genome maternal plasma DNA analysis via shotgun next-generation sequencing	DNA fragments libraries are generated from maternal plasma and then sequenced	Can be used routinely for the noninvasive detection of aneuploidies involving	Because circulating cffDNA derives from the trophoblast, confined placenta

TABLE 2 The Advances of Genomic Testing in Prenatal Medicine
(Part 2)—cont'd

Technique	Description	Diseases Detected	Limitations
	randomly to generate very large numbers of sequence tags that are then aligned computationally to the human genome. Perfectly aligned matching tags are then quantified in a chromosome-specific or region-specific	chromosomes 21, 18, and 13 and other disorders involving sex chromosome copy number anomalies	mosaicism for a tested chromosomal abnormality, present in about 1%, may result in a positive cffDNA test but the fetus is unaffected. cffDNA screening for microdeletions has not yet been clinically validated
Sequencing targeted chromosomes	Target chromosomes are sequenced to determine copy number in a fashion that is analogous to whole-genome approach with reduced content	Detection of aneuploidy or copy numbers anomaly in specific chromosome	You can only study the targeted chromosome
Highly multiple PCR approach with >11.000 single-nucleotide polymorphisms (SNP)	Sequence data is analyzed in a sophisticated manner such that it hypothesized that the fetus is monosomic, disomic, or trisomic	Able to identify the presence of fetal aneuploidy, triploidy, or uniparental disomy. Detects also nonpaternity or consanguinity	The limited scope of variant detection confines the analysis to preselected points in the genome. Further, most SNP-based diagnostic are probabilistic, nondeterministic, with variable degrees of clinical validity
Whole-exome sequencing (WES)	WES consists in sequencing complete coding regions from maternal plasma	Used in clinical diagnosis of individuals with undiagnosed diseases,	The use of WES in prenatal practice is nascent and WGS is more powerful than

Continued

TABLE 2 The Advances of Genomic Testing in Prenatal Medicine (Part 2)—cont'd

Technique	Description	Diseases Detected	Limitations
	genome or from fetal genome in fetal cells from amniocentesis or CVS	children with multiple congenital anomalies, intellectual disabilities, metabolic disorders, and mitochondrial diseases	WES for detecting exome variants
Targeted exome capture (TEC)	Sequencing specific targeted exome capture from maternal plasma or from whole fetal genome present in fetal cells from amniocentesis or CVS	Mitochondrial disorders and newborn screening for inborn errors of metabolism	The use of TEC in prenatal practice is nascent
Whole-genome sequencing (WGS)	WGS consist in sequencing whole-genome maternal plasma and recovery of the whole fetal genome, or directly sequencing the whole fetal genome present in fetal cells from amniocentesis or CVS	Diagnosis of severe conditions, variants of unknown significance, nonmedical genetic markers, carrier status, susceptibility genes, and genes expressing conditions with late onset	It has not entered in current clinical practice. Today, the function of more than 90% of annotated genes in the human genome is unknown, as is the function of the 98% present of the noncoding regions

Most recently, several clinical laboratories are now offering diagnostic whole-exome sequencing (WES) to search for mutations in the coding sequence of the 20,000 human genes (Yang et al., 2013, 2014; Eng et al., 2014). For WES, the majority of coding exons, which represent only 2% of the genome but contain 85% of disease-causing mutations, are sequenced. When fetal congenital abnormalities are identified on prenatal ultrasound, karyotype and CGH reveal a positive diagnosis in up to 20%–30% of cases (Van Den Veyver and Eng, 2015), depending on the type of structural defect. For the reminder, single-gene tests or gene panel may be useful, but very recent data suggest that diagnostic WES can provide answers in a substantial proportion of the remaining cases (Talkowski et al., 2012; Filges et al., 2014). The cost of generating genome information has shown a rapid decline. The high-throughout genomic technologies make it possible to sequence the whole genome of a person at a price that is affordable for some health-care systems. It will not be long before WGS information will be available for routine medical care on a widespread scale. This will further enable the practice of personalized medicine and prenatal care, which has the potential to reduce costs and improve the quality of care (Abrahams and Silver, 2009).

3. WHOLE-GENOME SEQUENCING AS A PRENATAL DIAGNOSTIC TOOL

3.1 What Is Whole-Genome Sequencing?

The two basic strategies for NGS of genomic DNA (Fig. 1) are a whole-genome approach and a more targeted approach (Bamshad et al., 2011; Biesecker and Green, 2014). In the first one (WGS), genomic DNA is fractionated into random small fragments. A sequencing library of all the fragments is constructed by adding linker sequences and is used as the template for sequencing by synthesis with fluorescent light-emitting nucleotides. The sequenced material originates from coding and noncoding regions of genes, from intergenic sequences and includes mitochondrial DNA. Each region is sequenced multiple times and the sequencing depth is defined as the number of copies of each region represented in the pool of fragments (Korf and Rehm, 2013). The Laboratory Quality Assurance Committee of the American College of Medical Genetics and Genomics (ACMG) has published standards for minimal coverage when NGS is used for diagnostic purposes. A minimum of 30-fold coverage is considered adequate for diagnosis WGS, and a minimum of 10-to 20-fold coverage of all bases is needed to make accurate diagnostic calls with targeted panels. As technology is improving, deeper sequencing at reasonable cost is becoming available in a more routine manner (Levy and Myers, 2016). WGS is much more expensive than WES because of the higher relative cost of sequencing technology, and because of its need for a more comprehensive infrastructure to store, manage, and analyze data. However, the sequencing technology for WGS has been improving at a very fast rate while reducing its costs. As the cost differential between WGS and WES diminishes, WGS will become the preferred clinical practice because it yields approximately 100 times more data than WES at the same coverage. This is a welcome development since WGS extends the variation research space to the whole genome and provides more uniform and better coverage depth and genotype quality (Belkadi et al., 2015).

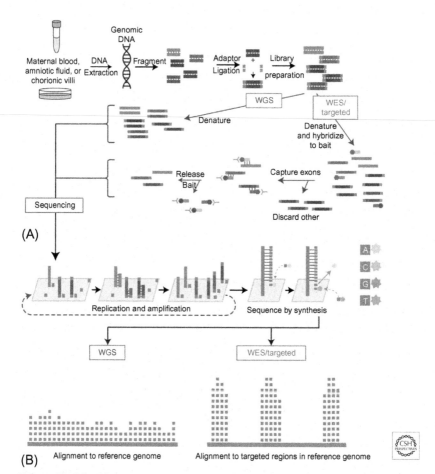

FIG. 1   Workflow for next-generation sequencing (NGS) (Van Den Veyver and Eng, 2015). **(A) Library preparation:** fetal genomic DNA is prepared from a prenatal sample (cffDNA from maternal blood, fetal cells from maternal blood, amniotic fluid, or chorionic villi) and fragmented. This is followed by adapter ligation and preparation of the sequencing library. When WGS is performed, the library is directly denatured for use as sequencing templates (*lines and arrows*). For targeted approaches such as WES (*lines and arrows*), an additional step to capture and enrich for fragments of interest (such as coding exons) for sequencing by hybridizing them to a library of baits with known sequence, followed by purification and bait release is required. **(B) Sequencing procedure:** fragments are immobilized by hybridization to linkers (here, represented on a solid surface, but other methods exist), followed by multiple rounds of replication and clonal amplification. Differently fluorescently labeled nucleotides are then added to the single-stranded templates and emitting light is used to identify the added nucleotide that is complementary to the template. The sequenced fragments are then aligned to the reference genome sequence in multiple copies (sequencing reads). In WGS, the alignment covers all regions for which sequencing was successfully sequenced. In targeted sequencing, such as WES, only those sequences represented in the baits are covered, typically with more reads for each.

Fetal DNA can be extracted directly from blood maternal plasma, from fetal cells present in blood maternal plasma, from fetal cells obtained from amniotic fluid or from fetal cells obtained from chorionic villi sample. In the case of fetal cells, they have to be cultured previously to the extraction of fetal DNA. *(Modified and reprinted by permission from Cold Spring Harbor. Copyright Cold Spring Harbor.)*

3.2   Whole-Genome Sequencing Applications in Prenatal Diagnosis

A greater advantage of WGS is the greater potential to identify the genetic component of health problems, at a lower cost compared to the current techniques. The sheer mass of data generated can reveal disease-causing alleles that could not be detected otherwise. WGS has the potential to identify all forms of genetic variations (Pang et al., 2014).

A recent paper showed the advantages of WGS over CGH as the standard of care for the clinical investigation of the etiology of congenital malformations and neurodevelopmental disorders (Stavropoulos et al., 2016). WGS identified all rare clinically NCVs that were detected by CMA. Clinical implementation of WGS as a primary test will provide a higher diagnostic yield than conventional genetic testing, such us CMA, and potentially reduce the time required to reach a genetic diagnosis.

There are different types of novel information that WGS produces.[2]

3.2.1   Variants of Unknown Significance

WGS will uncover many variants of unknown significance (VUS). Although often found in regions associated with important health functions, VUS are variants in a genetic sequence whose association with a particular disease is, by definition, unknown. Because the health-related impact of VUS cannot be stated with any degree of certainty, these variants do not yet reveal any medically important information. This inherent uncertainty is the main difference between VUS and the diagnostic information revealed through current prenatal genetic test.

3.2.2   Nonmedical Indicators

Nonmedical genetic markers will also be revealed through prenatal WGS. In addition to basic characteristics like eye color, parents might also have access to information about genes that can help predict nonmedical conditions, such as athletic ability, loyalty, criminality, and intelligence (Ostrander et al., 2009; Garcia et al., 2010). This kind of information stands in stark contrast to the diagnostic health information about serious disorders currently given to parents who engage in prenatal genetic testing.

3.2.3   Carrier Status

Another category of information that can be generated from prenatal WGS is a child's carrier status for genetic conditions. If a child is a carrier for a genetic condition, then once she/he reaches reproductive age and decides to have

---

2. We follow the analysis made by Donley et al. (2012).

children, her/his offspring would generally be at risk of having that condition only if her/his partner is also a carrier for the same condition. Carrier status does not usually reveal any health information relevant to the child's immediate health. For this reason, testing for carrier status in children is discouraged by nearly every professional organization.

3.2.4 Late-Onset Conditions

Highly or moderately penetrant late-onset conditions can also be discovered by prenatal WGS. Although these conditions often cause severe medical diseases, they generally do not affect a person's health until later in life. For instance, Alzheimer's and Parkinson's disease, although there are many other late-onset conditions that occur later in life having less devastating symptoms. The main difference between this category of information and the information currently found in targeted, risk-based tests is that late-onset conditions will not affect the person during childhood. As many of these late-onset conditions have limited treatments options, the information often merely provides a window into the end of one's life. This makes the information very powerful, and the reaction to it emotionally intense. Many adults decide not to be informed about this kind of finding. Predictive testing for late-onset conditions was uniformly discouraged in the pediatric setting for these reasons, but this has began to change and some societies began to recommended in favor, based on the parents' right to be informed (Hercher et al., 2016).[3]

3.2.5 Susceptible Genes

Prenatal WGS will also reveal many disease susceptibility genes; these are genetic markers of low or variable penetrance that suggest a genetic predisposition or increased statistical likelihood for developing a disease. This is totally different from the diagnostic information that is currently offered, which allows parents to make decisions based on information about the definitive presence of disease. Many chronic conditions belong to this category, for instance, genes indicating susceptibility for diabetes, mental health disorders, hearth conditions, and some cancers have already been discovered. The presence of susceptibility genes, however, does not mean that developing the disease is inevitable,

---

3. In November 2014, The National Society of Generic Counselors released a position statement on prenatal testing for adult-onset conditions as follows: "The National Society of Genetic Counselors (NSGC) encourages deferring prenatal genetic testing for adult-onset conditions if pregnancy management will not be affected. Prospective parents have the right to make fully informed and autonomous decisions about reproductive options and pregnancy management. However, prenatal testing for adult-onset conditions denies the future child the opportunity to make this decision for him/ herself as an adult. NSGC strongly recommends that prospective parents meet with a certified genetic counselor or other healthcare specialist with genetic counseling expertise for decision-making about prenatal testing." In 2016, the same society changed its statement position establishing that "any conflict between the right of prospective parents to obtain information and the right of the future child should generally be resolved in favor of the parents."

or perhaps even likely. Instead, the information will only give parents complex probabilities that their child could develop a medical condition. This probabilistic information is further complicated by the fact that environmental and behavioral factors play a large role in disease development. The number of susceptibility genes is very large and diverse and it will continue to grow (Govaerts et al., 2017).

3.2.6 Incidental Findings

WGS may reveal mutations in genes that are relevant for human health and disease, but that are unrelated to the initial reason for WGS test in 1%–6.5% of the cases (Ding et al., 2014). Incidental findings occur in all health-care fields (Borgelt et al., 2013; Yeh et al., 2013), including standard genetic testing, but their clinical and ethical complexity and frequency are much higher with WGS. How to handle these findings is widely debated (Green et al., 2013; Mcguire et al., 2013; Ross et al., 2013; Wolf et al., 2013; Eng et al., 2014). In 2013, both the European Society of Human Genetics (ESHG) and the American College of Medical Genetics (ACMG) published recommendations on how incidental findings should be approached (Van El et al., 2013; Kalia et al., 2016). The ESHG recommended that bioinformatics pipelines exclude genetic variants in genes unrelated to primary indication, in order to minimize the discovery of incidental findings. They recommend that if an unsolicited genetic variant is detected despite the initial filtering of the data and is indicative of a serious health problem, either in the person tested or a close relative, health professionals should report such variants. In contrast, the ACMG took a more prescriptive approach (Townsend et al., 2013). The ACMG subsequently modified their recommendations to permit patients to opt out of receiving incidental findings from clinical WGS. Whereas issues surrounding the ethics of presymptomatic testing of children for adult-onset conditions extend prenatally, the recent ACMG guidelines on incidental findings specifically exclude prenatally obtained WGS results, leaving a gap in guidance on how these incidental findings should be handled if detected prenatally in fetal samples or in simultaneously sequenced parental genomes (Kalia et al., 2016).

3.3 Limitations of Whole-Genome Sequencing

Although laboratories have begun to add screening for other aneuploidies, such as microdeletions and duplications, there is significant concern as more rare conditions are included without adequate clinical validation, high cumulative false positive rates, resulting unnecessary diagnostic procedures, and high false negative rates resulting in missed genetic diagnoses. Awareness of these issues by providers and patients is incomplete and marketing of WGS is highly focused on avoidance of the risk of diagnostic procedures. Although proof-of-principle studies have shown that is technically feasible to noninvasively sequence the entire fetal genome, this is not currently achievable in a time- and cost-effective manner. Thus, until the noninvasive analysis of cffDNA improves to the point that it will have the same accuracy as that of karyotyping and CMA on amniotic fluid or CVS samples, genetic counseling should objectively present the limitations and benefits of all currently available approaches in the context of the individual woman's a priori risk, her desire for genetic knowledge about her pregnancy, and personalized risk-benefit considerations.

4.   ETHICAL PROBLEMS SURROUNDING WHOLE-GENOME SEQUENCING

The promise of personalized medicine on prenatal care will make parents want to use the WGS test as soon as possible. The factors that will drive the use of the WGS will be mainly economic benefits and the parents themselves will be pressured to use the WGS. However, the WGS may not be more useful, compared to other tests, since most of the information does not have robust scientific evidence about its meaning, due to the large number of VUS that will be detected. Some of the ethical problems around WGS are explained in the following sections.

4.1   Large Amounts of Information

The future expanding of WGS could generate large amounts of information regarding late onset diseases (e.g., Huntington's), predispositions to severe and common diseases (e.g., breast cancer and diabetes), minor abnormalities and even nonmedical information (e.g., paternity), and physical traits (e.g., eye color).

Most of this information that WGS generates would probably not be as helpful for parents choosing this method, compared to the information obtained through the current methods of prenatal genetic testing. Using this uncertain information produced by WGS could be problematic in different ways. The first one is an avalanche of genetic information that is difficult to interpret and explain, some of it with unknown significance, which could lead to confusion to the interested couples and health professionals. This anxiety and confusion could result in lowering the threshold of pregnancy termination (abortion), as well as conflicts between parents and providers over whether the information should be distributed and how it should be acted on.

The second one is that the broad knowledge about a child's genetic information could exacerbate an inappropriate belief in genetic determinism, which could affect how parents raise their children in potentially harmful way. The children future autonomy and the "right not to know" will be undermined without sufficient justification.

4.2   Anxiety and Confusion in Parents

The use of WGS will change our expectation in the use of genetic test. First of all, it will lower the threshold used by clinicians to recommend prenatal genetic testing. Currently, these recommendations are based on risk factors. Prenatal WGS will be more universally offered to all parents, independent of risk factors, because WGS has the potential to reveal relevant non-risk-related factors. Also, there is no risk of miscarriage when fetal genome recovered from maternal blood is used in WGS testing. Unless an extremely restrictive threshold is established, parents who undergo WGS will receive some medically relevant information about their future child, even though much of that information will reveal less serious, or less well-understood, genetic markers. This will lead the expectant parents to think about their child as a clean slate during pregnancy and the image of a normal, healthy baby will be dramatically affected by this technology. Substantial research is needed to understand the psychosocial impact of this information and how it will affect parent's perceptions before birth.

Without education and genetic counseling, parents could overestimate or over value the genetic information in some way that unfiltered information will run the risk of confusing parents and create an intense concern. The idea of a "perfect child" will evolve and the qualities and quantities of new information generated by WGS will augment the anxiety that parents feel about the child during pregnancy.

This anxiety will be unnecessary in many cases because in late-onset diseases, the information will only serve to let parents know that their child will suffer a particular condition during its life but that they cannot help them. Nonmedical markers, such as markers related with violence or aggressive behavior, might also run the risk of causing unnecessary anxiety and even changes in parenting style. This could change the understanding of expectant parents about what is a healthy and normal baby.

4.3   Parents' Reproductive Choices, Expectations, and Genetic Determinism

Another problem relates to parents' reproductive choices. The decision to terminate a pregnancy based on a positive result of a prenatal genetic testing is often taken by the parents, because of the devastating nature or their child disease. "If WGS helps parents make these difficult choices by giving them access to more information, then it is a useful new reproductive technology. On the other hand, to the extent that some decisions to terminate pregnancies based on this kind of information represent confusion or a lack of complete understanding, the dissemination of this information would be problematic" (Donley et al., 2012). Increased information might ultimately cause an increased number of terminations (Lewis et al., 2013; Menezes et al., 2013;

Verweij et al., 2013), sometimes for minor or unimportant reasons (Benn and Chapman, 2010; King, 2011; Yagel, 2013).

WGS could also have an impact on child rearing. For instance, in the case of nonmedical genetic markers, learning a child's predicted intelligence (IQ) before birth may affect familial expectations, which in turn could influence how the child is raised. If intellectual expectations are low, the parent might become more tolerant of poor academic outcomes. The resulting lack of the parents encouragement and support could cause the child to fall below what his or her performance might have been without knowledge of these genetic associations. Because children and parents today are largely unaware of a child's genetic aptitude for many concealed traits, some potential deficiencies may never recognizably manifest. Knowledge of this information could lead to an erroneous acceptance of genetic determinism. Because the information prenatal WGS generates might produce both harms and benefits in the context of child rearing, adequate justification is needed for providing parents with certain types of information.

4.4   Right of the Parents to Information vs Right of the Child to "Not to Know"

WGS information can affect the child's adult life. There is a right of the child (adult) not to know a certain type of its own genetic information. This establishes a tension between the right to information on the part of parents and the child's right "not to know." There are clear differences between the protocols referring to genetic tests in children, which recommend keeping certain types of information safe until the child is an adult, and prenatal diagnosis protocols where these issues appear in a more blurred manner. It is necessary to limit the paternal access to information in order to respect the possibility that children can decide, as adults, whether or not to be informed. There is a clear risk of loss of the child's autonomy as a result of this knowledge. The parents' desire to perform genetic tests for their children should be taken seriously and should be discouraged if it does not give an actual benefit to the fetus or prevent further medical complications (Deans et al., 2015). "Parents should not be precluded from requesting any category of information generated by prenatal WGS, but a choice to deviate from standard practices should be accompanied by careful counseling to mitigate the potential harms" (Donley et al., 2012).

Most guides or recommendations of the professional societies on prenatal genetic tests do not take into account the future autonomy of the child. This should not be surprising since these guidelines are intended for prenatal genetic tests to diagnose mostly early-onset diseases and whose treatment will begin in childhood. In order to adequately implement the WGS prenatal analysis, it will be necessary to adapt the guidelines and protocols of the different medical associations so that they may include protection for the future autonomy of the child.

5.   THE IMPORTANCE OF PRE-WGS TEST AND POST-WGS DIAGNOSTIC COUNSELING

5.1   Counseling in WGS

If WGS does become commonplace, providing some genetic counseling to expectant parents will be crucial. Understanding complex and sometimes ambiguous genetic information is no easy task. Many people do not even know the difference between "risk" and "diagnosis." Sometimes, the quantity of information that parents receive is so large that they cannot process it all, and thus they look for health-care providers to guide them through understanding the information. Parents' autonomy depends on the development of a truly informed consent. WGS could easily compromise such consent. It is practically impossible to provide detailed information on all possible results and genetic conditions that a diagnostic WGS test can yield, and patients are only be informed about the broad categories of results that can be obtained from diagnostic WGS. For all these reasons, the role of physicians together with a well implemented pre-WGS test and post-WGS diagnostic counseling will have a crucial importance.

5.2   Pre-WGS Test Counseling

Traditionally, pretest counseling focuses on two elements: the explanation of invasive prenatal diagnosis procedures, and the achievement of informed consent for execution of the diagnostic procedure. If the risk of fetal loss during the diagnostic test is eliminated, it would seem that pretest counseling could be limited simply to obtaining informed consent for completion of prenatal WGS. However, the choice whether or not to conduct the WGS test itself should be made by the parents only after receiving relevant information about the implications of the test. It is important to highlight that the physician or genetic counselor must not impose a choice on the parents, but only help them to make a decision, avoiding any form of pressure that might determine their choice. For this, it must be clearly communicated in the course of the counseling session what difficulties the parents will encounter, the methodology of the test used and the degree of certainty of the results (Table 3).

At an ethical level, the duties of professional deontology require:

(1)   informed consent prior to WGS, which necessitates, as we have indicated, a complete and rigorous disclosure of information; and
(2)   strict quality control at the level of the techniques used by the laboratories involved in WGS. There is currently a wide range of reliability in the tests depending on the use of one technique or another as well as the use of one or multiple genetic markers.

TABLE 3 Pre-WSG Test Counseling Points (Sachs et al., 2015)

Points	Description
Testing is optional	Inform patients that all screening and prenatal diagnostic testing is optional. Women should be given an opportunity to consider the potential implications of WGS test results
Define screening	Clarify that WGS is a genetic test. Testing is performed using fetal DNA obtained from maternal blood sample, or amniocentesis or chorionic villi sample
Review clinical features and variability of conditions	Describe the types of clinical conditions and genetic diseases that can be detected using a sensitive and neutral language. If further information about these conditions is requested prior to testing, a referral to a qualified provider for more detailed discussion is optimal
Describe technology	The specific testing methods will vary by laboratory
Reporting format	Women should be told when and how their results will be provided
Sensitivity	Patients should be informed about the sensitivity of WGS. Specific detection rates will depend on the laboratory or sequenced used
False positive rate and confirmation of abnormal results	Clarify that WGS using fetal DNA from maternal blood sample could have a false positive result. There are several biological and nonbiological explanations for false positive WGS results. Therefore, a WGS using fetal DNA from cultured chorionic villi cells or amniocytes, it will be necessary for confirmation of an abnormal WGS result
Variants of unknown significance (VUS)	VUS are variants in a genetic sequence whose association with disease is, by definition, unknown. Because the health-related impact of VUS cannot be stated with any degree of certain, the variants do not yet reveal any medically important information
Limitation	Until WGS improve to the point that it will have the same accuracy as that of karyotyping and CMA on amniotic fluid or CVS samples, genetic counseling should objectively present the limitations and benefits of all currently available approaches

TABLE 3 Pre-WSG Test Counseling Points—cont'd

Points	Description
Incidental findings	Explain to patients that WGS may raise suspicion for maternal condition, paternity of the child, or other incidental finding during WGS
Timing	WGS can be performed at any time during pregnancy after 10 weeks of gestation. Discuss laboratory specific turnaround time for results in the context of the patient's gestational age and the local legal limit for abortion. Include in this discussion the turnaround time for amniocentesis results in the even that confirmatory diagnostic testing is performed

5.3 Post-WGS Diagnostic Counseling

Some studies show that information currently given to mothers upon the detection of trisomy 21, particularly during the first trimester of pregnancy, does not offer the pregnant woman a good understanding of the current or future health prospects of her child (Favre et al., 2007). Only with such information, however, can the parents exercise real autonomy and make reproductive decisions that respect ethical principles. At the same time, as the techniques of genetic testing advance, there is a need for skilled and qualified counselors to transmit the resulting genetic information to parents (Grody et al., 2013).

In most cases, when the WGS test returns negative results for the detection of known genetic diseases, it brings a sense of psychological relief for the future mother. However, when the test returns positive, meaning the fetus has a genetic problem, the health professionals are in a unique position to assist the woman. In such situations, the counselor cannot allow a woman to become dependent on his advice, but must present pertinent information so that a responsible decision can be made. The counselor must sincerely endeavor to inform the woman, without exaggerating or diminishing the severity of the disease's anticipated impact on the life of the family. Counselors must give the woman the time necessary for a calm reflection, providing answers for her thought-out questions, and helping her toward alleviating her feelings of anxiety. There are signs that, by improving the scientific and human quality of information during the decision-making process about a genetic disease, there is a statistically significant trend in recent years, for the woman to elect continuation of pregnancy and to reject abortion (Christian et al., 2000). E. Parens and A. Asch have pointed out this issue by stating, "Ignorance is one of the primary sources of the discrimination suffered by people with disabilities" (Parens and Asch, 2000).

5.4   Ethical Reflection on WGS Test in Prenatal Diagnosis

The efforts of science and technology to find genetic markers of diseases and to develop techniques that can make a method of prenatal diagnosis more sensitive and specific are, in themselves, good. As such, the technique of WGS may be evaluated positively. We believe that it is moral to do research in order to find genetic markers that would establish a correct and efficient method for the prenatal diagnosis of fetuses carrying a particular genetic disorder. We believe it is moral to make WGS test available to all pregnant women who are in a high-risk situation, defined on the basis of medical and ethical criteria, without compromising the integrity of the fetus, because it will allow the treatment of those disorders and, therefore, the pursuit of the good of the child. Still, *the positive ethical assessment of the clinical application* of this new diagnostic technique, satisfying all the requirements for the successful execution of the method, *depends on the conditions that precede and accompany it.*

The first condition is that *the purpose* for which the method is applied *has to be good.* Unfortunately, in much of the research there is evidence, albeit not always explicit, that the endpoint of this procedure could be eugenic, namely not allowing the birth of subjects with genetic profiles deemed as not optimal. This eugenic aim is supported by reports emphasizing that the implementation of WGS could increase the detection and abortion of fetuses with nonsevere diseases.

The second condition is that *it has to do no harm.* Three types of subjects can potentially suffer the consequences of this new diagnostic method: prospective parents, children, and society. For the prospective parents who are told that they have a high risk of having a child with a genetic disease, and to whom prenatal diagnosis is proposed, a period of profound psychological suffering begins that will affect the whole family. Especially distressing is the waiting period after completion of amniocentesis for the required growth of cell cultures. This period of anxiety will be significantly reduced with the use of new WGS techniques, where results may be available within 24 to 48 hours. In most cases, they will get a negative result. However, this does not exclude the possibility that other genetic diseases for which the appropriate genes have not yet been identified might affect the child.

For the child in the womb, the use of prenatal WGS techniques eliminates the risk of indirect abortion, safeguarding the child from the harm associated with invasive tests. However, the use of WGS technology in the context of a liberal eugenic mentality will reinforce the image of the disabled person as an individual to be excluded from society, thereby causing harm to the child. Furthermore, the earlier diagnosis promised by WGS is considered as a factor that will enable a greater number of direct eugenic abortions.

For society, WGS can lead to serious harm owing to its ability to reinforce an underlying societal mentality of genetic determinism. As disability advocates will note, prenatal diagnosis is ethically problematic for at least two reasons.

First, selective abortion expresses negative or discriminatory attitudes not only about the disease but also about those who have it. And second, prenatal diagnosis establishes intolerance toward diversity, not only in society but also in the family and, ultimately, this can affect the attitudes of parents toward their children. In fact, the number of geographical areas or social strata in which, for example, there are no children with Down syndrome is increasing (Quinones and Arijeta, 2017).[4] Social pressure against the genetically damaged may not represent an ideology of racial superiority, but it does reveal several tendencies of the pragmatic and hedonistic culture in which we live. The social reluctance to accept visible genetic disabilities, the aspiration for the perfect child, the intolerance shown to the suffering of self and others, and the economic rationality that rejects the added cost of a genetic disease, are some of the symptoms of the discriminatory attitudes in our Western societies.

6. CONCLUSION

The rapid development of genetic engineering and the globalization of information should prompt us to reflect more thoroughly on the ways in which new techniques of prenatal diagnosis are applied in the daily practice of medicine. The use of techniques of recovery of cffDNA present in maternal peripheral blood for WGS is a reality in clinical practice in the case of certain diseases. In the coming years, it will become part of routine fetal screening. In this paper, we have offered a bioethical reflection about the possible difficulties and problems of the use of these techniques (Table 4).

The widespread use of WGS in prenatal diagnosis could empower parents by offering knowledge of their child's diseases and time to prepare, particularly in the setting of a genetic disease of considerable impact. Many parents who discover that they have a child affected by a genetic disease choose not to terminate, but are empowered by the information and the time for preparation that early screening gives them. In this sense, WGS has a positive moral assessment.

It is the relation between prenatal diagnosis and eugenic abortions that converts WGS into a problematic tool from an ethical point of view. The widespread use of WGS could, in fact, decrease the autonomy of women in the decision-making process (Lewis et al., 2013; Seavilleklein, 2009). The use of WGS techniques will result in reduced costs of screening, an increase in the number of disabled fetuses detected, and a decrease in the number of indirect abortions caused by invasive techniques. These properties of WGS are well known by those health authorities that may use policies of extensive WGS screening as an instrument of eugenic prevention of genetic diseases. WGS cannot be used as a justification that leads parents to decide whether the expected child is worthy to live or not.

---

4. Iceland is the first country that recognizes 100% abortion of children with Down syndrome prenatal diagnosed (Quinones and Arijeta, 2017).

TABLE 4 Recommendations for the Implementation of WGS in Prenatal Diagnosis

Recommendation	Description
1. Implementation of WGS	WGS offers exciting new opportunities, and can empower families with increased knowledge about their reproductive risks and with decision-making autonomy. For this reason, **WGS has to be carefully introduced** in an evidence-based and ethically responsible manner and monitored after implementation in health-care systems
2. WGS and nonmedical providers	Considering that many of the innovations introduced by WGS are driven by for-profit companies, **medical associations will play an increasingly important role** in providing objective guidance to patients and providers. For example, in such basic questions as who will be the doctor responsible for giving and explaining the genetic information obtained from WGS to the parents
3. Education in WGS	Professional societies also have an active role in **educating physicians** on the differences between WGS and other prenatal genetic test as well as **educating parents** about the problems involved in carrying out WGS test
4. Genetic determinism	The knowledge of WGS information could affect the future of the child in the form of parental expectations and self-fulfilling prophecies. The idea that the future of **a person can be reduced to his/her genetic information** could be passed to society by using WGS
5. Pre-WGS test counseling	Pre-WGS test counseling is necessary because he duties of professional deontology require **informed consent prior to WGS**, which necessitates a complete and rigorous disclosure of information; and a strict quality control at the level of the techniques used by the laboratories involved in WGS
6. Post-WGS diagnostic counseling	There is a need for skilled and **qualified counselors to transmit the resulting WGS information** to parents. Counselors must give the woman the time necessary for calm reflection, providing answers for her thought-out questions, and helping her toward alleviating her feelings of anxiety
7. Autonomy of the child	It will be necessary to adapt the guidelines and protocols of the different professional societies and medical associations so that they may include **protection for the future autonomy of the child**
8. WGS information	The child's right to "not to know" can only be broken if the information is clearly useful for the parents or it can improve the health conditions of the child

TABLE 4 Recommendations for the Implementation of WGS in Prenatal Diagnosis—cont'd

Recommendation	Description
9. WGS and eugenics	Concerns are therefore expressed about WGS paving the road to a "quest for a perfect baby," progressively creating **new forms of eugenics in society**. Western societies already have numerous policies and regulations—for instance, relating to the rights of disabled individuals—that can help prevent a radical eugenic drift
10. Storage of WGS information	It will also be necessary to establish "who" can access to the WGS information for that child. **A blockchain technology system would be necessary** in such a way that a certain physician could only access to that part of the child's WGS information that it will be necessary for diagnosis and treatments of diseases related to his/her specialty

It is clear that "dignity" cannot be established at molecular level. As time unfolds, the detection methods of genetic abnormalities from WGS will become more effective. Given that many of the diagnosed diseases do not have a therapeutic treatment, abortion is often considered as a *preventive treatment*. Eugenics wrongly identifies the disease with the patient's life, the part with the whole. While a disease does not deserve to be respected, this rejection must never be erroneously manifested in social rejection of the patients suffering the disease, extolling abortion as a priority solution and abandoning the challenge of trying to find a cure consistent with the value of the life of the person. The use of WGS in this direction is in contrast to the first principle of medical ethics and, indeed, all ethics: *primum non nocere*, first do no harm. WGS could reinforce the image of the disabled person as an individual to be excluded from society. For this reason, physicians must play an important role in the process of pre-WGS test and post-WGS diagnostic counseling. WGS will change our medical routine over the next few years, and it will depend on the medical community's utilization of WGS as to whether it will remain a diagnostic tool or become a eugenic one.

In conclusion, we can say that the use of WGS techniques *to diagnose the existence of genetic diseases in the fetus in order to decide*, when the result is positive, *whether or not to perform an abortion*, implies and includes in itself the conditions that characterize a *negative ethical assessment*. On the other hand, when WGS is used to diagnose genetic diseases that do have treatments (whether immediately after birth or even in utero), and without an abortive intention in the event of a positive test result, i.e., when it is *for the best interest of the child*, the moral assessment is positive.

REFERENCES

Abrahams, E., Silver, M., 2009. The case for personalized medicine. J. Diabetes Sci. Technol. 3, 680–684.

ACMG Board of Directors, 2012. Points to consider in the clinical application of genomic sequencing. Genet. Med. 14, 759.

Ashoor, G., Syngelaki, A., Wagner, M., Birdir, C., Nicolaides, K.H., 2012. Chromosome-selective sequencing of maternal plasma cell free DNA for first-trimester detection of trisomy 21 and trisomy 18. Am. J. Obstet. Gynecol. 206. 322.e1-322.e5.

Ashoor, G., Syngelaki, A., Wang, E., Struble, C., Oliphant, A., Song, K., Nicolaides, K.H., 2013. Trisomy 13 detection in the first trimester of pregnancy using a chromosome-selective cell-free DNA analysis method. Ultrasound Obstet. Gynecol. 41, 21–25.

Bamshad, M.J., Ng, S.B., Bigham, A.W., Tabor, H.K., Emond, M.J., Nickerson, D.A., Shendure, J., 2011. Exome sequencing as a tool for Mendelian disease gene discovery. Nat. Rev. Genet. 12, 745.

Battaglia, A., Doccini, V., Bernardini, L., Novelli, A., Loddo, S., Capalbo, A., Filippi, T., Carey, J.C., 2013. Confirmation of chromosomal microarray as a first-tier clinical diagnostic test for individuals with developmental delay, intellectual disability, autism spectrum disorders and dysmorphic features. Eur. J. Paediatr. Neurol. 17, 589–599.

Belkadi, A., Bolze, A., Itan, Y., Cobat, A., Vincent, Q.B., Antipenko, A., Shang, L., Boisson, B., Casanova, J.-L., Abel, L., 2015. Whole-genome sequencing is more powerful than whole-exome sequencing for detecting exome variants. Proc. Natl. Acad. Sci. U. S. A. 112, 5473–5478.

Benn, P., Borrell, A., Chiu, R.W.K., Cuckle, H., Dugoff, L., Faas, B., Gross, S., Huang, T., Johnson, J., Maymon, R., Norton, M., Odibo, A., Schielen, P., Spencer, K., Wright, D., Yaron, Y., 2015. Position statement from the Chromosome Abnormality Screening Committee on behalf of the Board of the International Society for Prenatal Diagnosis. Prenat. Diagn. 35, 725–734.

Benn, P.A., Chapman, A.R., 2010. Ethical challenges in providing noninvasive prenatal diagnosis. Curr. Opin. Obstet. Gynecol. 22, 128–134.

Berg, J.S., Khoury, M.J., Evans, J.P., 2011. Deploying whole genome sequencing in clinical practice and public health: meeting the challenge one bin at a time. Genet. Med. 13, 499.

Bianchi, D.W., Parker, R.L., Wentworth, J., Madankumar, R., Saffer, C., Das, A.F., Craig, J.A., Chudova, D.I., Devers, P.L., Jones, K.W., Oliver, K., Rava, R.P., Sehnert, A.J., 2014. DNA sequencing versus standard prenatal aneuploidy screening. N. Engl. J. Med. 370, 799–808.

Bianchi, D.W., Platt, L.D., Goldberg, J.D., Abuhamad, A.Z., Sehnert, A.J., Rava, R.P., 2012. Genome-wide fetal aneuploidy detection by maternal plasma DNA sequencing. Obstet. Gynecol. 119, 890–901.

Bianchi, D.W., Simpson, J.L., Jackson, L.G., Elias, S., Holzgreve, W., Evans, M.I., Dukes, K.A., Sullivan, L.M., Klinger, K.W., Bischoff, F.Z., Hahn, S., Johnson, K.L., Lewis, D., Wapner, R.J., Cruz, F.D.L., 2002. Fetal gender and aneuploidy detection using fetal cells in maternal blood: analysis of NIFTY I data. Prenat. Diagn. 22, 609–615.

Biesecker, L.G., Green, R.C., 2014. Diagnostic clinical genome and exome sequencing. N. Engl. J. Med. 370, 2418–2425.

Borgelt, E., Anderson, J.A., Illes, J., 2013. Managing incidental findings: lessons from neuroimaging. Am. J. Bioethics 13, 46–47.

Boyer, S., Noyes, A., Boyer, M., 1976. Enrichment of erythrocytes of fetal origin from adult-fetal blood mixtures via selective hemolysis of adult blood cells: an aid to antenatal diagnosis of hemoglobinopathies. Blood 47, 883–897.

Chiu, R.W.K., Akolekar, R., Zheng, Y.W.L., Leung, T.Y., Sun, H., Chan, K.C.A., Lun, F.M.F., Go, A.T.J.I., Lau, E.T., To, W.W.K., Leung, W.C., Tang, R.Y.K., Au-Yeung, S.K.C., Lam, H., Kung, Y.Y., Zhang, X., Van Vugt, J.M.G., Minekawa, R., Tang, M.H.Y., Wang, J., Oudejans, C.B.M., Lau, T.K., Nicolaides, K.H., Lo, Y.M.D., 2011. Non-invasive prenatal assessment of trisomy 21 by multiplexed maternal plasma DNA sequencing: large scale validity study. BMJ 342, c7401.

Chiu, R.W.K., Chan, K.C.A., Gao, Y., Lau, V.Y.M., Zheng, W., Leung, T.Y., Foo, C.H.F., Xie, B., Tsui, N.B.Y., Lun, F.M.F., Zee, B.C.Y., Lau, T.K., Cantor, C.R., Lo, Y.M.D., 2008. Noninvasive prenatal diagnosis of fetal chromosomal aneuploidy by massively parallel genomic sequencing of DNA in maternal plasma. Proc. Natl. Acad. Sci. U. S. A. 105, 20458–20463.

Christian, S.M., Koehn, D., Pillay, R., Macdougall, A., Wilson, R.D., 2000. Parental decisions following prenatal diagnosis of sex chromosome aneuploidy: a trend over time. Prenat. Diagn. 20, 37–40.

Chu, T., Bunce, K., Hogge, W.A., Peters, D.G., 2009. Statistical model for whole genome sequencing and its application to minimally invasive diagnosis of fetal genetic disease. Bioinformatics 25, 1244–1250.

Collins F, Wall Street Journal, 2014. https://www.wsj.com/articles/francis-collins-says-medicine-in-the-future-will-be-tailored-to-your-genes-1404763139 (accessed 4 august 2017).

Deans, Z., Clarke, A.J., Newson, A.J., 2015. For your interest? The ethical acceptability of using non-invasive prenatal testing to test 'purely for information'. Bioethics 29, 19–25.

Ding, L.E., Burnett, L., Chesher, D., 2014. The impact of reporting incidental findings from exome and whole-genome sequencing: predicted frequencies based on modeling. Genet. Med. 17, 197.

Donley, G., Hull, S.C., Berkman, B.E., 2012. Prenatal whole genome sequencing: just because we can, should we? Hastings Cent. Rep. 42, 28–40.

Eng, C.M., Yang, Y., Plon, S.E., 2014. Genetic diagnosis through whole-exome sequencing. N. Engl. J. Med. 370, 1067–1069.

Fan, H.C., Blumenfeld, Y.J., Chitkara, U., Hudgins, L., Quake, S.R., 2008. Noninvasive diagnosis of fetal aneuploidy by shotgun sequencing DNA from maternal blood. Proc. Nat. Acad. Sci. 105, 16266–16271.

Fan, H.C., Gu, W., Wang, J., Blumenfeld, Y.J., El-Sayed, Y.Y., Quake, S.R., 2012. Noninvasive prenatal measurement of the fetal genome. Nature 487, 320–324.

Favre, R., Duchange, N., Vayssière, C., Kohler, M., Bouffard, N., Hunsinger, M.C., Kohler, A., Mager, C., Neumann, M., Vayssière, C., Viville, B., Hervé, C., Moutel, G., 2007. How important is consent in maternal serum screening for Down syndrome in France? Information and consent evaluation in maternal serum screening for Down syndrome: a French study. Prenat. Diagn. 27, 197–205.

Filges, I., Nosova, E., Bruder, E., Tercanli, S., Townsend, K., Gibson, W.T., Röthlisberger, B., Heinimann, K., Hall, J.G., Gregory-Evans, C.Y., Wasserman, W.W., Miny, P., Friedman, J.M., 2014. Exome sequencing identifies mutations in KIF14 as a novel cause of an autosomal recessive lethal fetal ciliopathy phenotype. Clin. Genet. 86, 220–228.

Garcia, J.R., Mackillop, J., Aller, E.L., Merriwether, A.M., Wilson, D.S., Lum, J.K., 2010. Associations between dopamine D4 receptor gene variation with both infidelity and sexual promiscuity. PLoS One 5, e14162.

Gilissen, C., Hoischen, A., Brunner, H.G., Veltman, J.A., 2011. Unlocking Mendelian disease using exome sequencing. Genome Biol. 12, 228.

Gonzaga-Jauregui, C., Lupski, J.R., Gibbs, R.A., 2012. Human genome sequencing in health and disease. Annu. Rev. Med. 63, 35–61.

Govaerts, L., Srebniak, M., Diderich, K., Joosten, M., Riedijk, S., Knapen, M., Go, A., Papatsonis, D., De Graaf, K., Toolenaar, T., Van Der Steen, S., Huijbregts, G., Knijnenburg, J., De Vries, F., Van Opstal, D., Galjaard, R.J., 2017. Prenatal diagnosis of susceptibility loci for neurodevelopmental disorders—genetic counseling and pregnancy outcome in 57 cases. Prenat. Diagn. 37, 73–80.

Green, R.C., Lupski, J.R., Biesecker, L.G., 2013. Reporting genomic sequencing results to ordering clinicians: incidental, but not exceptional. JAMA 310, 365–366.

Grody, W.W., Thompson, B.H., Hudgins, L., 2013. Whole-exome/genome sequencing and genomics. Pediatrics 132, S211–S215.

Hercher, L., Uhlmann, W.R., Hoffman, E.P., Gustafson, S., Chen, K.M., 2016. Prenatal testing for adult-onset conditions: the position of the national society of genetic counselors. J. Genet. Counsel. 25, 1139–1145.

Herper, M. 2017. Illumina promises to sequence human genome for $100—but not quite yet [Online]. Available: https://www.forbes.com/sites/matthewherper/2017/01/09/illumina-promises-to-sequence-human-genome-for-100-but-not-quite-yet/#3ec7cce2386d (accessed 4 September 2017).

Hillman, S.C., Willams, D., Carss, K.J., Mcmullan, D.J., Hurles, M.E., Kilby, M.D., 2015. Prenatal exome sequencing for fetuses with structural abnormalities: the next step. Ultrasound Obstet. Gynecol. 45, 4–9.

Horn, R., Parker, M., 2017. Opening Pandora's box?: ethical issues in prenatal whole genome and exome sequencing. Prenat. Diagn. 37, 1–6.

Huang, R., Barber, T.A., Schmidt, M.A., Tompkins, R.G., Toner, M., Bianchi, D.W., Kapur, R., Flejter, W.L., 2008. A microfluidics approach for the isolation of nucleated red blood cells (NRBCs) from the peripheral blood of pregnant women. Prenat. Diagn. 28, 892–899.

Iverson, G.M., Bianchi, D.W., Cann, H.M., Herzenberg, L.A., 1981. Detection and isolation of fetal cells from maternal blood using the flourescence-activated cell sorter (FACS). Prenat. Diagn. 1, 61–73.

Kalia, S.S., Adelman, K., Bale, S.J., Chung, W.K., Eng, C., Evans, J.P., Herman, G.E., Hufnagel, S.B., Klein, T.E., Korf, B.R., Mckelvey, K.D., Ormond, K.E., Richards, C.S., Vlangos, C.N., Watson, M., Martin, C.L., Miller, D.T., 2016. Recommendations for reporting of secondary findings in clinical exome and genome sequencing, 2016 update (ACMG SF v2.0): a policy statement of the American College of Medical Genetics and Genomics. Genet. Med. 19, 249.

King, J.S., 2011. And genetic testing for all… The coming revolution in non-invasive prenatal genetic testing. Rutgers Law J. 42, 599–819.

Kitzman, J.O., Snyder, M.W., Ventura, M., Lewis, A.P., Qiu, R., Simmons, L.E., Gammill, H.S., Rubens, C.E., Santillan, D.A., Murray, J.C., Tabor, H.K., Bamshad, M.J., Eichler, E.E., Shendure, J., 2012. Non-invasive whole genome sequencing of a human fetus. Sci. Transl. Med. 4. 137ra76-137ra76.

Knoppers, B.M., Zawati, M.H., Senecal, K., 2015. Return of genetic testing results in the era of whole-genome sequencing. Nat. Rev. Genet. 16, 553–559.

Korf, B.R., Rehm, H.L., 2013. New approaches to molecular diagnosis. JAMA 309, 1511–1521.

Levy, S.E., Myers, R.M., 2016. Advancements in next-generation sequencing. Annu. Rev. Genom. Hum. Genet. 17, 95–115.

Lewis, C., Silcock, C., Chitty, L.S., 2013. Non-invasive prenatal testing for Down's syndrome: pregnant women's views and likely uptake. Public Health Genom. 16, 223–232.

Lo, Y.M., Chan, K.C.A., Sun, H., Chen, E.Z., Jiang, P., Lun, F.M., Zheng, Y.W., Leung, T.Y., Lau, T.K., Cantor, C.R., Chiu, R.W., 2010. Maternal plasma DNA sequencing reveals the genome-wide genetic and mutational profile of the fetus. Sci. Transl. Med. 2. 61ra91-61ra91.

Lo, Y.M., Corbetta, N., Chamberlain, P.F., Rai, V., Sargent, I.L., Redman, C.W., Wainscoat, J.S., 1997. Presence of fetal DNA in maternal plasma and serum. The Lancet 350, 485–487.

Lo, Y.M., Tein, M.S.C., Lau, T.K., Haines, C.J., Leung, T.N., Poon, P.M., Wainscoat, J.S., Johnson, P.J., Chang, A.M., Hjelm, N.M., 1998. Quantitative analysis of fetal DNA in maternal plasma and serum: implications for noninvasive prenatal diagnosis. Am. J. Hum. Genet. 62, 768–775.

Mazloom, A.R., Džakula, Ž., Oeth, P., Wang, H., Jensen, T., Tynan, J., Mccullough, R., Saldivar, J.S., Ehrich, M., Van Den Boom, D., Bombard, A.T., Maeder, M., Mclennan, G., Meschino, W., Palomaki, G.E., Canick, J.A., Deciu, C., 2013. Noninvasive prenatal detection of sex chromosomal aneuploidies by sequencing circulating cell-free DNA from maternal plasma. Prenat. Diagn. 33, 591–597.

Mcguire, A.L., Joffe, S., Koenig, B.A., Biesecker, B.B., Mccullough, L.B., Blumenthal-Barby, J.S., Caulfield, T., Terry, S.F., Green, R.C., 2013. Point-Counterpoint. Ethics and genomic incidental findings. Science 340, 1047–1048.

Menezes, M., Meagher, S., Costa, F.D.S., 2013. Ethical considerations when offering noninvasive prenatal testing. Rev. Bras. Ginecol. Obstet. 35, 195–198.

Miller, D.T., Adam, M.P., Aradhya, S., Biesecker, L.G., Brothman, A.R., Carter, N.P., Church, D.M., Crolla, J.A., Eichler, E.E., Epstein, C.J., Faucett, W.A., Feuk, L., Friedman, J.M., Hamosh, A., Jackson, L., Kaminsky, E.B., Kok, K., Krantz, I.D., Kuhn, R.M., Lee, C., Ostell, J.M., Rosenberg, C., Scherer, S.W., Spinner, N.B., Stavropoulos, D.J., Tepperberg, J.H., Thorland, E.C., Vermeesch, J.R., Waggoner, D.J., Watson, M.S., Martin, C.L., Ledbetter, D.H., 2010. Consensus statement: chromosomal microarray is a first-tier clinical diagnostic test for individuals with developmental disabilities or congenital anomalies. Am. J. Hum. Genet. 86, 749–764.

Mouawia, H., Saker, A., Jais, J.P., Benachi, A., Bussières, L., Lacour, B., Bonnefont, J.P., Frydman, R., Simpson, J.L., Paterlini-Brechot, P., 2012. Circulating trophoblastic cells provide genetic diagnosis in 63 fetuses at risk for cystic fibrosis or spinal muscular atrophy. Reprod. BioMed. Online 25, 508–520.

Nicolaides, K.H., Syngelaki, A., Ashoor, G., Birdir, C., Touzet, G., 2012. Noninvasive prenatal testing for fetal trisomies in a routinely screened first-trimester population. Am. J. Obstet. Gynecol. 207. 374.e1-374.e6.

Nicolaides, K.H., Syngelaki, A., Gil, M., Atanasova, V., Markova, D., 2013. Validation of targeted sequencing of single-nucleotide polymorphisms for non-invasive prenatal detection of aneuploidy of chromosomes 13, 18, 21, X, and Y. Prenat. Diagn. 33, 575–579.

Ostrander, E.A., Huson, H.J., Ostrander, G.K., 2009. Genetics of athletic performance. Annu. Rev. Genom. Hum. Genet. 10, 407–429.

Palomaki, G.E., Deciu, C., Kloza, E.M., Lambert-Messerlian, G.M., Haddow, J.E., Neveux, L.M., Ehrich, M., Van Den Boom, D., Bombard, A.T., Grody, W.W., Nelson, S.F., Canick, J.A., 2012. DNA sequencing of maternal plasma reliably identifies trisomy 18 and trisomy 13 as well as Down syndrome: an international collaborative study. Genet. Med. 14, 296–305.

Palomaki, G.E., Kloza, E.M., Lambert-Messerlian, G.M., Haddow, J.E., Neveux, L.M., Ehrich, M., Van Den Boom, D., Bombard, A.T., Deciu, C., Grody, W.W., Nelson, S.F., Canick, J.A., 2011. DNA sequencing of maternal plasma to detect Down syndrome: an international clinical validation study. Genet Med. 13, 913–920.

Pang, A.W., Macdonald, J.R., Yuen, R.K., Hayes, V.M., Scherer, S.W., 2014. Performance of high-throughput sequencing for the discovery of genetic variation across the complete size spectrum. G3: Genes|Genomes|Genetics 4, 63–65.

Parens, E., Asch, A. (Eds.), 2000. Prenatal testing and disability rights. Georgetown University Press, Washington, DC.

Pergament, E., Cuckle, H., Zimmermann, B., Banjevic, M., Sigurjonsson, S., Ryan, A., Hall, M. P., Dodd, M., Lacroute, P., Stosic, M., Chopra, N., Hunkapiller, N., Prosen, D. E., Mcadoo, S., Demko, Z., Siddiqui, A., Hill, M., Rabinowitz, M., 2014. Single-nucleotide polymorphism–based noninvasive prenatal screening in a high-risk and low-risk cohort. Obstet. Gynecol., 124, 210-218.

Peters, D.G., Yatsenko, S.A., Surti, U., Rajkovic, A., 2015. Recent advances of genomic testing in perinatal medicine. Semin. Perinatol. 39, 44–54.

Porreco, R.P., Garite, T.J., Maurel, K., Marusiak, B., Obstetrix Collaborative Research Network, Ehrich, M., van den Boom, D., Deciu, C., Bombard, A., 2014. Noninvasive prenatal screening for fetal trisomies 21, 18, 13 and the common sex chromosome aneuploidies from maternal blood using massively parallel genomic sequencing of DNA. Am. J. Obstet. Gynecol. 211, 711–712.

Quinones, J, Arijeta, L., 2017. "What kind of society do you want to live in?": inside the country where Down syndrome is disappearing. Aug 15, 2017 ed.: CBS NEWS. [https://www.cbsnews.com/news/down-syndrome-iceland/] (Accessed Jan 27, 2018).

Ross, L., Rothstein, M.A., Clayton, E., 2013. Mandatory extended searches in all genome sequencing: "incidental findings," patient autonomy, and shared decision making. JAMA 310, 367–368.

Sachs, A., Blanchard, L., Buchanan, A., Norwitz, E., Bianchi, D.W., 2015. Recommended pre-test counseling points for noninvasive prenatal testing using cell-free DNA: a 2015 perspective. Prenat. Diagn. 35, 968–971.

Seavilleklein, V., 2009. Challenging the rhetoric of choice in prenatal screening. Bioethics 23, 68–77.

Song, Y., Liu, C., Qi, H., Zhang, Y., Bian, X., Liu, J., 2013. Noninvasive prenatal testing of fetal aneuploidies by massively parallel sequencing in a prospective Chinese population. Prenat. Diagn. 33, 700–706.

Stavropoulos, D.J., Merico, D., Jobling, R., Bowdin, S., Monfared, N., Thiruvahindrapuram, B., Nalpathamkalam, T., Pellecchia, G., Yuen, R.K.C., Szego, M.J., Hayeems, R.Z., Shaul, R.Z., Brudno, M., Girdea, M., Frey, B., Alipanahi, B., Ahmed, S., Babul-Hirji, R., Porras, R.B., Carter, M.T., Chad, L., Chaudhry, A., Chitayat, D., Doust, S.J., Cytrynbaum, C., Dupuis, L., Ejaz, R., Fishman, L., Guerin, A., Hashemi, B., Helal, M., Hewson, S., Inbar-Feigenberg, M., Kannu, P., Karp, N., Kim, R.H., Kronick, J., Liston, E., Macdonald, H., Mercimek-Mahmutoglu, S., Mendoza-Londono, R., Nasr, E., Nimmo, G., Parkinson, N., Quercia, N., Raiman, J., Roifman, M., Schulze, A., Shugar, A., Shuman, C., Sinajon, P., Siriwardena, K., Weksberg, R., Yoon, G., Carew, C., Erickson, R., Leach, R.A., Klein, R., Ray, P.N., Meyn, M.S., Scherer, S.W., Cohn, R.D., Marshall, C.R., 2016. Whole-genome sequencing expands diagnostic utility and improves clinical management in paediatric medicine. Npj Genomic Med. 1, 15012.

Stumm, M., Entezami, M., Haug, K., Blank, C., Wüstemann, M., Schulze, B., Raabe-Meyer, G., Hempel, M., Schelling, M., Ostermayer, E., Langer-Freitag, S., Burkhardt, T., Zimmermann, R., Schleicher, T., Weil, B., Schöck, U., Smerdka, P., Grömminger, S., Kumar, Y., Hofmann, W., 2014. Diagnostic accuracy of random massively parallel sequencing for non-invasive prenatal detection of common autosomal aneuploidies: a collaborative study in Europe. Prenat. Diagn. 34, 185–191.

Talkowski, M.E., Ordulu, Z., Pillalamarri, V., Benson, C.B., Blumenthal, I., Connolly, S., Hanscom, C., Hussain, N., Pereira, S., Picker, J., Rosenfeld, J.A., Shaffer, L.G., Wilkins-Haug, L.E., Gusella, J.F., Morton, C.C., 2012. Clinical diagnosis by whole-genome sequencing of a prenatal sample. N. Engl. J. Med. 367, 2226–2232.

The American College of Obstetrics and Gynecology, 2013. Committee opinion n°. 581: the use of chromosomal microarray analysis in prenatal diagnosis. Obstet. Gynecol. 122, 1374–1377.

Townsend, A., Adam, S., Birch, P.H., Friedman, J.M., 2013. Paternalism and the ACMG recommendations on genomic incidental findings: patients seen but not heard. Genet. Med. 15, 751.

Van Den Veyver, I.B., Eng, C.M., 2015. Genome-wide sequencing for prenatal detection of fetal single-gene disorders. Cold Spring Harb. Perspect. Med. 5.

Van El, C.G., Cornel, M.C., Borry, P., Hastings, R.J., Fellmann, F., Hodgson, S.V., Howard, H.C., Cambon-Thomsen, A., Knoppers, B.M., Meijers-Heijboer, H., Scheffer, H., Tranebjaerg, L., Dondorp, W., De Wert, G.M.W.R., 2013. Whole-genome sequencing in health care: recommendations of the European Society of Human Genetics. Eur. J. Hum. Genet. 21, S1–S5.

Verweij, E.J., Jacobsson, B., Van Scheltema, P.A., De Boer, M.A., Hoffer, M.J.V., Hollemon, D., Westgren, M., Song, K., Oepkes, D., 2013. European Non-Invasive Trisomy Evaluation (EUNITE) study: a multicenter prospective cohort study for non-invasive fetal trisomy 21 testing. Prenat. Diagn. 33, 996–1001.

Verweij, E.J., Oepkes, D., De Boer, M.A., 2013. Changing attitudes towards termination of pregnancy for trisomy 21 with non-invasive prenatal trisomy testing: a population-based study in Dutch pregnant women. Prenat. Diagn. 33, 397–399.

Wolf, S.M., Annas, G.J., Elias, S., 2013. Patient autonomy and incidental findings in clinical genomics. Science 340, 1049–1050.

Xue, Y., Ankala, A., Wilcox, W.R., Hegde, M.R., 2014. Solving the molecular diagnostic testing conundrum for Mendelian disorders in the era of next-generation sequencing: single-gene, gene panel, or exome/genome sequencing. Genet. Med. 17, 444.

Yagel, S., 2013. Non-invasive prenatal testing: more questions than answers. Ultrasound Obstet. Gynecol. 42, 369–372.

Yang, Y., Muzny, D.M., Reid, J.G., Bainbridge, M.N., Willis, A., Ward, P.A., Braxton, A., Beuten, J., Xia, F., Niu, Z., Hardison, M., Person, R., Bekheirnia, M.R., Leduc, M.S., Kirby, A., Pham, P., Scull, J., Wang, M., Ding, Y., Plon, S.E., Lupski, J.R., Beaudet, A.L., Gibbs, R.A., Eng, C.M., 2013. Clinical whole-exome sequencing for the diagnosis of Mendelian disorders. N. Engl. J. Med. 369, 1502–1511.

Yang, Y., Muzny, D.M., Xia, F., Et, A.L., 2014. Molecular findings among patients referred for clinical whole-exome sequencing. JAMA 312, 1870–1879.

Yeh, D.D., Imam, A.M., Truong, S.H., Mclaughlin, E.L., Klein, E.N., Avery, L.L., Velmahos, G.C., 2013. Incidental findings in trauma patients: dedicated communication with the primary care physician ensures adequate follow-up. World J. Surg. 37, 2081–2085.

Zimmermann, B., Hill, M., Gemelos, G., Demko, Z., Banjevic, M., Baner, J., Ryan, A., Sigurjonsson, S., Chopra, N., Dodd, M., Levy, B., Rabinowitz, M., 2012. Noninvasive prenatal aneuploidy testing of chromosomes 13, 18, 21, X, and Y, using targeted sequencing of polymorphic loci. Prenatal Diagn. 32, 1233–1241.

Zimmermann, S., Hollmann, C., Stachelhaus, S.A., 2013. Unique monoclonal antibodies specifically bind surface structures on human fetal erythroid blood cells. Exp. Cell Res. 319, 2700–2707.

Chapter 13

Noninvasive Prenatal Genetic Diagnosis

Sorin Hostiuc
Department of Legal Medicine and Bioethics, Faculty of Dental Medicine, "Carol Davila" University of Medicine and Pharmacy, Bucharest, Romania

1. INTRODUCTION

Prenatal testing aims to identify various morphological or physiological disorders of the zygote, embryo, or fetus, either before implantation (preimplantation genetic diagnosis), or early during the gestational period, and usually includes two main steps: screening and diagnosis. Prenatal screening tests identify medical issues as cheaply as possible, making it available to large population groups, and usually have a high sensitivity and a lower specificity. Prenatal diagnosis aims to certify the presence of a medical issue and characterize it as well as possible, having a much higher specificity compared to prenatal screening procedures. The most commonly used prenatal tests include amniocentesis, chorionic villi sampling, ultrasonography, serum biomarker testing, and genetic screening (either pre- or postconception). Some tests, such as chorionic villi sampling, are invasive, and associated with significant risks for either the mother or the baby; others, such as ultrasonography, are noninvasive, and therefore do not bear additional medical risks. Some are performed on fetal tissue, while other on the mother; some are used to detect fetal abnormalities, while others, such as the test for pregnancy-associated plasma protein A, are done to detect conditions that could pose a risk to either the pregnancy or the mother. Some tests are performed to detect genetic abnormalities, and other are used for sex evaluation. The steps of prenatal testing are in general well regulated, many tests are given free of charge to prospective mothers in different countries, and the recommendations often include counseling for related issues or ethical issues. One approach, published by Skirton et al. (2014) described the steps needed for offering prenatal diagnostic tests for women at a high genetic risk pregnancies, namely: (1) the presence of a significant family history prior to the pregnancy, (2) the women were identified during pregnancy as having a fetus at risk for a genetic conditions (e.g., family history, genetic tests), or

Clinical Ethics at the Crossroads of Genetic and Reproductive Technologies.
https://doi.org/10.1016/B978-0-12-813764-2.00013-1

293

(3) the fetus was found to be at risk due to an abnormal ultrasound. The guideline has 11 recommendations, including:

1. The presence of a designated professional who should act as a coordinator of the medical management team, and as a liaison with the patient.
2. Proper counseling should be offered by trained professionals, irrespective of the decision of the parent/parents.
3. Independent interpreting services should be offered if needed.
4. Consequences for other family members should be presented and discussed with the prospective parents.
5. Disclosure should be performed only according to national regulation, only if required, and with the consent of the parents.
6. If the diagnosis is uncertain, and the parents decide to terminate the pregnancy, they should be made aware of the presence of additional medical investigations, which could clarify the underlying cause of the fetal abnormality, even is this would lead to a slight delay of the abortive procedure.
7. Prenatal diagnosis should be made by a trained professional, and the testing should be done in an appropriate laboratory.
8. Trained professionals should present the results of the test, and posttest counseling should be offered if needed.
9. If prenatal diagnosis is declined or is not considered useful, the parents should be referred to specialists for the management of the pregnancy and for after birth support.
10. Additional counseling should be offered after birth, to discuss future reproductive choices and consequences for the parents and, if needed, other family members.
11. Consent should be obtained for any tissue-related sampling, storage, and analysis (Skirton et al., 2014).

Prenatal testing for genetic conditions usually tries to determine the presence of monogenic nuclear disorders, aneuploidies, or other chromosomal abnormalities (such as microdeletions), mitochondrial diseases and, as of recently, the identification of genes associated with multigenic/multifactorial conditions (Korpi-Steiner et al., 2017).

Monogenic disorders are hereditary conditions in which mutations are affecting one or both pairs of a specific gene and can be classified as autosomal dominant, autosomal recessive, and sex-linked. The most frequent monogenic disease in the Caucasian population is cystic fibrosis, affecting around 0.04% of all live births (Cram and Zhou, 2016). Overall, the prevalence of the most frequent monogenic disorders is presented in Table 1.

For some monogenic disorders that are more prevalent in the general population, specific prenatal tests for their identification were developed, including for cystic fibrosis, beta-thalassemia, sickle-cell anemia, spinal muscular atrophy, fragile-X syndrome, Duchenne muscular dystrophy, or hemophilia (Traeger-Synodinos, 2006).

TABLE 1 Various Types of Genetic Disorders and Their Estimated Prevalence (Pooled From Multiple Sources)

Disease	Prevalence (Live Births) (%)	Type
Familial hypercholesterolemia	0.25	AD
Klinefelter syndrome	0.2	Aneuploidy
Sickle cell anemia	0.16	AR
Down syndrome	0.1	Aneuploidy
Polycystic renal disease	0.08	AD
Type I Neurofibromatosis	0.04	AD
Turner syndrome	0.04	Aneuploidy
Patau syndrome	0.04	Aneuploidy
Cystic fibrosis	0.04–0.05	AR
Tay-Sachs disease	0.03	AR
Marfan syndrome	0.025	AD
DiGeorge syndrome	0.025	Microdeletion
Hereditary spherocytosis	0.02	AD
Edwards syndrome	0.02	Aneuploidy
Duchenne muscular dystrophy	0.014	X-linked
Hemophilia	0.01	X-linked
Phenylketonuria	0.008	AR
Huntington's disease	0.007	AD

Aneuploidies are pathological conditions defined by the presence of an abnormal number of chromosomes in a cell. Here are included some of the most common genetic abnormalities identified in newborns, such as Down syndrome (trisomy 21), Turner syndrome (45X0), Klinefelter syndrome (47XXY), and cri-du-chat (chromosome 5p deletion syndrome).

Noninvasive prenatal diagnosis or testing (NIPD, NIPT) is a screening test involving the analysis of cell-free fetal DNA (cffDNA) from maternal blood during pregnancy. cffDNA are DNA fragments, usually around 200 pair bases, entering the mother's circulatory system from the trophoblasts. They may be identified from as early as a 5 weeks gestation (Lo et al., 1998), and their amount increases progressively during pregnancy. cffDNA can be used from weeks 7–10 to test the presence of single gene disorders (Hill et al., 2011), much earlier compared to other, currently used methods, such as amniocentesis or chorionic villi sampling. cffDNA is routinely used in some countries to test for aneuploidies, such as Edward's, Patau, Down, Turner, Klinefelter, and for Rh compatibility testing (De Jong et al., 2011). However, the usefulness of cffDNA is continuously expanding, more genetic tests being developed each year. For example, the commercially available test Vistara(R) from Natera identifies the following monogenic disorders (the tested gene is presented in parenthesis): achondroplasia (FGFR3), Alagille syndrome (JAG1), Antley Bixler syndrome (FGFR2), Apert syndrome (FGFR2), cardiofaciocutaneous syndrome 1,3,4 (BRAF, MAP2K1, MAP2K2), CATSHL syndrome (FGFR3), CHARGE syndrome (CHD7), Cornelia de Lange syndrome 1,2,3,4,5 (NIPBL, SMC1A, SMC3, RAD21, HDAC8), Costello syndrome (HRAS), Crouzon syndrome (FGHR2, FGFR3), Ehlers-Danlos syndrome, classic, type VIIA, cardiac valvular form, type VIIB (COL1A1, COL1A2), epileptic encephalopathy, early infantile 2 (CDKL5), hypochondroplasia (FGFR3), intellectual disability (SYNGAP1), Jackson Weiss syndrome (FGFR2); juvenile myelomonocytic leukemia (PTPN11), LEOPARD syndrome (Noonan syndrome with multiple lentigines (PTPN11, RAF1), Muenke syndrome (FGFR3), Noonan syndrome 1,3,4,5,6,8 (PTPN1, SOS1, RAF1, KRAS, NRAS, SHOC2, BRAF, MAP2K1, HRAS, CBL), osteogenesis imperfecta type I, II, III, IV (COL1A1, COL1A2), Pfeiffer syndrome type 1,2,3 (FGFR2), Rett syndrome (MECP2), Sotos syndrome 1 (NSD10, thanatophoric dysplasia I, II (FGFR3), and tuberous sclerosis 1,2 (TSC1, TSC2) (Natera, 2017).

The accuracy of cffDNA testing is increasing each year. However, for some disorders, the accuracy is still lower compared to more invasive prenatal diagnosis procedures, which have, however, a higher miscarriage risk. There are some studies, mainly regarding Down's syndrome, suggesting that women would often prefer, if they have a higher risk, to suffer a diagnosis-related miscarriage to having a baby with this disease (Kuppermann et al., 2000). If the testing is done, the patients would accept a decreased accuracy, provided the risk associated with invasive prenatal procedures is nullified. For example, Hill et al. found that patients would accept a noninvasive procedure having an accuracy decreased by 16% (2013). Chan et al. (2014) showed that a noninvasive test that detects Down's syndrome with a 95% accuracy would be rendered acceptable by the prospective parents. Gil et al. (2014) in a recent metaanalysis, found the weighted pooled detection rates, and false-positive rates for trisomy 21 to be 99% and 0.08%, for trisomy 18—96.8% and 0.15%, 92.1% and 0.2% for trisomy 13, 88.6% and 0.12% for monosomy X, and 93.8% and 0.12% for sex-chromosome aneuploidies other than monosomy X. The main reason for a decreased accuracy (and subsequently the false positives) of the results of cffDNA testing is represented by the fact that the sequenced DNA contains both maternal and fetal cell-free DNA. Therefore a positive test may be caused by other factors that the actual presence of an abnormal genotype, including placental mosaicism, a vanishing twin, or even a maternal tumor (Bianchi and Wilkins-Haug, 2014).

If, however, we look at the predictive value, which also takes into account the fact that these conditions have a low prevalence in the general population, the results are much less impressive (Dondorp et al., 2015). For decision-making solely based on this procedure, the positive predictive value is more important because it tells us not only the sensibility but also the actual chance that, in case of a positive test, the fetus will have the genetic trait. For example, a study performed by Bianchi et al. showed that the positive predictive values for cffDNA versus standard testing in aneuploidy were 45.5% versus 4.2% for trisomy 21, and 40% versus 8.3% for trisomy 18 (Bianchi et al., 2014b). Many Mendelian disorders for which were developed tests based on cffDNA have a decreased specificity and a very low prevalence in the general population. For example, for Costello syndrome, a conservative computation, taking into account a specificity of 92% (detection rate according to Viastra(R)), a theoretical specificity of 99.9% and a prevalence of 1/300,000 would yield a positive predictive value of 0.3%. Therefore even if the test detects 92 out of 100 cases of Costello syndrome, from the 100 positive cases, only 0.3 would have the disease. These results have tremendous importance for decision-making in prenatal testing. If a parent finds out that his fetus tested positive, in this case for Costello's disease, she/he is faced with difficult decisions regarding the continuation of the pregnancy, especially if the pregnancy is close to the cut-off from which abortion is not allowed at the request of the mother. In these instances, correct information regarding the actual chances of the child having the disease is paramount, and confirmatory testing should be highly recommended to prove the presence of the disease.

2. HIGH-RISK VERSUS LOW-RISK POPULATIONS

Currently, the joint ESHG/ASHG position document with recommendations regarding responsible innovation in prenatal screening with NIPT includes different scenarios for using NIPT in common autosomal aneuploidies (Dondorp et al., 2015), namely:

(1) NIPT as the second test after combined first-trimester screening (cFTS) using current risk cut-offs. The use of NIPT as a second test decreases the need for invasive follow-up testing, reducing the risks associated with more invasive procedures. However, its use does not increase the detection rate significantly; this would make the association cFTS+NIPT to have an overall decreased detection rate compared to cFTS+invasive procedures. Additionally, some patients would also need additional invasive confirmation procedures, increasing the required steps to reach a positive result (Dondorp et al., 2015).

(2) NIPT as a replacement for cFTS. This approach has as the main advantages an increased detection rate for the considered aneuploidies, a decreased number of false positives, and an earlier response (NIPT can be done from

9 to 10 weeks). As disadvantages: (1) due to a low prevalence of these diseases in the general population, the tests have little predictive value, therefore needing, for confirmation, an invasive analysis; (2) cFTS also give information about other potential diseases, information that cannot be, at this moment, adequately supplied by the NIPT; (3) the cost of NIPT is much higher, causing a decreased addressability (if paid by the patients), or increased expenditure for the health-care system (which at the moment is not entirely justified) (Dondorp et al., 2015).

(3) NIPT as a second test after lowering cFTS cut-off value. This approach will increase detection rate associated with a decrease in the overall costs of introducing NIPT (Dondorp et al., 2015).

A similar approach can be used for less rare monogenic disorders, with the mention that both sensitivity and specificity should be close to 100%, as is currently the case for some aneuploidies such as Down's syndrome. Otherwise, due to the low predictive value, the usefulness of the test would be limited, even as a screening procedure, making it difficult to recommend due to the cost/benefit ratio.

Bianchi et al. (2014a) tried to summarize the main arguments for and against using NIPT in low-risk populations, with an emphasis on testing for Down's syndrome, the leading arguments being presented in the Table 2 (modified). Details about prenatal testing in low-risk populations will be given in another chapter of this book.

3.   CONSENT AND DECISION-SHARING

Consent in prenatal diagnosis should encompass at least three types of information: (1) the description of the analyzed disorders, (2) a description of the characteristics of the prenatal test, and (3) potential implications of the results (Newson, 2008). The primary risk associated with invasive, fetal diagnostic procedures is represented by the threat of miscarriage (van den Heuvel et al., 2010). As NIPT removes this risk, many physicians tend to superficialize the content of the informed consent, failing to provide relevant data about other potential risks, such as the possibility of terminating the pregnancy if the testing is positive (van den Heuvel et al., 2010), the actual meaning of a positive test, and what it actually means living with a child with that particular genetic condition. Various studies suggested or proved a critical disparity between the main components of making an informed choice, namely obtaining relevant information, understanding it, and making an informed decision (Green, 2004; van den Berg et al., 2005; van den Heuvel et al., 2010; Newson, 2008). van den Berg et al. (2005), for example, showed that even though a high percentage of subjects from their study group considered to have made an informed choice regarding prenatal screening, the decisions were uninformed due to value inconsistency, insufficient knowledge or both. A systematic review testing the understanding

TABLE 2 Arguments for and Against Using NIPT in Low-Risk Populations

Arguments Favoring NIPT in Low-Risk Populations	Arguments Against NIPT in Low-Risk Populations
• Prenatal testing is already acceptable in many countries; NIPT is just another type of prenatal testing	• It may increase the anxiety of parents, which could be unjustified in the case of a false positive result (more frequent in low-risk populations, and in diseases with a very low prevalence); therefore, it might cause maleficence
• It allows the respect for procreative autonomy	• There are still some unresolved issues regarding using NIPT in low-risk populations, such as an analysis of test failure rates, samples with low fetal percentage, and so on. Before these issues are addressed, we cannot be entirely sure of the actual usefulness of NIPT in low-risk populations
• It respects the principle of beneficence (in relation to the mother), as it will enable the birth of nondiseased children	
• Invasive procedures are associated with various risks, such as miscarriage, infection, and so on. These odds are decreased by NIPT (therefore respecting the principle of nonmaleficence). The risks are not entirely removed, as a positive result usually needs confirmation through invasive measures (see above)	• The testing methodology is not yet standardized—e.g., different companies offer different tests performed using various techniques, such as whole genome sequencing, selective amplification before sequencing, with distinct advantages and disadvantages. This makes extremely difficult to give parents proper information, and therefore might alter the decision-making process
• The testing can allow an abortion earlier compared to many invasive diagnostic methods	• There are no extensive cost-benefit analyses performed on low-risk populations. Therefore, the feasibility of using NIPT as a screening procedure cannot be assessed from a resource allocation perspective
• For some disorders, it has sensitivity and specificity close to 100% (especially for common aneuploidies)	• If the technique is not reimbursed, due to its high costs, it could generate injustice, as the method could be accessed only by the wealthier; they would have a decreased risk of having children with fetal abnormalities, or a lower risk of miscarriage compared to those not so fortunate
	• As NIPT is a genetic test, pre- and posttest counseling should be offered, which is quite tricky taking into account the rapid pace of development in the field, the exponential increase in usage, and a relatively low number of genetic counselors (an issue present inclusive in more developed countries)
	• Based on the results of the NIPT screening, and taking into account information and understanding generated difficulties, many parents would base their decision to abort based on this initial screening, therefore existing the risk of losing a high number of nondiseased fetuses

Modified from Bianchi, D.W., Oepkes, D., Ghidini, A., 2014a. Current controversies in prenatal diagnosis 1: should noninvasive DNA testing be the standard screening test for down syndrome in all pregnant women? Prenat. Diagn. 34, 6–11.

regarding the participation in genetic screening of pregnant women and newborns suggested that often the levels of expertise needed for adequate decision-making were not achieved; therefore, even if information was provided correctly, it was not adequately understood by the parents, being only superficially acquired and not retained (Green, 2004). Therefore, to allow a truly informed choice, physicians (and counselors) should augment the level of understanding of the patients. Some methods, such as an extensive use of leaflets or videos seem to help (Green, 2004), but even they do not allow a proper, relevant understanding required to make a choice that is truly informed. Additionally, NIPT poses significant difficulties in providing correct information to the patients; Minear et al. (2015) identified three main issues, namely: (1) giving reliable and accurate information is tough, taking into account the rapid pace of progress, and the fact that NIPT began to be also used in lower risk population and for additional diseases; (2) both patient and physician education regarding NIPT (and especially recent developments) needs proper materials/educators, which are difficult to find; and (3) rapid commercialization of NIPTs have augmented the challenges posed to informed decision-making, some tests being made readily available to prospective customers before the clinical validation of the data.

The informed consent form should include at least the information presented in Table 3.

4. COUNSELING FOR NIPT

There are two main types of counseling related to NIPT, which will be analyzed separately: pre- and posttest counseling.

Pretest counseling. The primary goal of pretest counseling is to give the parents enough relevant information to make an informed decision regarding the decision of whether to accept or not the testing. Additionally, counselors should make sure that the parents correctly understood relevant information, and made the decision accordingly. Pretest counseling contains overlapping data with the one needed before signing the informed consent. After the initial positive results with NIPT, physicians believed that this technique would ease the information process, as they would not have to explain to patients all the possible screening tests and their usefulness, which could be replaced with simple messages such as: "If you want to know about trisomy 21, we take a tube of blood and let you know in a week or so whether your baby is affected" (Oepkes et al., 2014). This approach was soon proven to be overly optimistic, as NIPT complexity increased significantly both regarding the actual technique, and in the number of disorders potentially identified (Oepkes et al., 2014). Some common questions arising from pretest counseling for trisomy 21 are presented in Table 4 (with permission from Oepkes et al., 2014).

Posttest counseling should focus on issues such as the possibility of a false positive result and what would it mean to the patient, and the fact that a positive

TABLE 3 Informed Consent Form for NIPT

Type of Information	Details
Information about NIPT	<ul><li>It is a screening test. A positive result should be followed by a confirmation test</li><li>Can be performed after 9 weeks (approximately, depending on the technique)</li><li>An interpretation of the results, which should not suggest the fact that it diagnoses a condition, but rather that shows a risk profile</li><li>Details about the actual technique used (e.g., whole genome sequencing), and a summary of the method (including the type of biological sampling, where will the samples be analyzed and interpreted)</li><li>Details about the limits of the technique, for example, will only detect certain genetic conditions, will not recognize balanced chromosomal rearrangements, other genetic disorders, birth defects, will yield no result if the quantity of free fetal DNA is too low, it might not detect mosaicism, the fact that a triploidy might not be discovered in case of a vanishing fetus or a twin gestation</li><li>It could yield information about the genetic profile of the father or other relatives</li><li>The costs (if incurred by the parents)</li></ul>
Information about tested disorders	<ul><li>Which disorders are screened, and a summary of the main signs/symptoms</li><li>Sensitivity, specificity, and positive predictive value (mandatory)</li><li>Information about disclosing the genetic sex of the fetus. Moreover, if the prospective parents choose not to receive this information, but the test identifies a sex-chromosome condition, she/he should be informed that data about the sex should be brought to their attention</li></ul>
Information about alternatives	<ul><li>Other screening procedures</li><li>Confirmatory procedures could be needed (and which are they)</li><li>The possibility of not performing the procedure</li></ul>
Information about the potential implications of the results	<ul><li>A positive result needs a confirmatory procedure, associated with risks such as miscarriage, infection, infertility, and so on</li><li>Information about posttest counseling, support groups, and so on</li><li>Other test related risks</li><li>Management of unexpected/incidental findings</li></ul>
Data confidentiality issues	<ul><li>How is the data stored, who has access to it, what secondary uses of the data are allowed (e.g., Scientific research), and in which conditions</li></ul>
Information about biological sampling storage	<ul><li>How will the samples be stored, for how much time, how will they be disposed of, and so on</li></ul>

TABLE 4 Common Pretest Counseling Questions Related to Trisomy 21

- What are trisomies 21, 18, and 13?
- What are risks for these anomalies in general?
- What are the a priori risks for trisomy for this individual patient?
- In case of a fetal trisomy, what are the chances of spontaneous miscarriage/perinatal demise from the time of testing onwards?
- What are the remaining risks for other significant fetal anomalies when trisomies are excluded?
- What are the chances of detecting other fetal anomalies using (routine) ultrasound in the first and second trimesters?
- From what testing options can the pregnant woman choose?
- How accurate is each of these tests; What are the benefits, limitations, and risks?
- What are the chances of a failed test, of the need for redraw and a uninterpretable result?
- When is the result available?
- How is the result communicated (negative and positive, high/low probability, risk score)?
- What does it mean when the result is positive?
- What are the odds of being affected given a positive result?
- What are the options when the test is positive?
- What are the options when the diagnosis of trisomy is confirmed?
- What is the remaining risk for trisomy when NIPT is negative?
- Is the sex of the fetus tested, is the result communicated, and how reliable is this?
- Can other abnormalities besides trisomies 21, 18, and 13 be found by NIPT and, if so, will they be communicated?
- Can abnormalities in maternal DNA be detected and, if so, will they be notified?

NIPT should always be followed by an invasive, diagnostic procedure. The parents should never be offered the option of termination without a confirmation from a positive diagnostic test. Moreover, we would argue that termination should be recommended only in very particular circumstances, such as the presence of a high risk for the mother, and not for the existence of congenital pathologies of the fetus. The main reason for this recommendation resides in the fact that the physician/counselor initiates a primary relationship with the parent/ parents, but secondary, and intrinsically, with the fetus (Dal Pozzo and Marsh, 1987). Even if the primary duties of the physicians are toward the primary "client," they also have an indirect responsibility to protect the fetus, based on principles such as respect for the embryo (Warnock, 1987) and/or later ontogenetically, on the right to be born of the fetus, and to respect the telos of the fetus. Based on this secondary relationship, physicians should either refrain in recommending termination, adopting a nondirectiveness approach (irrespective of the severity of the genetic condition), or recommend continuation of the pregnancy (only if the congenital disease has a minor to moderate severity (such as polydactyly, Down syndrome, single congenital kidney, and so on) (Hostiuc et al., 2013). To be noted that a few studies showed that both patients and physicians would prefer a shared decision-making approach, at least for some conditions. For example, Légaré et al. (2011) in a survey comprising 109 pregnant women and 41 family physicians, showed that, overall, both groups wished to engage in shared decision-making regarding prenatal Down syndrome screening. However, this approach is most likely generated by issues that undermine the capacity of the patient to act voluntarily, such as an extreme anxiety associated with a potentially severe diagnosis, or a lack of adequately understanding complex medical information by the parents.

5. SHOULD PARENTS BE ALLOWED TO GET TESTED "FOR INFORMATION ONLY"?

There are many instances in which parents do not want to terminate the pregnancy if the tests come positive for a particular disease. Some parents would like to know the gender of the baby or have time to prepare for a potentially harmful outcome, or just to screen the baby for as many potentially adverse conditions as possible (Deans et al., 2013). The "get tested for information only" is more frequently associated with NIPT compared to other methods, due to its sensitivity and lack of medical risks for the mother and the fetus. Nowadays, NIPT is mainly used in clinical practice for testing chromosomal abnormalities and some monogenic disorders. However, with the recent developments of genome-wide sequencing techniques, in a not so distant future, the parents may have the genetic profile of the fetus from the first weeks of life, and could choose to terminate the pregnancy not only based on the presence of severe diseases, but also on the risk profiles for multigenic disorders, or even depending on the presence/absence of physical, intellectual, or psychological traits, in which case we should question whether this could be morally acceptable. The main argument for allowing it would be procreative autonomy—the parents are the ones who should choose the conditions in which to have and to continue a pregnancy. Even nowadays, giving birth is a process of selection—selecting a suitable partner, with specific genetic traits such as hair color, constitution, intelligence, selecting only some of the embryos during in vitro fertilization techniques, or selecting to terminate a pregnancy if some severe diseases are suspected/confirmed. Why not allow the parents to make a truly informed decision regarding the genetic profile of their infant? Another argument favoring the procedure is beneficence. By allowing parents to select a child with a favorable genetic profile, he/she will have a decreased risk of getting sick, or will be more resistant to environmental factors, or to infectious agents; she/he will get better grades, go to a better university, or get wealthier. As this approach would decrease the number of persons with severe diseases or would cause them to appear later in life, or have less severe phenotypes, it could also decrease health care-related costs in the long term, making it much more cost-efficient when compared to other preventive measures. Even if it has potentially significant short-term and long-term benefits, testing children prenatally for a wide array of disorders/risk factors has three main negative consequences. They could lead to negative discrimination, could alter the telos of the infant, and can be considered a eugenic procedure.

6.   A TELEOLOGICAL APPROACH TO NIPT

Teleology is the science (logos) of the end goals. According to Aristotle, every natural entity has a final, intrinsic purpose. For example, the fundamental purpose of an acorn is to become a majestic oak tree (Ginsborg, 2004). Telos, or finality, from a Kantian perspective, should be generated intrinsically, and any action diverting a person from reaching the wished telos is immoral (the moral agents will not be any more ends in themselves, but instead means to other ends). Telos in prenatal bioethics is inherently linked with the moral status of the embryo and subsequently is analyzed in relation to the ethical acceptability of action directed against the well-being of the embryo (Hostiuc, 2014). For example, Moreno, on a critique of a paper written by Hyun and Jung, when discussing the use of human blastocysts created by somatic cell nuclear transfer (SCNT), said: "since what it is to be an embryo includes a certain 'teleological trajectory,' which includes being organized to develop further—and since the products of SCNT appear to lack this telos, they are not really human embryos. In fact, they really should not even be called 'cloned human embryos,' insofar as that suggests that they have the same telic properties as the original cell" (Moreno, 2006). Pellegrino, while performing an analysis of the moral acceptability of in vitro fertilization, argued that "Calling the embryo by another name does not alter the objective fact of interrupting a built-in process of development whose telos is a full-grown human being" (Pellegrino, 1999). If a being has moral status (or, by using the Kantian terminology, is a moral agent), we have to respect its right to act autonomously (in the present or the future). Regarding the moment when an embryo acquires a moral status, there are more than a few theories (for details see the corresponding chapter of this book, or Hostiuc, 2014). The Kantian theory states that the moral status is a fundamental attribute of any moral agent (Hostiuc, 2014)—according to Kant "respect is always directed toward persons and never toward things" (Kant and Braileanu, 1930). Moral agents, according to the Kantian theory, are represented by normal adults, capable of self-governing in moral issues, or autonomous (Schneewind, 1992). Autonomy contains two fundamental elements: (1) a lack of external controls, telling us what morality is, and (2) the fact that through self-governance we can control ourselves (Schneewind, 1992). The Kantian theory grants all adults, irrespective of their gender, race, or other qualities, full autonomy, a full moral status. However, as it links it to rationality, it considers that all persons not capable of moral reasoning to be outside this concept, included here being infants, embryos, animals, or persons with mental/intellectual deficits that render them unable to reason. However, even if we were not to consider embryos as moral agents, and we do not have the moral duties generated by this status, we should still respect them, based on what Steinbock calls "Kantian respect" (Steinbock, 2007), which is "made up of constraints on our behaviors toward others that spring from our recognition of others as mature agents on an equal moral footing with ourselves" (Quinn, 1984). This Kantian respect should allow the possibility of the future moral agent to choose his telos—some may want to be aviators, or physicians, or musicians. Nobody should interfere with his or her free choice, based on this principle of Kantian respect. However, when we start using prenatal testing techniques, we are doing just that, especially if we were to extend the ethical acceptability of prenatal testing to nondisease-related characteristics. By providing our child with a selected genetic profile, we are actually interfering with his/her free choice. If we give him "good" athletes' genes, he could be less likely to become a scientist or a musician. Parents, on the other side, wish to have the best babies they possibly can, and their reproductive autonomy should also be respected.

We must take into account that our views about what is right for the child and what is not, what traits are beneficial and what is not, is mainly a social construct, based on probabilities and stereotypes. We believe that is better to be tall, or intelligent, or have blue eyes because this is what society tells us, or because we saw people with specific characteristics be better in life. However, each life is unique, and we should not make decisions regarding a particular individual solely based on probabilities.

By choosing to act (by selecting embryos through IVF procedures, or by performing abortive procedures in case of undesired genetic profiles), we generate a conflict between procreative autonomy and fetal telos. This conflict could be solved by using the following algorithm: (1) if prenatal testing shows significant, health-related issues, procreative autonomy should prevail, and parents should be given the choice to act according to their wishes, fully knowing the medical details (physicians should however refrain in recommending abortion at any costs); (2) if prenatal testing shows significant risks for severe child-onset diseases, procreative autonomy should prevail, and parents should be given the choice to act according to their wishes; however, physicians could recommend nonabortion if a shared decision-making process is permissible in that particular physician-patient relationship; (3) if prenatal testing shows significant risk for severe adult-onset diseases (such as Huntington's disease), procreative autonomy should still prevail, but physicians could refuse to recommend or act toward a negative course of action for the baby (such as abortion), based on the conscience clause; (4) if prenatal testing shows mild-moderate risks for either child-onset or adult-onset disorders, or if parents want information that could be used for eugenic/enhancement purposes, telos should prevail over procreative autonomy, and physicians should refuse to perform the test—a selective selection of an embryo, or an abortion based on this information would certainly generate maleficence, and could significantly alter the telos, with physicians being directly involved in the selection process.

7.   NIPT AND DISCRIMINATION

NIPT can cause negative discrimination both through its use and its lack of use. By performing NIPT, the parents can select to continue or not the pregnancy, depending on various genetic traits. Termination of pregnancy caused by the presence of genetic characteristics has been considered discriminatory, based on two main arguments: the expressivist argument, and the parental attitude argument. According to the expressivist argument, prenatal testing aimed at disabling characteristics generates a negative view of persons living with that disability. Asch said, for example, that "As with discrimination more generally, with prenatal diagnosis, a single trait stands in for the whole, the trait obliterates the whole. With both discrimination and prenatal diagnosis, nobody finds out about the rest. The tests send the message that there's no need to find out about the rest" (2000). Mary Johnson, similarly, argued that "A decision to abort based on the fact that the child is going to have specific individual characteristics such as mental retardation, or in the case of cystic fibrosis, a build-up of mucus in the lungs, says that those characteristics take precedence over living itself, that they are so important and so negative, that they overpower any positive qualities there might be in being alive" (1990). There are many counterarguments to the expressivist view, including: (1) the fact that prospective parents may not have the financial or emotional means of supporting a child with a disability (Parens and Asch, 2000), (2) there could be other family members with the same disability, further stretching the emotional and financial burdens (Parens and Asch, 2000), (3) one or both parents could have the same disability, and do not want their offsprings to live with it (Parens and Asch, 2000); (4) the any-particular distinction, according to which, if parents are allowed to make abortion-related choices based on nonmedical reasons (the prospective mother is too young, the prospective parents are emotionally immature, they have other obligations), why would one not be allowed an abortion for a medical, objective reason? (Parens and Asch, 2003), and (5) if other proactive measures, aimed at minimizing the risk of giving birth to diseased children, such as not taking Thalidomide, or taking folic acid are allowed, and not seen as morally problematic (one could argue that, on the contrary, performing such tasks would be considered morally problematic), why should prenatal testing be considered morally problematic? (Parens and Asch, 2003). According to the parental attitude argument, using prenatal testing as a selection method by the parents is discriminatory based on a specific parental attitude, according to which they want to have a perfect child (Parens and Asch, 2003). Again Asch stated that: "When prospective parents select against a fetus because of predicted disability, they are making an unfortunate, often misinformed decision that a disabled child will not fulfill what most people seek in child rearing, namely, 'to give ourselves to a new being who starts out with the best we can give, and who will enrich us, gladden others, contribute to the world, and make us proud'" (2002). The genetic traits of the fetus are however not the only element shaping the development of children—education, parental care, and social environment play essential roles as well. This is especially important if discussing NIPT for multifactorial disorders, in which interactions between genetics and the environment are essential in the phenotypic manifestation of a particular trait.

Another critical issue that has to be noted is that a positive NIPT result needs to be confirmed through more invasive measures, which sometimes can only be done after the legal limit for abortion at the request of the mother. In this case, many legislatures would allow termination of the disease if severe, or significant, but often without clearly establishing the limit. Without clear regulations, the analysis of seriousness of the disorder is left at the disposal of the physicians, who view it very differently. Savulescu, for example, showed that some Australian physicians view cleft palate as a major disorder, which could warrant termination, while others saw it as a minor one (Savulescu, 2001a). Another study, performed in Romania, showed that, when asked to give an acceptability score to various abortion-related procedures, physicians gave an average score (scale 1–10, 1 meaning least severe and 10—most severe) of 1.55 if prenatal testing has identified morphological abnormalities, such as polydactyly, 6.55 for moderate disorders such as Down syndrome, and 9.34 for extremely severe disorders, such as anencephaly (Hostiuc et al., 2013). If for both minor and extremely severe abnormalities the results were unequivocal, this was not the case for moderate disorders. For Down syndrome, the same study showed that 25% of all respondents gave an acceptability score below or equal to 4 for the termination, suggesting a low acceptability of the procedure, and another 25% a score above 8, suggesting a high acceptability of the procedure (Hostiuc et al., 2013). This variability could generate inequalities in access to a medical procedure, which would contradict the ethical principle of justice. To minimize this problem, the seriousness of the disorder, which by itself would allow termination of pregnancy for medical reasons, should be better defined. Again Savulescu discussed, in a highly debated article, the existence of "a right not to be born": "There may be a few conditions which afflict humans which are so terrible that they are comparable to death. People who are permanently unconscious in a persistent vegetative state have life-prolonging medical treatment withdrawn on the basis that continued treatment is not in their interests. Permanent unconsciousness is like death. If one had a life of constant pain, limitation of movement and sensation, and severe impairment of the ability to interact with other people the world might be worse than death. But mild to moderate intellectual disability is not such a condition" (2002). According to this approach, if the survival or quality of life would be minimal at best, termination of pregnancy could be allowed, for medical reasons. However, in this category should not be included diseases causing mild–moderate disabilities (such as Down syndrome), or morphological disorders that could be treated surgically (such as cleft palate).

Another facet of NIPT-associated discrimination is generated by the high costs (still) of the procedure, especially when testing for a broader array of disorders. As the costs are high, and in many instances, they are not reimbursed, the availability of the procedure is limited. Therefore people who can afford the technique will have access to it and have more data regarding the embryo/fetus based on which to enforce their procreative autonomy. They will be able to make more informed choices regarding the continuation/termination of pregnancy, will have, as a mean, fewer children with significant disabilities or even will have the choice to select children with desired traits, which could generate further social inequalities.

8. EUGENICS, PROCREATIVE BENEFICENCE, AND NIPT

According to the principle of procreative beneficence, "couples (or single reproducers) should select the child, of the possible children they could have, who is expected to have the best life, or at least as good a life as the others, based on the relevant, available information"(Savulescu, 2001b). Savulescu considers that the selection for the best possible genetic profile of the offspring is morally required, based on this principle, and moral persuasion could be allowed in order to direct parents toward this approach, respecting three fundamental principles: information (obtained through prenatal genetic diagnosis and prenatal testing), free choice of which child to have, and noncoercive advice regarding which child will be expected to be born depending on the best chances of having the best life (Savulescu, 2001b). The author has tried to differentiate procreative beneficence from eugenics, by saying: "Procreative beneficence is different to eugenics. Eugenics is selective breeding to produce a better *population*. A *public interest* justification for interfering in reproduction is different from procreative beneficence, which aims at producing the best child, of the possible children couple could have. That is an essentially private enterprise. It was the eugenics movement itself that sought to influence reproduction, through involuntary sterilization, to promote social goods" (Savulescu, 2001b). Other authors have even extended this principle, emphasizing the social value of the children. For example, Elster considered that procreative beneficence should be replaced with a more general principle, entitled "General Procreative Beneficence," according to whom "If couples (or single reproducers) have decided to have a child, and selection is possible, then they have a significant moral reason to select the child, of the possible children they could have, whose life will maximize the expected overall value in the world" (Elster, 2011).

However, procreative beneficence received counterarguments from many parts, mainly because it was considered to be strongly eugenic. Bennet, for example, argued that "Based on this definition eugenics is a 'public interest justification for interfering with reproduction' 'to promote social goods.' As we have seen the establishment of a moral obligation to bring to birth the best child we can is not built on the private interests of the prospective parents (…) *(but is)* built on is an idea of making the world a better place than it could otherwise have been, not in terms of any individual person's welfare, but in terms of creating the greatest total score for what is regarded as the goods of life. If a project is not interested in the welfare of particular people but creating what those proposing this project believe is the best world possible, then this is exactly what eugenics is—promoting social and not personal goods. It is true that those proposing this obligation to create only the best children do not advocate coercion at the level of forced sterilizations in order to achieve this end but it is not true that no coercion at all is implied. As we have seen establishing a strong case for a moral obligation to bring to birth the best children we can inevitably puts pressure on people to this end" (2009). Stoller argued that Savulescu's approach is strictly consequentialist and that other approaches to morality, such as deontological or virtue ethics will lead to different conclusions: "In the area of procreation, then, when we focus solely on consequences, we ignore key components of our sense of morality. In most contexts, when dealing with a person whose identity is already determined, consequentialist accounts may provide moral answers and possibly encompass deontological and virtue ethics concerns along the way. When, however, we engage in the activity of creating life itself, and the question is not how to treat a particular person but which person if any to create, consequentialist thinking fails to capture all the relevant facets of the issue. Indeed, in the area of procreation, acting beneficently may not require the consequentialist demand of creating the best children; it may instead demand recognition of deontological and virtue ethics as an appropriate moral lens" (2008). Hotke refutes the principle of procreative beneficence based on its premise, namely that morality requires us to act rationally: "It is untenable to claim that morality requires us to do what we have most reason to do. If this were true, then many acts we believe to be permitted are in fact morally wrong. For any act that we do, if there is some other act that there was more reason to do, then the act that we do is morally wrong. For instance, if I choose to forgo breakfast rather than eat breakfast, then assuming there was more reason to eat breakfast than not to eat breakfast, I have acted morally wrongly; I ought to have eaten breakfast. This conclusion is absurd" (2014). As seen above, both parties have important arguments for and against procreative beneficence. However, from a clinician's point of view, things are much simpler. As stated before, we believe that physicians should fully respect the procreative autonomy of the prospective parents. This respect should allow a proper information of all relevant issues, and nondirectiveness, unless there are significant, or serious potential consequences generated by going this way. They should also respect the right of the child to an open future, and not interfere with its potential telos. If the prospective parents would like to act upon the principle of procreative beneficence, it is their right, based on procreative autonomy (within some limits, such as selecting the sex of the embryo, which should only be allowed if gender-related genetic conditions are identified). But physicians should not be involved in it, except from a purely technical point of view—they can perform procedures, such as NIPT, and can give objective, nondirective information, but should never recommend termination of pregnancy based on the principle of procreative beneficence. Even more, the conscience clause should include prenatal testing, and physicians should refuse NIPT, based on it, if they think that prospective parents interfere significantly, and based on nonmedical reasons, with the telos of the fetus.

9. CONCLUSIONS

NIPT has the potential to become the de facto standard in prenatal testing in the next few years. However, there are still a lot of unresolved issues that needs to be addressed before allowing its indiscriminate use in clinical practice.

REFERENCES

Asch, A., 2002. Disability equality and prenatal testing: Contradictory or compatible. Fla. St. Univ. Law Rev. 30, 315.

Bennett, R., 2009. The fallacy of the principle of procreative beneficence. Bioethics 23, 265–273.

Bianchi, D.W., Wilkins-Haug, L., 2014. Integration of noninvasive Dna testing for aneuploidy into prenatal care: what has happened since the rubber met the road? Clin. Chem. 60, 78–87.

Bianchi, D.W., Oepkes, D., Ghidini, A., 2014a. Current controversies in prenatal diagnosis 1: should noninvasive DNA testing be the standard screening test for down syndrome in all pregnant women? Prenat. Diagn. 34, 6–11.

Bianchi, D.W., Parker, R.L., Wentworth, J., Madankumar, R., Saffer, C., Das, A.F., Craig, J.A., Chudova, D.I., Devers, P.L., Jones, K.W., 2014b. DNA sequencing versus standard prenatal aneuploidy screening. N. Engl. J. Med. 370, 799–808.

Chan, Y.M., Leung, T.Y., Chan, O.K.C., Cheng, Y.K.Y., Sahota, D.S., 2014. Patient's choice between a non-invasive prenatal test and invasive prenatal diagnosis based on test accuracy. Fetal Diagn. Ther. 35, 193–198.

Cram, D.S., Zhou, D., 2016. Next generation sequencing: coping with rare genetic diseases in China. Intractable Rare Dis. Res. 5, 140–144.

Dal Pozzo, E.E., Marsh, F.H., 1987. Psychosis and pregnancy: some new ethical and legal dilemmas for the physician. Am. J. Obstet. Gynecol. 156, 425–427.

De Jong, A., Dondorp, W.J., Frints, S.G., De Die-Smulders, C.E., De Wert, G.M., 2011. Advances in prenatal screening: the ethical dimension. Nat. Rev. Genet. 12, 657.

Deans, Z., Hill, M., Chitty, L.S., Lewis, C., 2013. Non-invasive prenatal testing for single gene disorders: exploring the ethics. Eur. J. Hum. Genet. 21, 713–718.

Dondorp, W., De Wert, G., Bombard, Y., Bianchi, D.W., Bergmann, C., Borry, P., Chitty, L.S., Fellmann, F., Forzano, F., Hall, A., 2015. Non-invasive prenatal testing for aneuploidy and beyond: challenges of responsible innovation in prenatal screening. Eur. J. Hum. Genet. 23, 1438.

Elster, J., 2011. Procreative beneficence–Cui Bono? Bioethics 25, 482–488.

Gil, M., Akolekar, R., Quezada, M., Bregant, B., Nicolaides, K., 2014. Analysis of cell-free DNA in maternal blood in screening for aneuploidies: meta-analysis. Fetal Diagn. Ther. 35, 156–173.

Ginsborg, H., 2004. Two kinds of mechanical inexplicability in Kant and Aristotle. J. Hist. Philos. 42, 33–65.

Green, J.M., 2004. Psychosocial Aspects of Genetic Screening of Pregnant Women and Newborns: A Systematic Review. York Publishing on behalf of NCCHTA, York.

Hill, M., Finning, K., Martin, P., Hogg, J., Meaney, C., Norbury, G., Daniels, G., Chitty, L., 2011. Non-invasive prenatal determination of fetal sex: translating research into clinical practice. Clin. Genet. 80, 68–75.

Hill, M., Fisher, J., Chitty, L.S., Morris, S., 2013. Women's and health professionals' preferences for prenatal tests for down syndrome: a discrete choice experiment to contrast noninvasive prenatal diagnosis with current invasive tests. Obstet. Gynecol. Surv. 68, 171–173.

Hostiuc, S., 2014. Moral status of the embryo. Clinical and legal consequences. Gineco. eu 10, 102–104.

Hostiuc, S., Buda, O., Hostiuc, M., 2013. Late abortion. Attitudes amongst young physicians in Romania. Arch. Gynecol. Obstet. 288, 431–437.

Hotke, A., 2014. The principle of procreative beneficence: Old arguments and a new challenge. Bioethics 28, 255–262.

Johnson, M., 1990. Aborting defective foetuses-what will it do. Link Disabil. J. 14, 172–175.

Kant, I., Braileanu, T., 1930. Critica Ratiunii Pure. Editura Casei Şcoalelor.

Korpi-Steiner, N., Chiu, R.W.K., Chandrasekharan, S., Chitty, L.S., Evans, M.I., Jackson, J.A., Palomaki, G.E., 2017. Emerging considerations for noninvasive prenatal testing. Clin. Chem. 63, 946–953.

Kuppermann, M., Nease, R.F., Learman, L.A., Gates, E., Blumberg, B., Washington, A.E., 2000. Procedure-related miscarriages and Down syndrome–affected births: implications for prenatal testing based on women's preferences. Obstet. Gynecol. 96, 511–516.

Légaré, F., St-Jacques, S., Gagnon, S., Njoya, M., Brisson, M., Frémont, P., Rousseau, F., 2011. Prenatal screening for down syndrome: a survey of willingness in women and family physicians to engage in shared decision-making. Prenat. Diagn. 31, 319–326.

Lo, Y.D., Tein, M.S., Lau, T.K., Haines, C.J., Leung, T.N., Poon, P.M., Wainscoat, J.S., Johnson, P.J., Chang, A.M., Hjelm, N.M., 1998. Quantitative analysis of fetal DNA in maternal plasma and serum: implications for noninvasive prenatal diagnosis. Am. J. Hum. Genet. 62, 768–775.

Minear, M.A., Alessi, S., Allyse, M., Michie, M., Chandrasekharan, S., 2015. Noninvasive prenatal genetic testing: current and emerging ethical, legal, and social issues. Annu. Rev. Genomics Hum. Genet. 16, 369–398.

Moreno, J.D., 2006. The name of the embryo. Hast. Cent. Rep. 36, 3.

Natera 2017 http://www.Natera.com/ 2017, Oct 11.

Newson, A.J., 2008. Ethical aspects arising from non-invasive fetal diagnosis. Semin. Fetal Neonatal Med. 13, 103–108. Elsevier.

Oepkes, D., Yaron, Y., Kozlowski, P., Rego De Sousa, M., Bartha, J., Van Den Akker, E., Dornan, S., Krampl-Bettelheim, E., Schmid, M., Wielgos, M., 2014. Counseling for non-invasive prenatal testing (NIPT): what pregnant women may want to know. Ultrasound Obstet. Gynecol. 44, 1–5.

Parens, E., Asch, A., 2000. Prenatal Testing and Disability Rights. Georgetown University Press, Washington, DC.

Parens, E., Asch, A., 2003. Disability rights critique of prenatal genetic testing: Reflections and recommendations. Ment. Retard. Dev. Disabil. Res. Rev. 9, 40–47.

Pellegrino, E.D., 1999. The pre-embryo: an illusory category of convenience. Pediatr. Rev. 20, E32–E34.

Quinn, W., 1984. Abortion: identity and loss. Philos. Public Aff. 13, 24–54.

Savulescu, J., 2001a. Is current practice around late termination of pregnancy eugenic and discriminatory? Maternal interests and abortion. J. Med. Ethics 27, 165–171.

Savulescu, J., 2001b. Procreative beneficence: why we should select the best children. Bioethics 15, 413–426.

Savulescu, J., 2002. Is there a "right not to be born"? Reproductive decision making, options and the right to information. Bmj Publishing Group.

Schneewind, J.B., 1992. Autonomy Obligation and Virtue: An Overview of Kant's Moral Philosophy. Cambridge University Press, Cambridge.

Skirton, H., Goldsmith, L., Jackson, L., Lewis, C., Chitty, L., 2014. Offering prenatal diagnostic tests: European guidelines for clinical practice. Eur. J. Hum. Genet. 22, 580–586.

Steinbock, B., 2007. Moral status, moral value, and human embryos: implications for stem cell research. In: Steinbock, B. (Ed.), The Oxford Handbook of Bioethics. Oxford University Press.

Stoller, S.E., 2008. Why we are not morally required to select the best children: a response to Savulescu. Bioethics 22, 364–369.

Traeger-Synodinos, J., 2006. Real-time PCR for prenatal and preimplantation genetic diagnosis of monogenic diseases. Mol. Asp. Med. 27, 176–191.

Van Den Berg, M., Timmermans, D.R., Ten Kate, L.P., Van Vugt, J.M., Van Der Wal, G., 2005. Are pregnant women making informed choices about prenatal screening? Genet. Med. 7, 332–338.

Van Den Heuvel, A., Chitty, L., Dormandy, E., Newson, A., Deans, Z., Attwood, S., Haynes, S., Marteau, T.M., 2010. Will the introduction of non-invasive prenatal diagnostic testing erode informed choices? An experimental study of health care professionals. Patient Educ. Couns. 78, 24–28.

Warnock, M., 1987. Do human cells have rights? Bioethics 1, 1–14.

Chapter 14

Prenatal Testing in Low-Risk Populations: A US Perspective

Jazmine L. Gabriel* and Lauren Diskin†
*WellSpan York Cancer Center, York, PA, United States, †Obstetrics and Gynecology, WellSpan York Hospital, York, PA, United States

1. INTRODUCTION

In 2011, cell-free fetal DNA testing, also called noninvasive prenatal testing or screening (NIPT or NIPS), became clinically available in the United States as a screening test for fetal aneuploidy (i.e., missing or extra chromosomes, such as in Down syndrome). This screening test makes use of the discovery that fragments of cell-free placental DNA can be found in the maternal blood stream (Lo et al., 1997). Since the placenta and fetus share the same genetic material, except in rare instances, this genetic material can be analyzed to detect fetal aneuploidy as early as 9–10 weeks into pregnancy [American College of Obstetricians and Gynecologists (ACOG) and Society for Maternal-Fetal Medicine, 2015, p. 2].

Several factors make the screening test appealing to both pregnant women and providers: ability to test early in pregnancy; high sensitivity and specificity, particularly for trisomy 21 and trisomy 18; lower false-positive rate than conventional screening methods for common aneuploidies, perceived ease of testing (blood test rather than invasive test); fast turn around time (7–10 days); and reliable sex determination early in pregnancy [American College of Obstetricians and Gynecologists (ACOG) and Society for Maternal-Fetal Medicine, 2015; Bianchi et al., 2015]. Many of the factors that make cell-free DNA screening appealing are, at the same time, reasons why the screening test has been the subject of much debate and ethical concern. For instance, comparing cell-free DNA *screening* to (invasive) diagnostic *testing*, a comparison which is implicit in the name *noninvasive prenatal testing (NIPT)*, may blur the distinction between a screening test, which can have false positives and false negatives, and a diagnostic test. This misunderstanding can have implications ranging from false reassurance from (false) negative results and unnecessary anxiety from (false) positive results to irreversible decisions, such as keeping

Clinical Ethics at the Crossroads of Genetic and Reproductive Technologies.
https://doi.org/10.1016/B978-0-12-813764-2.00014-3

or terminating a pregnancy based on false positives or false negatives. A number of the concerns raised by cell-free DNA screening can be addressed by careful attention to pre- and postcounseling; however, the complexity and quantity of information relevant to patient decision-making require a degree of provider time and genetics expertise that is often unfeasible in busy prenatal clinics.

Professional societies, such as the American College of Obstetricians and Gynecologists (ACOG) and the Society for Maternal-Fetal Medicine (SMFM), first recommended the test as a "screening option" for women at increased risk of fetal aneuploidy. Those at increased risk include pregnant women over age 35, those with ultrasound findings associated with aneuploidy, positive first-trimester or second-trimester screening results, history of pregnancies affected by aneuploidy, or a balanced Robertsonian translocation in one parent. By 2015, ACOG and SMFM revised their statements to include cell-free DNA as screening option for all women, regardless of aneuploidy risk.

Extending cell-free DNA screening as an option to the general obstetric population amplifies a number of the ethical concerns raised by offering the test to high-risk women, but some issues, for instance, providing adequate pre- and postcounseling and understanding test performance in low-risk populations, are so exacerbated as to become qualitatively different. In general, many of the concerns discussed below revolve around the information requirements for genuine informed consent and how this process can be achieved given the limited time of prenatal providers and the limited genetic counseling workforce. Other issues include the role of commercial laboratories in driving test adoption, access to testing and counseling, impact on the experience of pregnancy and parenting, and impact on people with disabilities.

2. BACKGROUND ON OFFERING THE TEST TO LOW-RISK WOMEN

In 2012, ACOG and SMFM issued a joint Committee Opinion on NIPS that concludes that cell-free DNA screening "can be offered" to women whose pregnancies are at increased risk for aneuploidy, but that it "should not be part of routine prenatal laboratory assessment, but should be an informed patient choice after counseling" (American College of Obstetricians and Gynecologists, 2012, p. 1). With respect to low-risk women and women with multiple gestations, ACOG and SMFM state clearly that "[c]ell free fetal DNA should not be offered" (American College of Obstetricians and Gynecologists, 2012, p. 1). In September 2015, ACOG and SMFM revised their positions to address the question of extending the test to all pregnant women. The American College of Medical Genetics (ACMG), on the other hand, says it "was careful not to restrict NIPS to specific patient groups" (Gregg et al., 2016, p. 3).

Since 2015, cell-free DNA screening has become increasingly commonplace in low-risk populations, intensifying attendant ethical issues. The increasing routinization of cell-free DNA testing for low-risk pregnancies, however, is not due to any *recommendation* on the part of ACOG or SMFM.[1] Their joint committee opinion of 2015, reaffirmed in 2017, states that providers should have a *discussion* with all patients about the "risks, benefits, and alternatives of various methods of prenatal screening and diagnostic testing, including the option of no testing" [American College of Obstetricians and Gynecologists (ACOG) and Society for Maternal-Fetal Medicine, 2015, p. 1]. While they say that "any patient may choose cell-free DNA analysis...regardless of her risk status," they emphasize the importance of counseling about limitations, benefits, and alternatives.

Neither of these carefully worded statements can be easily construed as a recommendation of cell-free DNA for all women, but the societies make their position even clearer: "Given the performance of conventional screening methods, the limitations of cell-free DNA performance, and the limited data on cost-effectiveness in the low-risk obstetric population, conventional screening methods remain the most appropriate choice for first-line screening for most women in the general obstetric population" (American College of Obstetricians and Gynecologists (ACOG) and Society for Maternal-Fetal Medicine, 2015, p. 1). Because their position continued to be misinterpreted, the SMFM issued a Special Report clarifying their recommendations. They state, "The purpose of this statement is to clarify that the SMFM *does not recommend* that cell-free DNA aneuploidy screening be offered to all pregnant women [....] However, SMFM believes, due to the ethics of patient autonomy, that *the option should be available* to women who request additional testing beyond what is currently recommended by professional societies" (SMFM Statement, 2015, p. 753, my italics).

The revision of the 2012 position to include the option of cell-free DNA for all pregnant women appears to stem primarily from a concern about "the ethics of actively withholding available tests from one group," rather than a changed opinion based on the review of new information (SMFM Statement, 2015, p. 753). In fact, the SMFM statement explicitly addresses the question of data on this issue and concludes that "[l]imited data at the present time on the effectiveness and clinical utility for improving patient outcomes *preclude a recommendation* that cfDNA be actively offered to all pregnant women" (p. 753, my italics). They cite a study that found that cell-free DNA was appropriate as a first-line screening for women over 40 years (SMFM Statement, 2015, p. 753; Kaimal et al., 2015).

In an earlier (2014) statement, issued prior to the decision to extend cell-free DNA to low-risk women, the SMFM addresses the question of data on the effectiveness and clinical utility of cell-free DNA for low-risk populations by

---

1. This paper focuses on ACOG and SMFM, but many other professional societies (American College of Medical Genetics, National Society of Genetic Counselors, the Royal College of Obstetricians and Gynecologists, International Society for Prenatal Diagnosis, etc.) have issued statements. Comparing the nuances of these statements is beyond the scope of this chapter.

discussing a 2014 study by Bianchi et al. (Society for Maternal-Fetal Medicine, 2014a,b). The study, published in the New England Journal of Medicine, aimed to address the dearth of literature, particularly in the United States, on the use of cell-free DNA in low-risk populations (Bianchi et al., 2014). The authors, whose declared conflict of interests include support from one of the major commercial laboratories offering cell-free DNA screening, conclude that "noninvasive prenatal cfDNA testing performed better than standard screening methods" and that it "merits serious consideration as a primary screening method for fetal autosomal aneuploidy" (Bianchi et al., 2014, pp. 806–807). The basis for their conclusion was a reduction in false positive rates for trisomies 21 and 18 and higher positive predictive for trisomies 21 and 18, compared to conventional screening.

In addition to the authors' conflict of interest, the SMFM statement notes several limitations of the study: (1) The study was underpowered to compare detection rates. (2) It is not valid to compare false-positive rates in isolation, e.g., false-positive rates of trisomy 21 and 18 only, rather than false-positive rates of the test in general, which includes trisomy 13 and the sex chromosome aneuploidies. (3) Nearly a third of blood samples were drawn in the third trimester, a time when test performance is improved but when screening is not generally done because the results are not clinically relevant. (4) The study compared cell-free DNA to several different conventional screening tests offered in the first and second trimesters. These tests have different performance characteristics. Furthermore, fewer than 3% of study participants had the screening test with the best performance, namely, integrated screening. (Malone et al., 2005, as cited in Society for Maternal-Fetal Medicine, 2014a,b). (5) Conventional screening tests detect abnormalities that cell-free DNA testing does not, and which may be of more relevance to a population at lower risk of trisomies 21, 18, and 13. (6) The incidence of Down syndrome in the study population was 1/381, approximately twice that of a genuinely low-risk population (1/700). (As discussed below, PPV increases with incidence of the condition). (7) The no-call rate (the number of patients for which the lab can make no determination of results) was not representative of the no-call rate typically reported (0.9% vs up to 12%). One study found that 22% of the 8% of no-call results had fetus affected with aneuploidy (Pergament et al., 2014, as cited in American College of Obstetricians and Gynecologists, 2016).

The decision a year later that cell-free DNA could be included among the screening choices offered to the general obstetric population does not appear to be based on new information about the validity of the test in low-risk populations. ACOG defines various levels of evidence on which recommendations are based, and this decision was based neither on the appearance of consistent scientific evidence that would have enabled a Level A recommendation, nor on the interpretation of limited or inconsistent scientific data that would have enabled a Level B recommendation. Rather, the decision was based on consensus and expert opinion. According the SMFM's Special Report clarifying recommendations about cell-free DNA, the consensus was about ethics not data: "[D]ue to the ethics of patient autonomy,…the option should be available…" (SMFM, 2015, p. 753).

However, the ethics of patient autonomy is about more than options; it is about informed decision-making. The SMFM simultaneously expresses concern that the majority of low-risk women do not undergo extensive counseling prior to testing and states that it be an "absolute requirement" that women understand the limitations and benefits of screening. Thus, in an effort to further patient autonomy, their decision to extend the cell-free DNA to a low-risk population may have inadvertently undermined the autonomy they sought to defend.

Furthermore, some have expressed concern that the commercial laboratories, rather than evidence of clinical utility, are driving the increase in cell-free DNA screening (Mozersky and Mennuti, 2013). The commercial laboratories themselves are often responsible for or involved with many of the studies, which can result in misleading information (Society for Maternal-Fetal Medicine, 2014a,b). Conflict of interest among study authors may or may not impact study results, but it does make careful analysis and interpretation of the results especially important.

3. OVERESTIMATION OF THE SIGNIFICANCE OF GENETIC INFORMATION

Communication about the limitations and benefits of cell-free DNA is challenging in part because of the quantity and complexity of information relevant to complete understanding. However, the challenge is increased by the simple fact that the word DNA has powerful connotations related to human identity that may be more resistant to correction than, for instance, "first-trimester screening." Regardless of debates about genetic exceptionalism, that is, whether genetic information is in actuality different than other types of medical information, genetic explanations continue to hold a certain allure for many (Alper and Natowicz, 1992; Green and Botkin, 2003; Remennick, 2006). It may be particularly challenging to understand that there can be ambiguity around genetic testing results or that a genetic test could have false positives and false negatives.

The coupling of the terms genetic and information further complicates this issue, as in colloquial speech "information" often connotes something determinate and knowable (Adriaans, 2013). Although there can be *ambiguous* information, "information" without a modifier is generally assumed to be true or false in colloquial speech. Thus, understanding how a genetic testing could be a screening test that reveals *risk* rather than yes or no answers may be especially challenging.

A related issue, to be addressed below in the discussion on disorders with reduced penetrance or variable expressivity, is the tendency of the public toward genetic determinism (Parrott et al., 2012; Kaphingst et al., 2016). The idea that something is "in my genes" does not generally admit of shades of gray. In one study, lower rates of health literacy were associated with lower genetic literacy, which is not surprising, but part of what this meant was that those with lower health literacy attributed greater importance to genetic information and less importance to family history information (Kaphingst et al., 2016). In other words, those with lower health literacy may be more inclined toward genetic determinism.

There is general agreement that genetic literacy could be improved (Kaphingst et al., 2016; Ravitsky et al., 2017). When one considers the complexity of the information relevant to patient decision-making, it is clear that most of us will need additional education to understand the benefits and limitations of genetic testing. With respect to cell-free DNA screening, the following genetic concepts are challenging and yet important for informed, that is, autonomous, decision-making.

4.   DIAGNOSTIC TEST VERSUS SCREENING TEST

Many NIPT companies advertise their tests as over 99% accurate, a number that is often quoted for the accuracy of diagnostic tests as well. Because the sensitivity (true positives/(true positives + false negatives) and specificity (true negatives/(true negatives + false positives), particularly for trisomy 21 (Down syndrome), are so high, some providers and patients have trouble understanding the test as a screening test rather than a diagnostic test (Benn and Chapman, 2016, p. 121; Daley, 2014). The problem may be partially attributable to misleading advertising by NIPT companies, including emphasis on test performance for common aneuploidies (even though tests now include a broader array of conditions with varying degrees of performance relative to each condition (Stoll and Lindh, 2015; Daley, 2014). However, the conflation of diagnostic and screening tests also stems from mistaking sensitivity and specificity for diagnostic reliability and failing to take into account the positive predictive value (PPV) of the test (Stoll, 2013).

PPV is the chance that a positive test result is a true positive (true positives/true positives + false positives). This statistic depends on the incidence of the condition in the population tested; the lower the incidence of the condition, the lower the PPV. This is because false positives rise, even in a test with excellent specificity, when the incidence of a disease is low in a given population. Consequently, understanding this concept is particularly important for counseling the general population, which by definition has a lower incidence of some of the conditions tested [American College of Obstetricians and Gynecologists (ACOG) and Society for Maternal-Fetal Medicine, 2015]. ACOG and the

SMFM encourage laboratories to report PPVs with patient results (for each aneuploidy tested) because of importance of this statistic for patient counseling [American College of Obstetricians and Gynecologists (ACOG) and Society for Maternal-Fetal Medicine, 2015, p. 2]. The ACMG, on the other hand, considers PPV "irrelevant to anyone not facing a positive test result"; including the PPV for each condition screened would result in "an excess of data cluttering a report" (Gregg et al., 2016, p. 7). While PPV might be irrelevant for negative results, explaining the PPV based upon pretest information is crucial for patients to understand both the value and risks of obtaining a test.

For instance, a 40-year-old woman, whose chance of having live-born infant with Down syndrome is approximately 1/100, the sensitivity and specificity of cell-free DNA screening are both over 99%. The PPV, or chance that a positive test is a true positive, is 93%. Put otherwise, there is a 7% chance that the test is a false positive. However, for a 20-year-old woman, whose chance of a live-born infant with Down syndrome is significantly lower, about 1/1400, the PPV, or chance that the positive test is a true positive, is 48%. In other words, for a 20-year-old woman whose cell-free DNA test is "positive" for Down syndrome, there is a 52% chance that the fetus is not affected. These numbers show how significantly different test performance can be for a "high-risk" pregnancy versus a "low-risk" pregnancy. Both the 40-year-old woman and the 20-year-old woman are likely to be told (by provider, commercial, or pamphlet) that the test is ">99% accurate." Strictly speaking, this may be true, but it is also misleading.

The chance of having a fetus with aneuploidy increases with maternal age. Some conditions, however, are rare regardless of age, and the PPV for these conditions reflects this. Commercial laboratories have expanded the number of conditions screened to include other common aneuploidies (trisomies 22 and 16, which do not result in live births), sex chromosome aneuploidies (e.g., Turner syndrome, Klinefelter syndrome) and some microdeletion and imprinting disorders (DiGeorge syndrome/22q11.2, Prader Willi syndrome, Angelman syndrome). Some laboratories now include whole genome analyses. One study found a PPV of 2%–4% for one of the more common conditions listed above, DiGeorge syndrome/22q11.2 (estimated prevalence of 1/2000–1/4000) (Begleiter and Finley, 2014). In other words, 96%–98% of women who received a positive screen for DiGeorge syndrome would not have a fetus affected with the condition. As more women in the general population undergo screening, more women will face results that need careful explanation by medical providers. Screening for microdeletions has not been clinically validated and is currently not recommended by ACOG and SMFM. (American College of Obstetricians and Gynecologists, 2016). For this reason, SMFM recommends that screening for microdeletions be "opt in," not automatically included with other conditions (Cell free DNA screening is not a simple blood test, p. 3).

5. REDUCED PENETRANCE AND VARIABLE EXPRESSIVITY

In addition to understanding that a positive result from a test that is 99% accurate may not indicate an affected fetus, patients may be faced with a true positive result for a condition with reduced penetrance or variable expressivity. Reduced penetrance means that not all who have a particular gene change will have a phenotype. For instance, a person can have a BRCA1 mutation and never get cancer. Variable expressivity means that not all with same genetic change will have the same severity of phenotype. For instance, particularly for sex chromosome aneuploidies, such as Turner syndrome, phenotype can vary considerably.

The ACMG recommends that pretest counseling include information about the chance, even following diagnostic confirmation, of variable prognosis. However, this is a challenging concept to understand. Further, to understand that a condition has variable expressivity or reduced penetrance is not the same as being to accept or integrate this information. Some have called this information "toxic knowledge" because many women wish, after the fact, that they did not know it (Bernhardt et al., 2013).

There may be an assumption on the part of parents that if a condition is screened for it must be relatively severe. Given that population-based screening programs with public support generally meet criteria that include significant severity, it is not unreasonable that parents would make such assumptions. Thus, even if cell-free DNA is not part of a public screening program, the subtlety of this distinction may be lost for many parents as cell-free DNA becomes a routine screening test for the general population. Providers will have the challenge of communicating that not all conditions screened are severe or predictable.

6. NEGATIVE TEST AND HEALTH BABY ASSUMPTIONS

Because cell-free DNA testing is often described as replacing other screening tests (and even diagnostic tests), parents may be under the false impression that the same conditions are screened by all tests. Further, because of the connotations carried by the term "genetic," there may be an assumption that the genetic test is even more comprehensive than first and second trimester screening. It may be especially challenging to communicate that cell-free DNA screening primarily gives information about the three common trisomies and that a negative result does not guarantee a healthy baby. While cell-free DNA is more accurate than conventional screening methods for trisomy 21, 18, and 13, conventional screening may pick up abnormalities that cell-free DNA would not.

The American College of Medical Genetics, in its updated position statement on "Noninvasive prenatal screening (NIPS) for fetal aneuploidy" (2016), notes that "[i]t is important to emphasize what NIPS does not provide to patients" (Gregg et al., 2016, p. 2). Cell-free DNA screening does not screen for single-gene disorders or open neural tube defects, and it does not screen for late pregnancy complications or replace routine fetal anatomic screening by ultrasound. This "emphasis" on what is not offered may be an acknowledgment of the ease with which patients might assume that cell free DNA screening is similar to conventional screening tests. Particularly in view of the fact that the ACMG describes this screening test as *"replacing conventional screening for Patau, Edwards, and Down syndromes across the maternal age spectrum"* (p. 1, my emphasis), it would be easy for someone not familiar with the extent of conditions and complications screened for conventional screening methods to assume that the one test replaces the other.

Furthermore, one aspect of the routinization of prenatal screening may be the assumption that the screening somehow helps the baby. That screening is a simple and routine part of pregnancy has been called the "collective fiction" by medical anthropologists (Press and Browner, 1994; Johnston et al., 2017). Press and Browner explain the source of this collective fiction in terms of a tension between the public's faith in the power and value of scientific knowledge and providers' (and pregnant women's) discomfort with discussing the full implications of prenatal testing of conditions for which there is no cure. Lurking in the background of this testing are issues people would rather deny or forget: selective abortion and eugenic selection. Consequently, a "collective fiction" and corresponding "collective silence" emerge in which prenatal testing is presented as a routine part of pregnancy that helps to ensure a healthy birth and healthy baby. There is a collective embrace of this fiction along with a collective silence around the attendant issue of abortion. Providers may thus avoid discussing the uncomfortable aspects of this testing, and pregnant women are "spared" having to confront difficult questions—spared, that is, until a positive test result forces them to confront questions for which they may be wholly unprepared.

The belief that one is doing something good for the baby by undergoing prenatal testing combined with the routinization of testing in prenatal clinics can make it difficult for women to opt out of such testing (Remennick, 2006). Remennick (2006) refers to "the emerging social pressure for comprehensive prenatal screening as an indispensable part of *good motherhood*" (p. 21). Women may feel a certain social accountability for bringing a child with a genetic condition into the world, particularly in a society with minimal support for people with disabilities (Charro and Rothenberg, 1994). Doctors may feel pressure to encourage, or even push, women to pursue screening because of concerns about birth outcomes and legal liability (Charro and Rothenberg, 1994; Press and Browner, 1994). What it means to claim the test is "optional" in the context of this social pressure and general routinization of screening is unclear. The emphasis on informed consent and women's autonomy may be misplaced when pressure from (and on) providers and society make prenatal screening hard to refuse.

7. INCIDENTAL FINDINGS: MATERNAL CONDITIONS

Another challenging, yet essential, aspect of informed consent for cell-free DNA screening is explaining to women that even though the focus of the test is the baby, maternal DNA is also examined. Thus, previously unknown maternal conditions may be discovered incidentally as a result of testing (Benn and Chapman, 2016, p. 121). However, the fact that a maternal condition rather than a fetal condition is responsible for the result may not be immediately clear. Parents may first get a positive screening result that indicates a likely affected fetus, only to learn that this is a false positive as a result of a maternal condition.

Detecting a maternal condition can be beneficial, for instance if there are important implications for the mother's health (Benn and Chapman, 2016, p. 121). For instance, if cell-free DNA detects maternal malignancy, a pregnant woman may begin treatment for cancer rather than having a cancer grow undetected. However, some maternal conditions may not have clinical implications. For instance, one might learn that a mother has undiagnosed 22q.11 (DiGeorge) syndrome. This could range from useful knowledge to disconcerting or stigmatizing knowledge. Perhaps a woman struggled in school and feels relieved or validated upon learning of this undiagnosed condition. Or maybe the woman has very few features of the condition, and learning of her diagnosis makes her feel that there is something wrong with her, where previously she had felt "normal."

In addition to finding a maternal condition with well-defined implications (e.g., cancer), there is also a chance of finding a copy number variation (CNV) that has been studied little, if at all. It is hard to predict what effect this information will have on people. People may interpret this to mean that "something is wrong with their DNA" or they may recognize that everyone is different and that it is unsurprising to find changes from person to person. It is possible that concept of "normality" or "health" may shift as more people find that they and/or their children have such variations (Parham et al., 2017, p. 49); however, this may take time and some people may be better equipped, either through education or temperament, to absorb this lesson than others. For many CNVs, researchers may not yet have information to determine clinical significance. It may be a lot to ask of people to take such findings with a grain of salt.

Problems of incidental findings are not unique to cell-free DNA screening. However, incidental findings relevant to the mother are essentially twice incidental. Classically, an incidental finding would be relative to expected findings for the tested individual (the fetus). A finding that is not relative to tested individual may be more challenging to communicate, particularly if parents do not understand that maternal DNA is tested along with fetal DNA. Indeed, the ACMG refers to these findings as "unanticipated findings" or "inadvertent discoveries," rather than incidental findings, and does recommend that providers and patients understand the potential for such findings before screening (Gregg et al., 2016, pp. 6–7).

8. ABORTION

Some have expressed concern that increasing the number of conditions screened and the number of individuals screened will increase the number of terminations (Parham et al., 2017, p. 47). For those who consider termination immoral regardless of timing and reason, this would mean that increasing the numbers of women screened with cell-free DNA is clearly problematic. However, for those who consider timing to be an important factor in the morality of termination, cell-free DNA screening might allow for terminations that fewer people find morally problematic (Parham et al., 2017, p. 48). Many who believe that termination is a personal, moral decision rightfully left to individual women, might still hold that it is better to terminate earlier in the pregnancy, if possible. Justifications for this position include increased maternal welfare (lower chance of complications), an intuition that the moral status of the fetus increases with gestational age, and the emotional and social difference of terminating before the pregnancy is visible or known to family, friends, and colleagues (Grimes, 1984; Lie et al., 2008; Van Schendel et al., 2014). The ACMG lists as a positive feature of cell-free DNA screening the possibility of earlier diagnosis following a positive screen. This allows for the "broadest range of reproductive options" (Gregg et al., 2016, p. 4). However, it may be unfair to characterize screening as leading to the broadest range of options, when termination is not a genuine option for many women, particularly those in lower income and rural locations.

Further, the possible moral benefit, if it can be called this, of earlier abortion is complicated by at least three factors: the ease with which screening results can be mistaken for diagnostic results, the range in severity of conditions screened, the use of diagnostic procedures and terminations for lethal chromosome abnormalities. First, while professional societies state explicitly that decisions about termination should not be based on cell-free DNA results, there are reports that decisions have been made without confirmation by diagnostic testing (Daley, 2014). Even when providers communicate that screening results need to be confirmed with diagnostic testing prior to termination decisions, some women make decisions against medical advice (Dar et al., 2014). Also, misinformation among providers abounds: a 2014 study found that 13% of MFMs offered cell-free DNA screening as diagnostic, despite professional guidelines, and 91% reported feeling confident in their knowledge about cell-free DNA (Haymon et al., 2014). Education gaps among other providers are likely to be even greater (Allyse et al., 2015).

The combination of greater numbers of low-risk women undergoing screening and commercial laboratories offering screening for an increasing number of (often quite rare) conditions makes for a PPV nightmare. More women screened means more will receive screen positive results, but because of the low PPV in low-risk populations, a significant proportion of them will not have affected fetuses. While cell-free DNA screening is touted as reducing invasive procedures, many women may end up undergoing additional testing and procedures who otherwise would not have. Terminating desired pregnancies because of misunderstood information is certainly problematic: an uninformed choice is not an autonomous choice, and there is the chance for harm to the woman during any medical procedure. However, increased termination of fetuses assumed to be imperfect may promote eugenic attitudes in society (Ravitsky et al., 2017).

Second, even if cell-free DNA is fully understood for the screening test that it is, the variable expressivity and reduced penetrance of some of the conditions screened means that women may find themselves faced with a decision about whether to keep a pregnancy without knowing how severely, if at all, the fetus may be affected. Despite their lack of severity in comparison with other genetic conditions, sex-chromosome aneuploidy diagnoses have been associated with high abortion rates, between 68% and 81% (Jeon et al., 2012). But the decision-making may be more complicated than terminating or continuing the pregnancy: Consider a woman carrying twins, one of whom screens positive and is then diagnosed by amniocentesis with a sex chromosome aneuploidy. The woman could decide to reduce the pregnancy, thus opening the possibility of losing both fetuses. She could keep the pregnancy and find that knowing the diagnosis may facilitate access to much-needed support or find that her relationship with her child is altered negatively by having this information. The child might have grown up without ever being diagnosed, if the condition is mild. Knowing that there is something "wrong" may alter the child's self-concept for the worse, or it may be welcome information to the child/future adult that explains certain qualities she has noticed in herself (Motulsky et al., 1994).

The answers to these questions cannot be known in advance. The ACMG's response to issues like these is to say that "clinical utility may vary between patients" and to counsel by considering "the patient's ability to accept uncertainty" (Gregg et al., 2016, p. 3). However, to emphasize some degree of relativism in reproductive decision-making does not alter the new moral challenges many women and families may face as a result of this new technology.

First, it is not always possible for people to know in advance how they will respond uncertain information. One study found that initially women wanted as much information as possible but when they were faced with challenging or uncertain results, they wished they did not have the information, which had become "toxic knowledge" (Bernhardt et al., 2013). Many others have written about how prenatal testing alters the experience of pregnancy, and even the experience of parenthood (Bernhardt et al., 2013; Rothman, 1986; Lippman, 1993).

Second, while it is true that parents will have different opinions about how severe is too severe, these attitudes are shaped by society and also by the information provided or omitted to patients. Decreasing numbers of individuals with various conditions can make it less socially acceptable to have a child with a given condition and can lead to decreased societal interventions that support parents and children with disabilities (Hickerton et al., 2012; Parham et al.,

2017; Parens and Asch, 2003; Remennick, 2006). Thus, offering the choice to parents does not enhance reproductive autonomy in a straightforward way, and may broaden eugenic attitudes to an increasing number of conditions as laboratories increase the number of conditions screened. If we do not first ask "which conditions should be screened?" pregnant women and their families will increasingly end up face challenging moral questions about which lives are worth living. These questions will have to be answered in the context of a society where a "preventative ethos" and technological imperative offer answers in advance: it is best to prevent whenever possible and make use medical technologies that allow prevention. For instance, one mother of a child with Down syndrome recalls being asked "Oh, did you find that out in time?" (Hickerton et al., 2012, p. 378).

Third, some labs are offering panels that include generally lethal chromosomal conditions, e.g., trisomy 16 and trisomy 22. If screen positive results are followed by diagnostic testing and/or termination, this could mean unnecessary testing and procedures for women and unnecessary costs for the healthcare system (Ravitsky et al., 2017). Further, if an increasing number of women choose diagnostic testing following a screen positive result for a lethal chromosome condition, then the number of fetal losses following diagnostic procedures may be increased, giving the false impression that the procedure caused the loss (Gregg et al., 2016, p. 4).

9.   ETHICS OF TESTING JUST FOR INFORMATION

Limiting cell-free DNA screening to pregnancies at high-risk means in part that the information sought is for medical reasons. Opening access to all pregnant women opens the possibility of testing "for information only" (Deans et al., 2014, p. 8). While drawing a line between testing for information only and for medical reason may be possible in theory only, the distinction can be used to raise several ethical issues that are not merely theoretical. One set of issues relates to the privacy and interests of the fetus as a future adult.

Deans et al. frame the issue by bracketing the question of termination on the basis of results. In other words, they ask about the ethics of testing just out of curiosity and a desire to have more information about the future child. Framed in this way, the question of the rights of the future adult, including preservation of privacy, can be raised as a potential objection, analogous to concerns about the right to open future raised by testing children for adult-onset disorders (Deans et al., 2014). Deans et al. also raise a concern about cell-free DNA screening objectifying the fetus and eventual child. While it may not be morally harmful to test out of curiosity, they write, it could be considered "distasteful" because the attitude evinced, namely valuing the traits of a child over "who that child is," is inconsistent with the behavior of a "virtuous parent" (Deans et al., 2014, p. 10).

With respect to testing for information about a baby's sex, one could argue that cell-free DNA screening raises no additional ethical concerns, since parents may already find out the sex during ultrasound. However, cell-free DNA screening would allow for earlier sex determination. Many parents find out the sex of their baby out of curiosity without intention to terminate based on the results. Deans et al. note that while information about the sex of the baby may be "trivial" in many cases, the larger social context could render the information morally significant. If, for instance, it is used to terminate female fetuses for discriminatory reasons, then access to this technology early in pregnancy could have social implications in addition to the moral concerns raised by terminating an individual fetus on the basis of sex.

10.   PRACTICAL ISSUES: WHO WILL COUNSEL PATIENTS?

The reasoning given by the SMFM for extending cell-free DNA screening as an option to all pregnant women was to respect women's autonomy. However, respect for autonomy entails more than options; it requires informed decision-making. Thus one of the central issues with extending cell-free DNA to the general obstetric population is a practical one: Who will counsel the patients? The ACMG considers pretest counseling "crucial" and suggests that patients should be able to make educated decisions about the use of cell-free DNA screening and its ramifications; however, their recommended topics of discussion appear to require a significant amount of time and education. It is not surprising that some consider one of the defining problems of extending cell-free DNA screening to average risk populations the "impossibility of providing specialty counseling and interpretation services to every woman considering prenatal screening and testing" (Parham et al., 2017, p. 46).

A related question is how or by whom will those responsible for counseling be educated? Many prenatal providers do not receive much training in genomics, and often receive information about cell-free DNA screening from the commercial laboratories (Dougherty et al., 2016; Parham et al., 2017; Farrell et al., 2016). Patients, as well, are receiving information from the laboratories themselves, whether by way of advertising online or pamphlets at doctor's offices (Parham et al., 2017; Farrell et al., 2016). In addition to the challenge of conveying complex information in a short period of time, there is also the challenge of communicating this information after patients have been exposed to advertising materials.

Previously, maternal-fetal medicine physicians or prenatal genetic counselors counseled high-risk women. This clearly cannot be the route for counseling the general obstetric population, as it would be neither practically possible nor theoretically desirable to have all low-risk women come to specialized high-risk clinics for a counseling about a screening test. Theoretically, prenatal genetic counselors could do the counseling, but practically speaking this is not currently possible. There are not enough genetic counselors to accomplish this task (Johnston et al., 2017). Furthermore, challenges around reimbursement mean that many general obstetric practices cannot afford a full-time genetic counselor (Parham et al., 2017). This leaves the majority of counseling to general obstetrics and obstetric nurses (Johnston et al., 2017).

11. OB/GYN COUNSELING

While most physicians are familiar with cell-free DNA screening, many may not have the education or training which would allow them to feel comfortable with such complex counseling. In 2015, a survey was completed at the Cleveland clinic to assess how physicians obtain information about cell-free DNA as well as how they incorporate cell-free DNA into their prenatal practice. The study, which included information based on physician self-report, found that providers initially acquire knowledge about cell-free DNA from a variety of resources including medical literature such as peer review journals, education didactic programs, as well as commercial labs that offer cell-free DNA (Farrell et al., 2016, p. 1). The survey also discussed what barriers physicians face with regards to cell-free DNA testing, and 41.9% responded that they had limited familiarity with cell-free DNA and 39.5% noted that patient health literacy is a major barrier to counseling. Both factors can impact the quality of informed consent for testing.

Furthermore, most countries do not have specific training programs in place for cell-free DNA counseling (Van Lith et al., 2015, p. 10). If cell-free DNA begins to become more integrated into routine prenatal care, it will be necessary to expand medical education for physicians. The Cleveland Clinic survey found that additional medical education efforts are needed to assist in the integration of cell-free DNA into private practice. It is essential that this information come from within the obstetrical profession as opposed to commercial labs, which would lead to ethical dilemmas involving biased informed consent (Farrell et al., 2016, p. 5). A prospective study by Macri and associates showed that integrating a designated genetics curriculum into resident education can improve physicians' knowledge and confidence regarding genetic counseling for patients (Macri et al., 2005). Integrating detailed information about NIPT into residency and continuing medical education curriculum for physicians could may be the first step in helping providers become more informed about new genetic testing methods and how to counsel patients properly. Additionally, updated educational resources for patients will be necessary to assist in patient understanding of this complex screening test.

Finally, an obstacle that many physicians face with integrating cell-free DNA for low-risk populations is how to incorporate this complex counseling into a busy office practice with a limited amount of time to spend with each patient. Obstetrical providers are responsible for extensively reviewing the patient's history and providing appropriate counseling to the patient during the first several prenatal visits. During the allotted time for the first prenatal appointment it is important for the provider to review the patient's past obstetrical history including number of pregnancies, type of delivery, and any complications during those pregnancies/deliveries. It is pertinent to obtain detailed information because this can impact management during the current pregnancy. It is also essential to review the patient's medical history in detail, including current medications to evaluate for conditions that may impact the pregnancy or which may become exacerbated during the pregnancy state. Patients must be counseled regarding tobacco abuse, alcohol consumption, and use of other drugs during the pregnancy. Additionally, pregnant patients must be counseled on appropriate diet and weight gain in pregnancy as well as the importance of regular and safe exercise while pregnant.

Providers must also screen for domestic violence as well as assess what support and resources the patient has in place for the pregnancy and after delivery. Other important discussions that must take place include the benefit of breast feeding and contraceptive counseling (Kirkham et al., 2005, pp. 1307–1313). In addition to these discussions, the provider must complete a detailed physical exam as well as schedule any necessary lab testing and ultrasounds. Furthermore, it is necessary to address any acute concerns the patient may have during these initial visits. Time constraints for appointments create a challenge for providers to adequately counsel patients on all of these important issues. To counsel about the nuances of cell-free DNA screening in addition to all this would require that providers have more time than is currently allotted, or feasible.

12.   INFORMATION ACCESS OUTSIDE OF THE UNITED STATES

The question of who will counsel low-risk patients about cell-free DNA is even more challenging to answer for women living in low- to moderate-income countries and in the more rural and resource-poor areas of high-income countries. Genetic specialists tend to cluster in urban centers, which means that testing and counseling are more accessible to the higher-income people who live in cities (Allyse et al., 2015). In areas with poor infrastructure and minimal access to general prenatal care, access to genetic testing for rare conditions, such as the aneuploidies, may seem be a low priority. This is particularly true given that there are more common genetic conditions for which testing would be (and has been) more useful. For instance, many low-income countries do have high rates of genetic conditions such as the thalassemias and sickle-cell disease (De Sanctis et al., 2017). Whether scarce health-care resources should be devoted to screening for much less common, nonheritable conditions is certainly questionable (Allyse et al., 2015).

Still, regardless of its clinical utility, cell-free DNA is being marketed across the globe. As of 2014, cell-free DNA companies were advertising in over

61 countries (Chandrasekharan et al., 2014). This testing is often ordered by providers with little education in genetics, and who are thus ill-equipped to provide pretest counseling and posttest interpretations of results (Allyse et al., 2015). Many women may end up screening without fully understanding, for instance, that the test is not diagnostic or that it does not look for more common and relevant genetic conditions, such as the thalassemias.

As screening becomes routinized, more women who would ordinarily not have been considered to have high-risk pregnancies, may experience the medicalization of pregnancy that characterizes the experience of women in many higher-income countries. In other words, pregnancy for the general and low-risk population may become a risky, medical event in which bonding with the fetus is increasingly deferred until the mother has results of tests (Rothman, 1986).

Despite the limitations of cell-free DNA screening, it is more accurate than maternal serum screening for certain conditions, and access to testing has the potential to decrease diagnostic procedures and the need for skilled personnel to perform such procedures; increase access to early abortion in states where it is legal; and increase access to resources for pregnancies identified to be at higher risk and for newborns with genetic conditions. Those who may benefit from extra time and attention devoted to counseling and informed consent, however, are less likely to receive this care.

13. PROBLEMATIC SOLUTIONS

One possible solution to the practical problems discussed above, namely the billing problem of employing a full time genetic counselor and the time problem of relying on general obstetricians and nurses, is to have genetic counselors employed by labs do the counseling. Indeed, commercial laboratories employ increasing numbers of genetic counselors (Stoll et al., 2017). While this solution addresses issues related to time and billing, it may create a problem even more challenging than those it solves: conflict of interest. While genetic counselors may view themselves as always placing patient needs above all else, there is no escaping the fact that employees are serving both patient/customers and their companies.

Furthermore, the conflict of interest exists regardless of whether professional judgment is influenced by the competing interest (Stoll et al., 2017). It is easy to tell ourselves that commercial interests will not influence us, but studies have shown that even small gifts to physicians can be influential (Lenzer, 2016, p. 2). There is no reason to believe genetic counselors, or any human beings, would be exempt from similar influence. Even if there is no evidence of genetic counselors placing company interests above patient interests, trust in the profession and in genetic testing more broadly may be undermined by relying on this framework (Stoll et al., 2017).

14.   LINEAR MODEL OF INFORMATION TRANSMISSION

The preceding discussion about logistics of counseling patients is based on the assumption that, if there were enough providers and enough time, patients could be counseled effectively and informed consent achieved. However, more information said aloud by a provider is not equivalent to more information understood and integrated by a patient. Informed consent and reproductive autonomy are not necessarily enhanced when more information is given. In fact, excessive amounts of information, particularly if it has no clear clinical utility, could hinder reproductive autonomy (Ravitsky et al., 2017, p. 40). The issue goes beyond the educational background of the patient. For instance, with respect to prenatal whole genome sequencing (NIPW) even providers trained in genomics have difficulty interpreting the results for affected patients; interpreting results for unaffected patients is substantially harder (Ravitsky et al., 2017). The shift toward a shared decision-making model for medical decision-making is an acknowledgment that patients need the opportunity to discuss challenging information in order to weigh its value. Merely listening or being exposed to information said aloud is not sufficient for decision-making.

15.   JUSTICE ISSUES: ACCESS TO TESTING

While insurance coverage of cell-free DNA in the United States is increasing, coverage remains uneven among both private insurers and state Medicaid programs. Thus access to cell-free DNA testing tends to be greater among those with higher incomes, exacerbating a "two-tier" system wherein those who can pay get access to a higher standard of care than those who cannot (Parham et al., 2017, p. 46).

In addition to access to screening, there is the issue of access to follow-up care. Those without health insurance or with inadequate health insurance may learn from testing that that their child will need extra care but be unable to access that care when the time comes.

The burden of caring for children with disabilities lies disproportionately among groups with fewer resources. For instance, the number of babies born with Down syndrome or other genetic conditions tends to be higher among groups with lower socioeconomic status, perhaps because of lower rates of abortion and/or prenatal testing (Resta, 2011, p. 1787). At the same time, there are fewer babies born with Down syndrome among higher socioeconomic groups, despite increasing maternal age and the concomitant increase in rates of aneuploidy.

The ACMG recommends that cell-free DNA screening be offered to all pregnant women who are not significantly obese, but emphasizes that pretest counseling is "crucial" and must go beyond discussion of the common trisomies (21,18, and 13). If one subtracts from the average risk obstetric population women whose providers cannot provide sufficient counseling and women who are significantly obese ($> 1/3$ of the population by CDC estimates), what percent of the average risk population is left? (Ogden et al., 2015). Given that rates of obesity are higher are higher among non-Hispanic black women, Hispanic women, and women who are not college educated, testing availability, which is already an issue for women of lower socioeconomic status, will be further limited and tend to break down along lines of class and skin color.

16. CONCLUSION

Expanding access to cell-free DNA screening raises a number of ethical issues, some of which are practical and thus potentially resolvable and others, which bear on our relationships to our bodies, our children, and our communities. Much attention has been paid to questions about autonomy, with some researchers arguing that limiting testing violates patient autonomy. Given that genuine autonomous choice requires adequate information and adequate opportunity to discuss and process this information, it seems that we should be less concerned about depriving people of testing and more concerned about how to address the practical problems of educating them about the meaning of this often-confusing information. More research is needed on how patients best integrate information about prenatal screening options and on the extent to which commercial laboratories, rather than clinical utility, is driving the push toward increasing access in the general population. As there is no cure for many of the conditions screened, patients and their families will need to be educated on the types of decisions they will face should results be positive.

REFERENCES

Adriaans, 2013. Pieter, "information" In: Zalta, E.N. (Ed.), The Stanford Encyclopedia of Philosophy. (fall 2013 edition). https://plato.stanford.edu/archives/fall2013/entries/information/.

Allyse, M., Minear, M.A., Berson, E., Sridhar, S., Rote, M., Hung, A., Chandrasekharan, S., 2015. Non-invasive prenatal testing: a review of international implementation and challenges. Int. J. Women's Health. 7, 133–126.

Alper, J.S., Natowicz, M.R., 1992. The allure of genetic explanations. BMJ 305 (6855), 666.

American College of Obstetricians and Gynecologists, 2016. Screening for fetal aneuploidy. ACOG practice bulletin no. 163. Obstet. Gynecol. 127 (5), e123–e137.

American College of Obstetricians and Gynecologists, 2012. Committee Opinion No. 545: Noninvasive prenatal testing for fetal aneuploidy. Obstet. Gynecol. 120 (6), 1532–1534.

American College of Obstetricians and Gynecologists (ACOG), Society for Maternal-Fetal Medicine, 2015. Committee opinion no. 640: Cell-free DNA screening for Fetal Aneuploidy. https://www.acog.org/Resources-And-Publications/Committee-Opinions/Committee-on-Genetics/Cell-free-DNA-Screening-for-Fetal-Aneuploidy.

Begleiter, M.L., Finley, B.E., 2014. Positive predictive value of cell free DNA analysis. Am. J. Obstet. Gynecol. 211 (1), 81.

Benn, P., Chapman, A.R., 2016. Ethical and practical challenges in providing noninvasive prenatal testing for chromosome abnormalities: an update. Curr. Opin. Obstet. Gynecol. 28 (2), 119–124.

Bernhardt, B.A., Soucier, D., Hanson, K., Savage, M.S., Jackson, L., Wapner, R.J., 2013. Women's experiences receiving abnormal prenatal chromosomal microarray testing results. Genet. Med. 15 (2), 139–145.

Bianchi, D.W., Parker, R.L., Wentworth, J., Madankumar, R., Saffer, C., Das, A.F., Craig, J.A., Chudova, D.I., Devers, P.L., Jones, K.W., Oliver, K., 2014. DNA sequencing versus standard prenatal aneuploidy screening. N. Engl. J. Med. 370 (9), 799–808.

Bianchi, D.W., Parsa, S., Bhatt, S., Halks-Miller, M., Kurtzman, K., Sehnert, A.J., Swanson, A., 2015. Fetal sex chromosome testing by maternal plasma DNA sequencing: clinical laboratory experience and biology. Obstet. Gynecol. 125 (2), 375–382.

Chandrasekharan, S., Minear, M.A., Hung, A., Allyse, M.A., 2014. Noninvasive prenatal testing goes global. Sci. Transl. Med. 6 (231).

Charo, R.A., Rothenberg, K.H., 1994. 'The good mother': the limits of reproductive accountability and genetic choice. In: Rothenberg, K.H., Thomson, E.J. (Eds.), Women and Prenatal Testing: Facing the Challenges of Genetic Technology. Ohio State University Press, Columbus.

Daley, B., 2014. Oversold Prenatal Tests Spur some to Choose Abortions. Boston Globe. Available at https://www.bostonglobe.com/metro/2014/12/14/oversold-and-unregulated-flawed-prenatal-tests-leading-abortions-healthy-fetuses/aKFAOCP5N0Kr8S1HirL7EN/story.html.

Dar, P., Curnow, K.J., Gross, S.J., Hall, M.P., Stosic, M., Demko, Z., Zimmermann, B., Hill, M., Sigurjonsson, S., Ryan, A., Banjevic, M., 2014. Clinical experience and follow-up with large scale single-nucleotide polymorphism-based noninvasive prenatal aneuploidy testing. Am. J. Obstet. Gynecol. 211 (5), 527–e1.

De Sanctis, V., Kattamis, C., Canatan, D., Soliman, A.T., Elsedfy, H., Karimi, M., Daar, S., Wali, Y., Yassin, M., Soliman, N., Sobti, P., 2017. β-thalassemia distribution in the old world: an ancient disease seen from a historical standpoint. Mediterr. J. Hematol. Infect. Dis. 9 (1), e2017018.

Deans, Z., Clarke, A.J., Newson, A.J., 2014. For your interest? The ethical acceptability of using non-invasive prenatal testing to test 'purely for information'. Bioethics 29 (1), 19–25.

Dougherty, M.J., Wicklund, C., Taber, K.A.J., 2016. Challenges and opportunities for genomics education: insights from an Institute of Medicine roundtable activity. J. Contin. Educ. Health Prof. 36 (1), 82–85.

Farrell, R.M., Agatisa, P.K., Mercer, M.B., Mitchum, A.G., Coleridge, M.B., 2016. The use of non-invasive prenatal testing in obstetric care: educational resources, practice patterns, and barriers reported by a national sample of clinicians. Prenat. Diagn. 36 (6), 499–506.

Green, M.J., Botkin, J.R., 2003. Genetic exceptionalism in medicine: clarifying the differences between genetic and nongenetic tests. Ann. Intern. Med. 138 (7), 571–575.

Gregg, A.R., Skotko, B.G., Benkendorf, J.L., Monaghan, K.G., Bajaj, K., Best, R.G., Klugman, S., Watson, M.S., 2016. Noninvasive prenatal screening for fetal aneuploidy, 2016 update: a position statement of the American College of Medical Genetics and Genomics. Genet. Med. 18, 10.

Grimes, D.A., 1984. Second-trimester abortions in the United States. Fam. Plann. Perspect. 260–266.

Haymon, L., Simi, E., Moyer, K., Aufox, S., Ouyang, D.W., 2014. Clinical implementation of noninvasive prenatal testing among maternal fetal medicine specialists. Prenat. Diagn. 34 (5), 416–423.

Hickerton, C.L., Aitken, M., Hodgson, J., Delatycki, M.B., 2012. "Did you find that out in time?": New life trajectories of parents who choose to continue a pregnancy where a genetic disorder is diagnosed or likely. Am. J. Med. Genet. A 158 (2), 373–383.

Jeon, K.C., Chen, L.S., Goodson, P., 2012. Decision to abort after a prenatal diagnosis of sex chromosome abnormality: a systematic review of the literature. Genet. Med. 14 (1), 27–38.

Johnston, J., Farrell, R.M., Parens, E., 2017. Supporting women's autonomy in prenatal testing. N. Engl. J. Med. 377 (6), 505–507.

Kaimal, A.J., Norton, M.E., Kuppermann, M., 2015. Prenatal testing in the genomic age: Clinical outcomes, quality of life, and costs. Obstet. Gynecol. 126 (4), 737–746.

Kaphingst, K.A., Blanchard, M., Milam, L., Pokharel, M., Elrick, A., Goodman, M.S., 2016. Relationships between health literacy and genomics-related knowledge, self-efficacy, perceived importance, and communication in a medically underserved population. J. Health Commun. 21 (sup1), 58–68.

Kirkham, C., Harris, S., Grzybowski, S., 2005. Evidence-based prenatal care: part I. General prenatal care and counseling issues. Am. Fam. Physician 71 (7), 1307–1316.

Lenzer, J., 2016. Two years of sunshine: has openness about payments reduced industry influence in healthcare? BMJ. 354. p.i4608.

Lie, M.L., Robson, S.C., May, C.R., 2008. Experiences of abortion: a narrative review of qualitative studies. BMC Health Serv. Res. 8 (1), 150.

Lippman, A., 1993. Prenatal genetic testing and geneticization: mother matters for all. Fetal Diagn. Ther. 8 (Suppl. 1), 175–188.

Lo, Y.D., Corbetta, N., Chamberlain, P.F., Rai, V., Sargent, I.L., Redman, C.W., Wainscoat, J.S., 1997. Presence of fetal DNA in maternal plasma and serum. Lancet 350 (9076), 485–487.

Macri, C.J., et al., 2005. Implementation and evaluation of a genetics curriculum to improve obstetrician-gynecologist residents' knowledge and skills in genetic diagnosis and counseling. Am. J. Obstet. Gynecol. 193 (5), 1794–1797.

Malone, F.D., Canick, J.A., Ball, R.H., et al., 2005. First-trimester or second-trimester screening, or both, for Down's syndrome. N. Engl. J. Med. 353, 2001–2011.

Motulsky, A.G., Holtzman, N.A., Fullarton, J.E., Andrews, L.B. (Eds.), 1994. Assessing Genetic Risks: Implications for Health and Social Policy. vol. 1. National Academies Press, Washington, DC.

Mozersky, J., Mennuti, M.T., 2013. Cell-free fetal DNA testing: who is driving implementation? Genet. Med. 15 (6), 433–434.

Ogden, C.L., Carroll, M.D., Fryar, C.D., Flegal, K.M., 2015. Prevalence of Obesity Among Adults and Youth: United States, 2011–2014. US Department of Health and Human Services, Centers for Disease Control and Prevention, National Center for Health Statistics, Hyattsville, MD, pp. 1–8.

Parens, E., Asch, A., 2003. Disability rights critique of prenatal genetic testing: reflections and recommendations. Dev. Disabil. Res. Rev. 9 (1), 40–47.

Parham, L., Michie, M., Allyse, M., 2017. Expanding use of cfDNA screening in pregnancy: current and emerging ethical, legal, and social issues. Curr. Genet. Med. Rep. 5 (1), 44–53.

Parrott, R., Kahl, M.L., Ndiaye, K., Traeder, T., 2012. Health communication, genetic determinism, and perceived control: the roles of beliefs about susceptibility and severity versus disease essentialism. J. Health Commun. 17 (7), 762–778.

Pergament, E., Cuckle, H., Zimmermann, B., et al., 2014. Single-nucleotide polymorphism-based noninvasive prenatal screening in a high-risk and low-risk cohort. Obstet. Gynecol. 124 (2 Pt 1), 210.

Press, N., Browner, C., 1994. Collective silences, collective fictions: how prenatal diagnostic testing became part of routine prenatal care. In: Rothenberg, K.H., Thomson, E.J. (Eds.), Women and Prenatal Testing: Facing the Challenges of Genetic Technology. Ohio State University Press, Columbus, OH, pp. 201–213.

Ravitsky, V., Rousseau, F., Laberge, A.M., 2017. Providing unrestricted access to prenatal testing does not translate to enhanced autonomy. Am. J. Bioeth. 17 (1), 39–41.

Remennick, L., 2006. The quest for the perfect baby: why do Israeli women seek prenatal genetic testing? Sociol. Health Illn. 28 (1), 21–53.

Resta, R., 2011. Are genetic counselors just misunderstood? Thoughts on "the relationship between the genetic counseling profession and the disability community: a commentary." Am. J. Med. Genet. A 155 (8), 1786–1787.

Rothman, B.K., 1986. The tentative pregnancy: Prenatal diagnosis and the future of motherhood. Viking Penguin, New York.

Society for Maternal-Fetal Medicine, 2014a. SMFM Statement: Maternal Serum Cell-Free DNA Screening in Low Risk Women. Available at https://www.smfm.org/publications/157-smfm-statement-maternal-serum-cell-free-dna-screening-in-low-risk-women.

Society for Maternal-Fetal Medicine, 2014b. Cell Free DNA Screening Is Not a Simple Blood Test. Available at https://www.smfm.org/publications/183-cell-free-dna-screening-is-not-a-simple-blood-test.

Society for Maternal-Fetal Medicine Publications Committee, 2015. SMFM statement: clarification of recommendations regarding cell-free DNA aneuploidy screening. Am. J. Obstet. Gynecol. 213 (6), 753–754.

Stoll, K., 2013. Guest post: NIPS is not diagnostic—convincing our patients and convincing ourselves. In: The DNA Exchange. Available at https://thednaexchange.com/2013/07/11/guest-post-nips-is-not-diagnostic-convincing-our-patients-and-convincing-ourselves/.

Stoll, K., Linkh, H., 2015. Guest post: PPV puffery? Sizing up NIPT statistics. In: The DNA Exchange. Available at https://thednaexchange.com/2015/05/04/guest-post-ppv-puffery-sizing-up-nipt-statistics/.

Stoll, K.A., Mackison, A., Allyse, M.A., Michie, M., 2017. Conflicts of interest in genetic counseling: acknowledging and accepting. Genet. Med.

Van Lith, J.M.M., Faas, B.H.W., Bianchi, D.W., 2015. Current controversies in prenatal diagnosis 1: NIPT for chromosome abnormalities should be offered to women with low a priori risk. Prenat. Diagn. 35 (1), 8–14.

Van Schendel, R.V., Kleinveld, J.H., Dondorp, W.J., Pajkrt, E., Timmermans, D.R., Holtkamp, K.C., Karsten, M., Vlietstra, A.L., Lachmeijer, A.M., Henneman, L., 2014. Attitudes of pregnant women and male partners towards non-invasive prenatal testing and widening the scope of prenatal screening. Eur. J. Hum. Genet. 22 (12), 1345–1350.

FURTHER READING

Gregg, A.R., Gross, S.J., Best, R.G., et al., 2013. ACMG statement on noninvasive prenatal screening for fetal aneuploidy. Genet. Med. 15, 395–398.

Gregg, A.R., Skotko, B.G., Benkendorf, J.L., Monaghan, K.G., Bajaj, K., Best, R.G., Klugman, S., Watson, M.S., 2016. Noninvasive prenatal screening for fetal aneuploidy, 2016 update: a position statement of the American College of Medical Genetics and Genomics. Genet. Med. Off. J. Am. Coll. Med. Genet. 18 (10), 1056–1065.

Haymon, L., Simi, E., Moyer, K., Aufox, S., Ouyang, D.W., 2014. Clinical implementation of noninvasive prenatal testing among maternal fetal medicine specialists. Prenat. Diagn. 34 (5), 416–423.

Jeon, K.C., Chen, L.S., Goodson, P., 2011. Decision to abort after a prenatal diagnosis of sex chromosome abnormality: a systematic review of the literature. Genet. Med. 14 (1), 27–38.

Chapter 15

Using Genetics for Enhancement (Liberal Eugenics)

Sonja Pavlovic*, Milena Ugrin*, Stefan Micic[†], Vladimir Gasic*, Jelena Dimitrijevic[†] and Ursela Barteczko[†]
*Institute of Molecular Genetics and Genetic Engineering, University of Belgrade, Belgrade, Serbia, [†]Center for the Study of Bioethics, University of Belgrade, Belgrade, Serbia

1. INTRODUCTION

Enhancement is an old idea in the history of human thought. From Icarus who wished to become more powerful than God planned humans to be, to Goethe's Faustus who chose to conspire with Satan in order to obtain the wisdom God did not envision for humans, throughout the history of mankind, people have made continuous efforts to improve themselves (Rakic, 2017). Personalized enhancement has become an emerging field of research in our time. The progress in genetics has significantly contributed to the growing aspiration that human traits can be improved. Since we have learned that our genetic and epigenetic signature is our main determinant, an idea that all types of personalized enhancement can be reached by genetic enhancement has become very strong. This article deals with the issue of using genetics for enhancement, seen through the eyes of both geneticists and bioethicists.

1.1 Definition of Enhancement

There are various approaches to the concept of enhancement. Most philosophers agree on the definition of enhancement as, "an improvement from which an individual should benefit" (Chan and Harris, 2007; Esposito, 2005). DeGrazia (2014) also defines enhancement in terms of improvement, "enhancements are interventions to improve the human form or function that do not respond to genuine medical needs."

Some researchers like Sarah Chan and John Harris (2007) have a wider definition of human enhancement. They consider dietary supplements, prosthetic limbs, vaccination, reading glasses, opera glasses, and hearing aids to be representative of various forms of enhancements. Others will narrow their

Clinical Ethics at the Crossroads of Genetic and Reproductive Technologies.
https://doi.org/10.1016/B978-0-12-813764-2.00015-5

definitions by excluding those as examples of treatment rather than enhancement because it helps to regain the normal status of some function of healthy human body. Thus, for Wolpe (Wolpe, 2002; Esposito, 2005) enhancement is the improvement of individuals who are not sick—those who are average or normal. This definition is problematic because there is no specific person whom we can consider perfectly average or normal (Esposito, 2005).

Ruth Chadwick (2008) analyzes four different understandings of enhancement, which roughly correspond to the following:

1. *Enhancement as a set of certain techniques developed for therapy, but used in a way that goes "beyond therapy"* in the sense of going beyond health-restoring treatments (the beyond therapy view). This approach is based on a distinction between treatment and enhancement. Eric T. Juengst (1998) defines enhancement as "interventions designed to improve human form or functioning beyond what is necessary to sustain or restore good health" (Savulescu, 2006).

2. *A quantitative understanding of enhancement*, seen as increasing or adding (to) a certain characteristic (the additionality view).

3. *A qualitative understanding of enhancement* namely making things better in some way (the improvement view). It can be the improvement of X, but it can also change X in such a way that we cannot regard it as X anymore. At species level, this means that if we change humans in a certain way, we might become a new species—posthumans. The problem with this definition is that it is quite uninformative.

4. *Enhancement as an umbrella term*, which covers a number of particular potential changes, such as, for example, extending the human life span or increasing the general immunity to pathogens.

Since it is hard to find a definition of "enhancement" that fits all cases that might fall under one label, we can use the term to refer to a wide variety of changes and techniques. It can include radical life extension; making people smarter, taller, or more emotionally stable; cosmetic changes; and pharmacological improvements. From these we can conclude that instead of differentiating enhancement from therapy and analyzing enhancement as a process, it might be more appropriate to analyze it on a case-by-case basis in which enhancement could be invoked (Gordijn and Chadwick, 2008).

There are also philosophers who think that the term "enhancement" is so fraught with erroneous assumptions and is so abused that we should not even use it (Savulescu, 2006). So, any exclusive enhancement definition fails, especially if we try to define it in terms of disease, normalcy, and health that are significantly culturally and historically bound. Thus the concept itself is a result of negotiated values (Wolpe, 2002). James Canton (2002) also claims that enhancement is a relative concept. Enhancement captures a certain value of discourses related to the human performance and it does not have a substantive transcultural meaning.

The fifth approach to the term of enhancement is the Functional Approach. It defines enhancement in terms of enhanced functions. The archetypal example of this approach is Douglas C. Engelbart's Augmenting Human Intellect (1962). Cognitive enhancement is defined here in terms of the improved general information processing abilities (Savulescu, 2006).

There is one more interesting approach that we ought to consider here, the Welfarist Approach to enhancement. According to this approach, human enhancement is any change in the biology or psychology of a person that increases his/her chances of leading a good life in a particular societal framework (Savulescu, 2006).

Also, it is useful if we consider the difference between positional and nonpositional goods (Buchanan, 2009). Positional goods mean a relative advantage, like height, and have value only if not everyone has them. Nonpositional goods are goods that are beneficial for people regardless of whether everyone has them, like the sense of self-identity (Savulescu, 2006).

1.2   Types of Enhancement

In order to reach to a better understanding of the concept of enhancement, we must also consider different types of enhancement: physical, cognitive, and moral. We need to stress that these categories are not strict, because (1) we can make a change that could generate benefits in more than one category and (2) it is plausible that we could not have true moral enhancement without cognitive enhancement, and true cognitive enhancement without moral enhancement. The reason why we discuss each category separately is because each of these categories has a specific, associated controversy.

1.2.1   Physical Enhancement

Some consider that physical enhancement will equate to cheating and will undermine the value of human effort (Kass, 2003). The most debated problem that is related to the physical enhancement is the ethics of sports. Allegedly, human physical enhancement undermines the notion of fair play. This objection is on target if we adopt the ancient Greek "Olympic" understanding of sports, in which the goal is to find the strongest, fastest, or most skilled man. But sport is not a genetic lottery. It is not equal to the dog race where we just mark physical capabilities. The winner is the person with combination of genetic potential, training, psychology, self-discipline, motivation, grit, and judgment. As Savulescu points out, the Olympic performance is a challenge for creativity and choice, not a very expensive horse race (Savulescu, 2006).

1.2.2   Cognitive Enhancement

As well as physical enhancement, cognitive enhancement also encounters the objection of cheating on the part of bioconservatives. Usually, it is a consequence of misunderstanding the specific aspects of enhancement. Cognitive enhancement, same as a high natural intelligence quotient (IQ), does not include any particular knowledge of itself. Knowledge, as well as any other success, still requires effort and learning.

Other objections to cognitive enhancement are related to the effects of the constant and fast development of our knowledge, which may result in inventions of various new and dangerous technologies. Without a proper moral enhancement, such a development could have catastrophic consequences, in particular from actions of the malicious agents who will misuse these dangerous technologies. Most influencing proponents of this line of argumentation are Savulescu and Persson (2008).

They consider that if the research of cognitive enhancement continues, as it is likely to, it must be accompanied by research into moral enhancement (Persson and Savulescu, 2008). To eliminate the risk that a small number of malevolent actors will cause the ultimate harm, cognitive enhancement would have to be accompanied by a moral enhancement that extends to all of us.

The example of martial arts is instructive in this respect. Martial arts, as their integral part of learning, have a moral codex that students must learn in order to apply them properly. In the same way, moral enhancement should accompany cognitive enhancement (Persson and Savulescu, 2008). Human morality must evolve beyond the morality that is concerned with those who are close to us in space, time, or genetic relations, the same way as our knowledge is evolving (Savulescu and Persson, 2012).

1.2.3 Moral Enhancement

This topic is obviously highly controversial. First of all, it is not clear enough how we can morally enhance humans? DeGrazia (2014) gives us some examples of how it could be achieved. For example, glucose increases the resistance to the temptation of doing something wrong, selective serotonin reuptake inhibitors can make us less inclined to assault people, and deep-brain stimulation (e.g., electrical stimulation of the amygdala) can reduce aggression.

DeGrazia distinguishes three different types of moral enhancement:

1. motivational improvement
2. improved insight or better understanding
3. behavioral improvement

However, some might argue that, even if successful, this would not be a genuine moral enhancement, since the moral act must come from one's own will and not from his genes/metabolism. Simkulet argues that these interventions fail to constitute genuine moral enhancement because they ignore a person's intentions and good intention is the true value of moral action (Simkulet, 2016). We can avoid this argument by pointing to the direct enhancement of motives, but then another problem will appear: a moral act, in order to be truly moral, must be free, and if we accept freedom as the possibility to act otherwise (Frankfurt, 1988), our will is not free.

This objection is questionable. First of all, it is not accepted that the only free action is the action in which we can do otherwise. Second, either some form of physical determinism is true or none is. If that determinism can be compatible with free will, then absolutely any form of determinism must be compatible as well. Thus if moral enhancement involves determinism, it is not necessarily antagonistic to freedom (Harris, 2013; Savulescu and Persson, 2012).

The level of oxytocin does influence on behavior, but it does not determine moral behavior. We cannot say that women are less free than men because they are less aggressive and more altruistic by their biological nature. Moral enhancement will not, by itself, produce moral or desirable behaviors. It will still require effort and learning (Savulescu and Persson, 2012). As DeGrazia (2014) points out, moral enhancement would not rob us of freedom, the same way as caffeine does not rob us the merits of writing a paper.

Another problem is that the moral deficiency of the human individual is not the only source of catastrophic threats, human stupidity is as well (Harris, 2011), as supported by near-miss nuclear accidents during the Cold War. We can also argue that without cognitive enhancement, moral enhancement is not possible, because a will to do what is good can never result in good acts without insight and understanding of what the good will be in that particular situation. This particularly applies to catastrophic threats of great inherent complexity, like the global climate change or bioterrorism.

It could be concluded that imposing obligatory moral enhancement, in order to edit out the immoral behavior and exclude a risk, is like hoping that gene editing will edit out the disability. Even if we succeed in that and even if all the babies are born perfectly able and moral, there will always be a risk of some accident of physical or psychological nature that will make one disabled or immoral. The risk of catastrophe made by just one immoral individual will always be a possibility that we have to deal with and take into account before introducing any new technology on the free market.

1.3 (How) Should We Distinguish Enhancement From Therapy?

Philosophers that define enhancement as something that goes beyond therapy have a problem of distinguishing enhancement from therapy. Some of those who propose this definition, argue that enhancement is immoral since it goes beyond normal. The problem with this definition is that, if we do not take it as convention and try to find the difference between enhancement and therapy, we find many obstacles. Here are some points of view on how this distinction can be made.

Resnik (2000) analyze different attempts to draw the line between therapy and enhancement. In the attempt to distinguish enhancement from therapy, the first distinction he draws is the line between somatic and germline interventions, based on which he classifies genetic interventions in somatic genetic therapy, germline genetic therapy, somatic genetic enhancement, and germline genetic enhancement. Somatic interventions are those that modify somatic cells and germline interventions are those which modify germ cells. Germline interventions are the ones that can affect future generations (Resnik, 2000). The gene therapy clinical trials that have been performed until recently were mostly performed on somatic cells (for an exception, see Tachibana et al., 2013).

The second distinguishing element, according to Resnik, is the aim of the intervention. The aim of gene therapy is to treat diseases. The aim of genetic enhancement is to perform other kinds of interventions, such as improving the human body. Since gene therapy serves morally legitimate goals, it has been considered morally acceptable, while enhancement has been considered something that serves morally illicit goals. If we differentiate therapy from enhancement based on the aims of medicine, we face two problems. The first one is that enhancement might improve our medical status (e.g., our immune system), and therefore cause beneficence based on the duty to intervene. Second, nowadays we have numerous medical procedures, such as: rhinoplasty, liposuction, and hair transplants, the aims of which are not curing diseases and which are still not considered immoral, based on the principle of beneficence as satisfaction.

Also, when we talk about goals and aims, we face the problem of intentions and expected outcomes (Caplan, 1992). It is problematic to argue that intentions alone are sufficient, if there is no evidence that a particular intervention is likely to have therapeutic benefits. But it is not clear that the therapeutic intent is a necessary condition for something to be considered a therapeutic intervention. An obvious example is placebo, which can be therapeutic even if not administered specifically for this purpose (Gordijn and Chadwick, 2008).

Proponents of this argumentation have to face two problems:

1. We do not have a clear and uncontroversial account of health and disease.
2. Why should other goals beside treating disease be morally illicit, when they serve for good in other realms (e.g., social acceptance)?

Health and disease are descriptive concepts that have an empirical, factual basis in human biology (Resnik, 2000). Boorse defended one of the most influential descriptive approaches to health and disease: a diseased organism lacks the functional abilities of a typical member of its species, which is considered healthy (Boorse, 1977). Of course, the "typicality" itself is a controversial issue. Is an inherited trait carried by 40% of the human population typical or not?

According to such a value-laden approach, our concepts of health and disease are based on social, moral, and cultural norms. In some cultures, schizophrenia is a disease, while in others it is a divine gift.

There are also practical problems related to this distinction. Interventions considered as enhancements may be brought about by using techniques that have originally been developed for therapeutic goals (Gordijn and Chadwick, 2008).

There were attempts in solving the problem of differentiating between enhancement and therapy by reference to normal. Therapy should restore the normal function. This raises the question of "normal," since there is no definition of "normal" that is culturally invariant. In addition, the concept of "normal" has changed across time. Numerous accounts with the aim to make a distinction between treatment and enhancement often ignore the fact that many interventions considered to be treatment actually alter typical functions of the body or psyche. At an individual level, therefore, the concept of normalcy lacks both precision and moral content (Chan and Harris, 2007).

Now let us turn to the other problem listed above. Since the concept of disease is descriptive, we cannot answer this question without making some normative assumptions. Since the descriptions have no normative import, the descriptive account of disease does not provide us with a way of drawing a sharp moral line between therapy and enhancement. We need to supplement it with normatively rich account, which will prove the wrongness of enhancement and the rightness of therapy.

Even if we had a clear distinction between enhancement and therapy, does this show why enhancement is morally wrong? Enhancement interventions could be morally acceptable if they do not violate other accepted moral norms and have independent advantages. There is nothing inherently wrong with enhancement. The rightness or wrongness of any concrete enhancement procedure will depend on its factual and normative aspects (Resnik, 2000); thus, we must analyze and evaluate every case individually.

Some philosophers defend a thesis that gene therapy and genetic enhancement are immoral because they change the human form (Resnik, 2000; Habermas, 2003). But this would be the problem only if we consider the human form as something that has inherent value. Besides, the human genome, hence the human form, can persist only through constant changes in order to adapt to the environment that is also constantly changing. And even if it could, considering the history of evolution, human form will change in the future by the entirely natural processes of natural selection and genetic drift anyway. So why would it be inherently more moral to leave such changes to irrational, chance-leaden process such as evolution, instead to rationally leaden plan for enhancement (suppose that we have such an aim to enhance human species as whole, for the sake of argument)?

It seems as though we are overestimating the blind watchmaker and underestimating human knowledge and risk management. The burden of proof to show what is morally wrong with enhancement is on the proponents of this thesis. There is nothing inherently immoral about enhancement as such. Some types of gene therapy are morally acceptable and some types of genetic enhancement is unacceptable (Resnik, 2000).

What is considered to be a change at species level, rather than at individual level? While this might not be immediately relevant for the most discussed methods of enhancement, it is important to keep in mind that for an enhancement to be considered an enhancement of the species, it would have to be transmissible down the germline. In other words, species enhancement does require the introduction of a change at the gene pool level (Gordijn and Chadwick, 2008). By its very nature, it will change the baseline for "typical" or "average" property of the population members. We can envision the species enhancement being an emergent consequence of individual enhancement choices and procedures, each of which will require resolving of their particular moral dilemmas. Therefore there is no need to invoke particular ethical weight to species enhancement, contra the usual arguments of bioconservatives (e.g., Fukuyama, 2002).

2. GENETIC ENHANCEMENT

The concept of genetic enhancement has been around for a few decades. What started in the early 1980s as a treatment for children who did not produce enough human growth hormone (hGH), developed into a possibility of genetic enhancement for healthy individuals. Innovations in recombinant deoxyribonucleic acid (DNA) technology led to the production of an hGH protein in host cells, such as bacteria *Escherichia coli (E. coli)*, without the risk of human pathogens transfer (Rezaei and Zarkesh-Esfahani, 2012). Through this "indirect" gene manipulation in the prokaryotic expression system, recombinant hGH could be safely produced in abundance and used for growth boost in athletes (Murray, 2002).

2.1 Genomics and Epigenomics as Base for Genetic Enhancement

Genetic information of each human being is stored in the DNA molecule. Certain portions of DNA are unique to each individual. Any two unrelated people are 99.9% identical at the genetic level, with 0.1% being different and making us all individual (genetic variation). It is now thought that genetic variation influences every aspect of human physiology, development, and adaptation. Consequently, an understanding of human genetic variation could play an important role in promoting health, combating disease, and implementing genetic enhancement.

The Human Genome Project has been completed in 2003, providing the first complete human genome sequence and thus enabling the development of genomics as a new discipline focused on the study of the structure and function of genomes (International Human Genome Sequencing Consortium, 2004). Genomics has enabled a first insight into the anatomy of the human genome and made a significant contribution to the understanding of the function of genes and their complex interactions. However, sequencing the entire human genome has brought new knowledge about the inventory and distribution of different sequence types in the genome. It was concluded that protein-coding genes represent only 1.2% of the genome. The Encyclopedia of DNA Elements (ENCODE) Project, which systematically delineated all functional elements encoded in the human genome, revealed that the vast majority of the human genome was involved in the regulation of gene expression—80.4% of the genome. It has become obvious that the number of genes for noncoding regulatory ribonucleic acids (RNAs) exceeds the number of protein-coding genes (ENCODE Project Consortium, 2012).

Additionally, it has become rapidly recognized that the DNA sequence (genome) by itself cannot explain the functionality and development of the living organism. The discovery of epigenetic mechanisms has introduced an epigenome whose variations could provide explanations for many cellular processes. Epigenetics studies genetic factors and mechanisms that regulate the expression of the genetic code without altering the DNA sequence. Epigenetic processes regulate the expression of genes by chemical modifications of DNA and chromatin structural proteins through DNA methylation and various forms of posttranslational modifications of histones and through the actions of small noncoding RNA molecules (Berger et al., 2009). The essential characteristic of the epigenome is that it can rapidly respond to environmental stimuli (Marsit, 2015).

Fascinating recent developments in molecular genetics, especially the improvement of modern technology for human genetic and epigenetic profiling, have led to increased knowledge regarding genetic base of health and disease.

2.2   High-Throughput Methodology for Genome Wide Genetic, Epigenetic, and Gene Expression Profiling

There are several approaches for the comprehensive analysis of the genetic and epigenetic profiles of a large number of people that have provided sufficient data on molecular markers that may be used in diagnosis, prognosis, and treatment of certain diseases, as well as for genetic enhancement. The best known are platforms for DNA analysis [DNA microarrays, genotyping arrays, single nucleotide polymorphism (SNP) arrays, next-generation sequencing] and hybridization platforms for the analysis of gene expression, or the amount of transcribed messenger RNA (hybridization microarrays, expression profiling). A special kind of study, GWAS (genome-wide association study) analyses a large number of genetic markers in different individuals suffering from the same disease, has contributed to the implementation of personalized medicine in clinical practice. GWAS analyses establish a relationship between molecular genetic markers and the pathological phenotype. Biomedical professionals are keen to understand the personal genetic profile of every individual. The complete genome sequencing is therefore becoming the ultimate genetic test. It can be performed once in a lifetime, as early as possible, and the data can be used throughout life, with the aim to achieve better health and longer life using principles of preventive and personalized medicine. It is important to note that for all these methods, bioinformatics data processing have a significant role (Pavlovic et al., 2014).

2.3 Personalized Medicine as a Model for the Implementation of Genetic Enhancement

Currently, genomics and epigenomics are fully integrated into medical practice. Personalized medicine, also called genome-based medicine, uses the knowledge of genetic basis of the disease to individualize treatment for each patient. A number of genetic and epigenetic variants, molecular genetic markers, are already in use in medical practice for diagnosis, prognosis, and follow-up of monogenic hereditary disorders, multifactorial or infectious diseases (Ugrin et al., 2017; Bojovic et al., 2017; Skodric-Trifunovic et al., 2015).

High throughput analysis of the whole genome (the complete set of DNA within a single cell of an organism), comprising DNA sequencing analysis and functional genomic analysis (mainly concerned with patterns of gene expression during various conditions), opened the door for personalized medicine. The application of genomics in clinical practice is the best example of successful translational research, the research that aims to move "from bench to bedside" or from laboratory experiments through clinical trials to point-of-care patient applications.

Genetic and epigenetic markers that will be used in genetic enhancement will be discovered using the same methodology that has led to personalized medicine. A growing number of whole genome sequencing and GWAS studies inevitably leads to the discovery of the markers associated with traits that could be enhanced.

2.4 From Predictive Genomics to Preventive Medicine and Genetic Enhancement

Predictive genetic testing represents the genetic analysis of healthy individuals in order to predict the risk for developing certain diseases before the appearance of the first symptoms (presymptomatic risk assessment). The aim of predictive genetics is to define predictive genetic risks factors and determinants of health and disease.

Predictive tests are not only performed when assessing the risk of developing a serious disease. Predictive genetic tests can indicate that a person needs to modify their diet so as not to have the harmful effects of nutrients, for example, gluten (for celiac disease), lactose (for adult hypolactasia), caffeine (for hypersensitivity), or fat (for obesity). Nutrigenomics and nutriepigenomics, based on the individual's genetic and epigenetic background, provides the ability to correct a congenital metabolic imbalance with a proper diet or certain food supplements (Pavlovic et al., 2014).

Predictive genomics is, at the moment, the field closest to genetic enhancement. Even now, some of the molecular markers used in predictive, especially nutrigenetic tests could be classified as genetic enhancement markers.

2.5 Ethical Issues in Personalized Medicine

Research and potential integration of personalized medicine into clinical practice has ethical, legal, and social (ELS) implications. It is indispensable to face ELS challenges raised by personalized medicine. There are two important ELS issues associated with personalized medicine: an increased amount of health information and the increase of disparities in health care (Brothers and Rothstein, 2015).

Advances in genomics and other "omics" technologies have led to substantial increase in individual health data. Several concerns related to the generation and storage of health "big data" are raised, among which privacy and discrimination are most important. Informational health privacy refers to protection of health data. Confidentiality should be guaranteed and patient's health data must not be disclosed unless authorized by the patient. Genetic discrimination is one of the biggest concerns in the postgenomic era. Genetic testing currently provides an indication that an individual has a predisposition to develop a certain illness or that an individual would not respond to standard therapy, increasing his/her morbidity and mortality risk. This information can cause discrimination in employment, insurance, mortgages, and other important activities. The importance of legislation regarding the prevention of genetic discrimination has been recognized in many countries (Otlowski et al., 2012). Personalized medicine will have a significant effect on the physician-patient relationship, not only because most physicians will not be trained enough to interpret "omics" findings, but also because of lack of time, since physicians will need more time to apply genomic insights in designing a treatment plan and, more importantly, to explain all aspects of the genetic testing and the obtained results to the patients (Marchant and Lindor, 2013). Accordingly, this will result in an increase of personal injury litigations. There are so many potential additional professional responsibilities brought to physicians by the implementation of personalized medicine, but the nature of these responsibilities remains uncertain. It is especially complicated to interpret and report the results of cancer predisposition screening and incidental findings revealed by whole-genome sequencing to patients. The possible duty of clinicians in these situations still remains undetermined and needs prompt discussion and decision from bioethicists and lawyers (Burke et al., 2013).

The idea of personalized medicine has brought a hope that it will contribute to the elimination of health disparities (Collins et al., 2003). However, inequalities in access to health-care services and information technologies have led to an increase of existing disparities in health care. Work in this area is also needed in order to provide fair access to health care globally (Brothers and Rothstein, 2015).

2.6 Gene Therapy, Molecular-Targeted Therapy, and Cellular Therapy

2.6.1 Gene Therapy

Gene therapy is a way of treating certain disorders by introducing genetic material into a cell to fight or prevent disease. Specific engineered genes are delivered to a patient's cells in order to target disease causing the genetic defect. For a successful gene therapy, it is necessary to design and create an effective therapeutic gene and a vector for its efficient delivery into a human cell (Pavlovic et al., 2015).

A therapeutic gene is not only a "healthy" copy of the gene that replaces a mutated gene, but also a genetic material deactivating a mutated gene that functions improperly, or any other genetic material that can fight a disease, when introduced in a cell or incorporated in a human genome. The genetic material can be delivered to a cell by using a "vector." The most commonly used vectors in gene therapy are viruses, since they are natural deliverers of genetic material (their own) into a human cell. Viral genome is altered in a manner to make a virus safe and noninfective, and to carry a therapeutic gene. Only a few gene therapy protocols have been very successful despite great efforts of many researchers (Anderson, 1992; Kastelein et al., 2013).

Recent revolutionary advances in gene editing now allow the introduction of a wide range of therapeutic genetic changes to cells with relative ease and accuracy, reducing the risk of unexpected side effects. Genome editing is a set of related techniques that make it possible for a scientist to introduce precise genetic changes to a specific section of the DNA. The last few years have seen a rapid change in ease, affordability, and accessibility of the genome editing technique.

Genome editing with engineered nucleases (GEEN) is a type of genetic engineering in which DNA is inserted, deleted, or replaced in the genome of a living organism using engineered nucleases, or "molecular scissors." Nine genome editors are currently in use. Among them, meganucleases, zinc finger nucleases (ZFNs), transcription activator-like effector-based nucleases (TALEN), and the CRISPR-Cas9 (Clustered Regularly Interspaced Short Palindromic Repeats; CRISPR associated protein 9) system are the most commonly used (Jinek et al., 2012; Le Cong et al., 2013).

2.6.2 Molecular-Targeted Therapy

Molecular-targeted therapy is the best example of accurate, causal therapy since the disease-causing molecular defect is a target of a drug. It is one of the cornerstones of personalized medicine. A therapeutic drug is often a small molecule drug that targets markers inside the cell or an antibody that attaches to specific targets on the outer surface of cells. Most molecular-targeted therapies are used in the treatment of cancer. They have contributed to the fact that several cancers are becoming a chronic disease (Pavlovic et al., 2017).

The most common mechanisms of epigenetic silencing (aberrant DNA methylation and modification of histones) have been used as targets for molecular therapeutics that are clinically applicable in the cancer treatment. These drugs could replace cytotoxic chemotherapy in near future (Pavlovic et al., 2017).

2.6.3 Cellular Therapy

Today, treatment of various diseases is performed by using the patient's own cells. That is one of the ways of applying the principals of personalized medicine in the treatment.

Stem cells have an extraordinary ability to produce all types of cells in the body. Stem cells are not only the characteristic of embryonic stage of development, but they are also present in adult tissues. In various tissues, adult stem cells represent the basis for the renewal of cells during life. An individual's stem cells could be used as an effective therapeutic method. This concept enables producing autologous stem cell lines of the patient, which could be used for transplantation of damaged tissues and organs.

Cell-based therapies have already employed genome editing to make customized blood cells to cure patients with severe forms of leukemia. This combines transplantation of patient's stem cells with genome editing to make those cells "a living drug" for the patient himself. In 2017, the Food and Drug Administration (FDA) approved the first ever cell-based gene therapy in the United States. The drug is a highly personalized cancer treatment called CAR T-cell therapy (CAR is short for chimeric antigen receptor) and is applied for treatment of acute lymphoblastic leukemia and other hematological malignancies (Porter et al., 2011).

2.7 Ethical Aspects of Gene Therapy, Molecular-Targeted Therapy, and Cellular Therapy

Ethicists are concerned about the potential consequences of gene therapy, such as: Who decides which traits are normal and which constitute a disability or disorder?, Will the high costs of gene therapy make it available only to the wealthy?, Could the widespread use of gene therapy make society less accepting of people who are different?, and May there be unforeseen dangerous consequences since our knowledge on genomics is not sufficient?

Even more ethical concerns are associated with germline gene transfer, genetic engineering, and genome editing at the gamete and zygote stages. The problems resulting from the current insufficient knowledge of the control of genetic mechanisms and limits of current technology could be severe, even lethal, or they even might be evident in adulthood, when the errors could be passed on to future generations. For these reasons, germline gene therapy has been considered ethically impermissible.

Beyond the medical risks, a number of long-standing ethical concerns exist regarding the possible practice of germline gene therapy and gene editing. Such modifications in human beings raise the possibility that we are changing not only a single individual but numerous future generations as well. Concerns involve issues ranging from the autonomy of future individuals to distributive justice, fairness, and the application of these technologies to "enhancement" rather than treating disease.

2.8 Gene Enhancement, Molecular-Targeted Enhancement, Cellular Enhancement

The same genetic engineering methodologies used in gene therapy can be applied for the introduction of changes into a genome to cause genetic enhancement. Also, the same approaches used in personalized medicine can lead to genetic enhancement. The introduction of the genetic material into the cell could produce the enhancement of human traits (gene enhancement). Knowledge the molecular structure of enhancement related genes could promote new drugs development (molecular-targeted enhancement) (Fig. 1). Finally, genome editing applied in stem cells (adult or embryonal) of a person could target a gene involved in enhancement of a certain trait (cellular enhancement) (Fig. 2).

Therefore, from a methodological point of view, genetic enhancement can be performed. However, there is not enough knowledge about the genetic basis

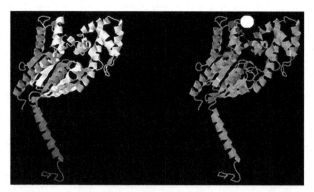

FIG. 1 Molecular-targeted enhancement through targeting mutated GNAQ protein. Three-dimensional models of a protein harboring a variant in an enhancement related gene *GNAQ* (*right*) targeted with a molecular drug (*white full circle*). The protein structure has been changed (*left*) with new potential for enhancement of smell and vision and the control of skin and hair color. For protein structure stability and modeling upon molecular-targeted enhancement, STRUM method and i-TASSER server were used.

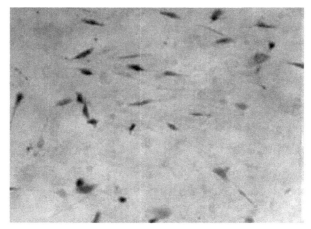

FIG. 2 Gene and cellular enhancement through transfer of gene for interleukin-10 (*IL-10*) to adipose-derived mesenchymal stem cells. Genetically engineered autologous stem cells can be administered to joints leading to prevention of rheumatoid manifestation characteristic for elderly population. Optical microscopic imaging and in situ staining of the vector with cloned *IL-10* gene confirms efficient gene transfer into stem cells.

of the traits that could be enhanced. Enhancement-related genes and genetic variants are to be discovered in order to be used in genetic enhancement.

However, there are some successful examples of genetic enhancement, demonstrating our knowledge of genetic basis of a nonpathological human traits leading to the improvement of athletic performance, such as muscle growth, blood production, endurance, oxygen dispersal, or pain perception. Animal models produced by genetic engineering and gene enhancement approach have been bred. Their nick names are "Schwarzenegger mice" and "marathon mice" and they represent a model for gene doping which could be applied in humans to enhance their athletic performance (Barton-Davis et al., 1998; McPherron et al., 1997).

Genetic enhancement could replace chemical and surgical interventions used for the purpose of self-improvement and could become the "beauty treatment" of the future.

One of the areas in which genetic enhancement might find initial applications is in sports, and the other one is the delay the aging process. Our knowledge of genetic enhancement markers related to those two fields are presented in the following text.

2.8.1   Genetic Enhancement of Athletic Performance

Improvement of athletic performance through the manipulation of performance-related genes (such as hGH), termed "gene doping," could result in increased muscle mass, muscle strength, or running endurance, as well as in promoting more rapid healing and reduce pain in sport injuries (Baoutina et al.,

2007). Several genes have already been considered as a target for athletic performance improvement among which are genes involved in endurance ability, muscle performance and power exercise, and susceptibility to injuries.

EPO

Erythropoietin (EPO) is a hormone responsible for the regulation of erythropoiesis. By increasing the EPO levels and, thus, the number of red blood cells from the circulatory system, the oxygen delivery to the tissues is enhanced and higher sport performance/endurance can be achieved (Gaudard et al. 2003). Aside from constant injections of the hormone, EPO levels could be raised in vivo, through intramuscular (IM) injection with a plasmid containing EPO gene (pEPO) (Fattori et al., 2005) and using ex vivo gene transfer approaches (Lippin et al., 2005). The increased sensitivity of erythroid progenitors to EPO and, therefore, the stimulation of erythropoiesis is also observed in individuals with mutations in the gene encoding the erythropoietin receptor (EPOR) making their endurance levels higher compared to the individuals without these mutations (Arcasoy et al., 2002).

VEGF Gene

The enhancement of physical endurance could also be achieved by increasing of number of blood vessels around each muscle fiber, which leads to an increase in gas, heat, and nutrient exchange between blood and working muscle fibers (Akhmetov et al., 2008). Although the formation of new blood vessels is stimulated by aerobic exercise training (Shono et al., 2002), individual differences in the degree of angiogenesis is determined by genetic factors (Rogers and D'Amato, 2006). One of the key regulators of angiogenesis is vascular endothelial growth factor (VEGF), the glycoprotein responsible for blood and lymphatic vessel cell proliferation (Bernatchez et al., 1999). Numerous studies have showed association of *VEGF* gene variants with physical performance of athletes. Namely, variants in regulatory region of *VEGF* gene affect its expression levels leading to the improvement of the maximal oxygen consumption (V_{O2max}) and thus could be considered as a potential target for gene doping (Akhmetov et al., 2008; Prior et al., 2006).

PPARGC1A, PPARA, and *PPARD* Genes

Peroxisome proliferator-activated receptors (PPARs) is a protein family whose members, PPARα, PPARγ, and PPARδ, regulate the expression of genes involved in lipid and carbohydrate metabolism (Akhmetov et al., 2007).

PPARA and PPARGC1A, coding PPARα and its coactivator PGC-1α, respectively, as well as PPARD, coding PPARδ, are highly expressed in tissues that catabolize fatty acids, such as skeletal muscles, and induce remodeling of their fiber composition (Akhmetov et al., 2007). It is now well established that the expression of these genes leads to a switch from "fast-twitch" glycolytic type IIb fibers to "slow-twitch" oxidative type I/IIa fibers, which results in higher endurance in sport performance (Baoutina et al., 2007). Although exercise increases the expression of *PPARA, PPARGC1A,* and *PPARD* genes and promotes the conversion to type I/IIa muscle fibers (Russell et al., 2003), a similar effect could be attained by the expression of these genes in the absence of training as well (Lin et al., 2002). Studies have shown that specific variants in *PPARA* and *PPARGC1A* genes could lead to their altered expression and are associated with and could be advantageous to top-level endurance athletes (Eynon et al., 2010). Also, carriers of the variant affecting exon 4 of the *PPARD* gene (rs2016520), associated with increased gene expression, have higher predisposition to endurance performance (Akhmetov et al., 2007).

ACE and ACTN3 Genes

Two genes strongly associated with physical performance in humans are the *ACE* and *ACTN3* genes, with a variant in *ACE* being the first genetic element shown to have a significant impact on athletic performance (Montgomery et al., 1998).

   ACE gene codes the angiotensin-converting enzyme (ACE), which degrades vasodilator kinins, and converts angiotensin I (ATI) to the vasoconstrictor angiotensin II (ATII). As part of the renin-angiotensin system, it is responsible for controlling blood pressure by regulating body fluid levels (Guth and Roth, 2013). It was shown that the insertion of a 287 bp Alu repeat sequence within intron 16 of the *ACE* gene is associated with lower plasma and tissue ACE activity and higher physical endurance (Puthucheary et al., 2011).

   ACTN3 codes the α-actinin-3 protein (ACTN3), a part of contractile apparatus in the "fast-twitch" muscle fibers that are responsible for the rapid muscle contractions and, thus, involved in strength- and power-orientated performance (MacArthur and North, 2007). Many studies have demonstrated that the common R577X variant within *ACTN3*, leading to the production of nonfunctional α-actinin-3 protein, could be associated with lower baseline muscle strength and poorer sprint performance (Clarkson et al., 2005; Niemi and Majamaa, 2005). However, the presence of R577X *ACTN3* variant alone does not ensure the predisposition for elite athletic performance since many of the α-actinin-3 functions can be compensated by other enzymes involved in the anaerobic muscle metabolism (Ginevičienė et al., 2011).

IGF-1 Gene

IGF-1 (insulin-like growth factor 1), encoded by *the IGF-1* gene, is a member of polypeptide hormones involved in tissue regeneration processes and, specifically, in exercise-associated muscle growth and development (Ben-Zaken et al., 2013). This role of IGF-1 made gene therapy using mIGF-1 (muscle-specific IGF-1), a promising approach for patients affected by some types of muscle disease, but also for athletes in need for extra physical strength. It was shown that variants within *IGF-1* promoter region could result in higher circulating IGF-1 levels, suggesting the importance of these variants in endurance and, especially, power sport performance (Ben-Zaken et al., 2013). Considering the limitations and side effects of the human recombinant IGF-1 injections, including the need for high doses, cardiac problems, and cancer progression, gene therapy represents a much safer and more effective way of providing muscles with high, stable concentrations of IGF-1 (Baoutina et al., 2007).

CK-MM Gene

As the energy supply for muscle activity represents one of the main factors in determining human physical performance, MM-CK (the muscle isoform of creatine kinase), one of the key enzymes responsible for maintaining this supply, could play a role in the enhancement of physical performances (Fedotovskaia et al., 2012). It was shown that a decrease in the activity of the MM-CK, encoded by the *CK-MM* gene, reduces the intensity of the muscle contraction and leads to an increased oxygen uptake (van Deursen et al., 1993). Variants located in the 3′-untranslated region of the *CK-MM* are believed to have an effect on the MM-CK activity in myocytes, leading to altered oxygen consumption, and subsequently making this gene a candidate for gene doping (Rivera et al., 1997).

TNC Gene

Athletic performance could also be enhanced by altering the injury-prone tissues such as ligaments or tendons to be more resistant and less likely to be damaged during physical performance. Tenascin-C, encoded by the *TNC* gene, binds components of the extracellular matrix and cell receptors, and regulates cell-matrix interactions and the tendon's response to mechanical load (Maffulli et al., 2013). The *TNC* gene is known to contain a GT dinucleotide repeat polymorphism within intron 17, which is shown to be associated with the chronic Achilles tendon injury. Although the effect of this polymorphism on the gene expression is not known, it was shown that alleles with 12 and 14 GT repeats are overrepresented in the subjects with this injury, while alleles with 13 and 17 GT repeats are underrepresented (Mokone et al., 2005).

COL1A1 and COL5A1

Collagen type I fibrils are a component of bone matrix and are involved in the formation of ligaments and tendons. They are formed from collagen Iα1 and Iα2 polypeptides, encoded by the *COL1A1* and *COL1A2* genes, respectively (Maffulli et al., 2013). *COL1A1* contains a G to T polymorphism in intron 1, which increases the affinity for the Sp1 transcription leading to elevated gene expression. This results in the formation of unstable collagen type I homotrimer consisting only of collagen Iα1 polypeptide chains (Mann et al., 2001). The results of several studies have implied that the rare TT genotype of the *COL1A1* protects the individual from acute soft tissue rupture and should be included in the risk models for athletes (Collins et al., 2010).

COL5A1 gene encodes α1 chain of type V collagen, which, although a minor fraction of the collagen content in ligaments and tendons, has a crucial role in collagen fibril assembly and fibrillogenesis (Wenstrup et al., 2004). A variant within the 3′-untranslated region of *COL5A1* (*Bst*UI restriction fragment length polymorphisms) could increase type V collagen production, decreased mean fibril diameter, and increased fibril density, which would lead to modifications to the mechanical properties of the musculoskeletal soft tissue and, hence, to the many phenotypes associated with these properties including Achilles tendon injuries (Mokone et al., 2006; Collins and Posthumus, 2011).

2.8.2 Genetic Enhancement of Athletic Performance—Ethical Issues

Similar to substance doping, gene doping has raised many ethical concerns. The first one is whether gene doping should be considered cheating and, as such, should it be prohibited? This question is based on the notion that sport should be a fair game and that no athlete should have an unfair advantage over another. What genetics has taught us is that no man is created equally and, by this definition, some of the professional athletes already have an innate genetic advantage. Does this mean that the athletes born with genetic modification that allows them to have greater performance abilities than others, should not be permitted to compete? Legalizing gene doping would remove the advantage held by a few, which would lead to competition based on equality. However, this presumption would only be true in an economically equal world. Given the economic difference between social classes and various countries, the wealthy would be able to purchase the gene enhancement technology more easily than the less fortunate ones, giving them the unfair advantage.

One of the argument against gene enhancement in sport is the fact that it could be harmful to an athlete, as well as for sport in general. Although the research on gene enhancement and its long-term effect on human health is still at the beginning, the term "harmful" seem to be a relative one. Namely, sports like boxing are harmful to the athletes by their nature but they are not being banned due to their harmfulness. It would appear that it all comes down to the degree of "harmfulness" which, until more research is done, cannot be determined. One should also consider the damaging effect gene doping could have on the quality of the sport. Many argue that by having athletes with the same genetic predisposition will make them less unique, thus, removing some of the advantages that lead to exciting plays. Although athletic performance, as many other physical characteristics and behavior, does not depend solely on a genetic composition of a person, this could be one more argument against gene doping (Miah, 2001).

Although the gene doping is believed to be inevitable, the ethical concerns regarding this type of gene enhancement could slow down its application in everyday life.

2.8.3 Aging-Related Genes—A Key for Antiaging and Longevity

Aside from athletic performance and performance-related genes, which genetic enhancement could find its initial application in, other genes could also be targets for gene manipulations aimed at improvement of nonpathological human traits. These include genes involved in age delaying, intelligence improvement, and skin color and elasticity, among others.

Due to great breakthroughs in biomedicine and molecular biology in the last century and the last decade of the current century, the proportion of the elderly population has increased substantially. Regrettably, many of their health problems have not been solved yet. The continuous loss of physiological integrity and increased susceptibility to death are still facets of the elderly life, no matter how many diseases are treatable. That is why the search for aging-related genes is a project on its own in contemporary biomedicine (Kwon et al., 2017).

Aging is a process that is ingrained in the genetic code of all human beings. For as long as mankind has existed, it has been treated as a given of existence, just like space and time. Until recently, only science fiction and fantasy have been going into the realm of possibility of healthy and vigorous century-old living beings. But now, aging-related genes may be the keys to a long and well life, even to the point of immortality.

Although the promise of defeating an ingrained process such as aging is inspiring, the entire endeavor has proven to be very complex. Many of the mechanisms of aging are unknown, so only through experiments, data mining, bioinformatics, and checking the conclusions of analyses can bring us even close to a solution. But how can aging-related genes be investigated in a reasonable amount of time?

Kwon and colleagues have used bioinformatics and literature mining for such an activity (2017). There is also a number of databases that deal with the mentioned genes. For example, Human Aging Genomic Resources (HAGR) is an online collection of research databases and tools for the biology and genetics of aging. It incorporates GenAge (a database of genes associated with aging in humans), AnAge (an extensive database of longevity records of vertebrate species), and GenDR (a newly incorporated database, containing both gene mutations that interfere with dietary restriction mediated lifespan extension and consistent gene expression changes induced by dietary restriction) (Tacutu et al., 2013).

Using HAGR, the targeted aging-related genes were picked if there were known drugs that affected the genes in question. It has been found that 522 out of 7146 genes have been identified as target genes for known drugs.

For example, one of the experiments on effect of metformin on fruit flies has proven to be promising, since it has increased the lifespan of the mentioned species (Anisimov, 2013). As far as it has been determined, metformin targets adenosine monophosphate (AMP)-activated protein kinase and the genes which determine it are the known aging-related genes *PRKAA1* and *PRKAA2*. The fruit flies that were given metformin had a double lifespan compared to the control group (Anisimov, 2013).

Another good example is the *TLR2* gene, which is targeted by a newly designed drug Erlotinib (Kwon et al., 2017), used for the treatment of Lyme disease. This gene is involved with pathways that affect longevity and aging, like the MAPK (mitogen-activated protein kinase) signaling pathway and PI3K-Akt [Phosphatidylinositol-4,5-bisphosphate 3-kinase—Protein kinase B (PKB), also known as Akt] signaling, as well as in a number of signaling pathways, which are connected to infections.

With the tools at our disposal, the journey is long, but it is possible. Regrettably, the problem of aging would not be solved in its entirety by this course of action alone. Considering that different cell types in the body age differently (Ori et al., 2015), new diseases associated with aging could spring up, creating new challenges for medicine and gene enhancement. Thus aging-related problems are not solvable with a single stroke.

Furthermore, with the drastic increase in the elderly population (He et al., 2015), the structure of the workforce all over the globe has become overburdened. People will be less capable of creating a workforce large enough to support the needs of the rest of the populace. These economic problems give rise to a dilemma of whether to increase the upper age limit of working or to create a selection of individuals who would be granted longevity.

3.   LIBERAL EUGENICS

The idea of liberal eugenics is fairly new, although its core notion, eugenics, is old and infamously known. Eugenics has been a matter of debate and dispute for centuries and it has been incorporated into political agendas and programs as an instrument for selection of desirable traits. But, what is eugenics all about? As an etymology suggests, it is concerned with good birth. The idea is mostly associated with Darwin's cousin Francis Galton who promoted the idea in works such as Hereditary Genius (1869), and Natural Inheritance (1894) but the idea has been debates, as we know of, since Plato. In the Book V of the Republic, Plato gives an analogy with dogs where he states the importance of good ancestry. The best bred dogs will produce the best offspring (Grube and Reeve, 1992). The same goes for people. The best members of the class of the guardians should procreate among themselves in order to ensure the production of the best children and, tomorrow, the best guardians. Plato believes that it is necessary to incentivize the procreation of the best and to disable the procreation of the worst members of society. These are some of the first written arguments that can be found in favor of eugenics.

On the same line of argumentation, centuries later, attempts were made to incorporate those ideas into the official political agenda. During the so-called progressive era in the United States in the beginning of the 20th century there was a practice to sterilize people with some intellectual and physical disabilities in order to stop their reproduction. It was promoted by having banners where it was written "Some people are born to be burden on the rest." Also, similar attempts were made in the Nazi Germany with Action T4, which required homosexuals, people with mental disabilities, Jews, and the Roma population to be forcefully sterilized or euthanized (Meyer, 1988).

But, does eugenics necessarily have to have such a bad reputation? Every normal parent has a strong desire to improve the well-being of its offspring and to maximize the possibilities for their children to live up to their full potential. Parents often try to secure best possible material conditions for their children and they try to provide the best possible education, making them capable of dealing with the extremely competitive surrounding. Mothers during pregnancy take vitamins and other supplements to ensure proper development of a baby and to enhance capacities as much as possible.

That concern for the benefit of the children has led some authors to develop the idea of liberal eugenics. Differentiating from the eugenic practice that was conducted at the beginning of the 20th century and promoting liberalism as a most desirable political ideology, they have come to the idea that parents should be ones who decide which traits their children will possess.

Liberal eugenics is a practice that allows the use of biotechnology in order to create enhanced children (Agar, 1998). The idea is that by using genetic engineering, genes can be manipulated in such a way to activate certain traits those genes are responsible for. Proponents of liberal eugenics value human life and freedom the most and they are trying to turn away from eugenics in the traditional sense where the state is the arbiter regarding what traits are desirable and the elimination of the ones who possess undesirable traits. Authoritarian eugenics is trying to design people by preestablished model while liberal eugenics believes that this right is reserved for parents only. Proponents accept that different people have different ideas of what makes a good life and what traits should their offspring have. Those ideals are guidelines for genetic engineering when those traits are to be chosen and the state must not have any influence on that matter.

However, it is a fact that future generations can, and probably will, have different ideals than their parents. Those are the boundaries imposed on liberal eugenics. Parents can influence desirable traits only until it does not jeopardize the potential life plans of future human beings. Nicholas Agar, the author who coined the term and one of the most famous proponents of liberal eugenics, stated: "I argue that the respect for the life plans of future persons can constrain paternal choice in a way that sharply distinguishes the new eugenics from its ugly ancestor. To help demonstrate this I compare genetic engineers access to life plans with their access to capacities, the properties of a person that help determine success in life plans. I suggest that a program of systematic life plan modification is beyond the research of genetic engineers and that this inability imposes restrictions on the other types of enhancement" (Agar, 1998).

The position the proponents are taking is that interference in the characteristics that children will have is formed to a certain extent and that extent is to allow children to choose, when the time comes, whatever the path they desire. Liberal eugenics should only make sure that their capacities are maximized for achieving that and not determining their telos. Although liberal eugenics is inclusive and tolerant for different world views, there are some boundaries that must be respected. "Liberal freedoms are always freedoms within limits; no freedom is absolute. The presumption in favor of the freedom of speech does not stop us from banning incitements to commit race crimes or false bomb alerts by passengers on 747s. Therefore, it would seem that liberal eugenicists should be open to the idea that some uses of enhancement technologies are just wrong and should be banned" (Agar, 2004).

The practice of liberal eugenics should serve only to those for whom it was conducted on for achieving their life plans and not parents, state, or someone else.

One of the problems that arises is what criteria we could use in order to evaluate new, unknown biotechnological interventions and what impact they will have?

We must secure moral institutions that would regulate the mentioned interventions by finding cases that are similar in a relevant way. Those cases must correspond to the unknown situations in a way that an analogy can be made that will point out the problematic moral questions common for them. Relevance is determined by a common denominator that shares the same properties in both cases that are creating the problem. Agar gives the idea of "moral images" as a filter for determining unknown situations in relation to known ones and he further states that we must establish moral intuitions about those cases (Agar, 2004). He ensures us that his goal is not to destroy moral principles but rather to guide us to the wide specter of potential moral concerns that will be caused by new biotechnologies.

One important distinction Agar points out is the distinction that separates the therapeutic benefits of genetic engineering from the eugenic benefits. Therapeutic benefits are directed toward the treatment of diseases. Cystic fibrosis is an example that Agar uses to illustrate what can be detected and treated in the prenatal period using prenatal screening (Agar, 1998). What he is trying to achieve with this example of genetic modification is the effort and possibility to reach the level that is considered normal, meaning to remove obstacles to the normal functioning of an individual.

That is an example of therapeutic usage of genetic engineering and there is an overall agreement that it should be done. On the other hand, the eugenics efforts of genetic engineering are to create people that will have abilities that go beyond the average person. For the critics of liberal eugenics, this latter use of genetic engineering is controversial, while the former is not the subject of debate.

Everyone agrees that pain should be avoided, but there is no consensus whether biotechnology should be used as a mean of improving the future person.

The main opponents are the authors known as bioconservatives. Their thesis is that any form of enhancement is morally unacceptable, since such procedures are "dehumanizing" and that it can actually be reduced to two things—that human nature has value in itself and the belief that the improvement will undermine the existing social order (Bostrom, 2007). The belief that human nature has value in itself does not have to rely on the perception of people as God's creatures, which would be an objection that could be expected of the people who are considering this issue from a religious point of view. This position is defended by Karl Rahner, one of the most famous Roman Catholic theologians of the 20th century. Other authors, such as Bill McKibben, believe that the arrangement of the genes given to people has value for themselves and that their modifications cannot be morally justified. There is no explicit mention of the "human nature," but the same argumentation can be found with those who use this notion, such as Leon Kass.

According to bioconservatives, the danger of enhancement can be summed up in two points, namely that the creation of posthumans (the term used to denote the ones upon the bio-enhancement was performed) is degrading to humans, and the next fear is that by the creation of the posthumans, humans are threatened to be extinct. This point of view states that genetic engineering will "produce" new people who will decisively depersonalize and dehumanize existing humans.

Fukuyama accepts the thesis of social Darwinism that people want to match with those who belong to a similar status like themselves. By creating genetically enhanced people, a new class is actually being created. This class would have similarities with the class of noblemen in feudal societies where a mere fact that a child is born to certain parents (having "blue blood") gives them certain rights and privileges that place them above the others. Fukuyama believes that modification of genes would have the same social effects (Fukuyama, 2002).

In this scenario, the value would not be judged by the ancestry but by the genetic structure those humans possess. Those with modified genes would be considered legitimate heirs of wealth, they should make decisions, and they would make the ruling class. This position of bioconservatives could lead to the strengthening of a deterministic position, social Darwinism, and thus to the elimination of other relevant factors such as environment (nurture), good fortune, and the efforts in achieving personal goals. According to Fukuyama, this would also lead to less sympathy for those who have a worse status. Such a situation would inevitably result in violent conflicts. He sees a way out of this state in two ways: either to ban all biotechnologies that make people better or make them accessible to everyone (Fukuyama, 2002).

It is important to notice that liberal eugenics does not mean obligatory genetic treatment nor complete freedom to choose whatever comes into one's mind. It is a method to aid offspring to have complete freedom to choose whatever path they want and to have the most useful characteristics to achieve that. The line that separates allowed and forbidden genetic modification is the freedom of future persons.

4.   REGULATION, HEALTH COVERAGE, AND PUBLIC OPINION

As the reality of germline cell modifying becomes more apparent, so do the policies concerning modification of future offspring become less and less useful in the climate, which is being formed. Germline cells are cells whose genetic material gets carried down to a new individual of a species, so it is natural for the state to regulate such an important part of mankind's existence. Unfortunately, policies on germline manipulation are polarized. Either they are rigorous to the point of banning or they do not exist (Isasi and Knoppers, 2015). France, Germany, Australia, and Canada are examples of countries that forbid germline modification. On the other hand, China, India, and Japan have sanctions that are not applicable (Isasi et al., 2016).

Policies concerning germline manipulation reflect the biases of the policy makers (Knoppers et al., 2017). The main reasons why policy makers do not make adequate policies concerning this subject are based on history and lack of distinct definitions of certain important aspects.

When it comes to history, gene therapy of humans, even for the sake of creating well-being, is associated with the core idea of eugenics, which aims at making the gene pool of humans better. Germany and India are distinct examples of fear of eugenics in their policies (Indian Council of Medical Research, 2000; Interdisciplinary Study Group "Gene Technology Report", 2008).

On the other hand, there is not a very clear distinction between enhancement and therapy (Committee on Human Gene Editing, and National Academies of Sciences, Engineering, and Medicine, 2017). This means that there is still a long way until the state forms a clear stance.

Insurance companies are not willing to subsidize unclear "treatments," which can be classified as "enhancements" as well (Buchanan et al., 2000). This is one of the direct consequences of polarized policies on gene therapy. For example, Glybera, one of the first approved drugs for gene therapy in Europe costs around 1.1 million euros, making it a very undesirable medical supply to reimburse. Therefore, it is clear why gene therapy is treated equally to cosmetic surgery. This treatment could also benefit insurance companies by labeling an expensive treatment as enhancement, especially if it is controversial in its implications (Mehlman, 1999).

Societal views are also an important factor. Public opinion can dictate the use of technology and the technologies being developed and the policies declared (McCaughey et al., 2016). An example of a technology which requires public opinion is the use of CRISPR/Cas9 in humans. According to the International Summit on Human Gene Editing, a broad social consensus is required before proceeding with germline editing (Baltimore et al., 2016). Despite this requirement, there have not been many surveys that could indicate the content of any general stance concerning the topic in question. There is only an indication that "enhancement" is viewed as an undesirable effect of the modifications (Blendon et al., 2016; Funk et al., 2016).

5.   CONCLUSION

Fascinating progress of genetics has strongly contributed to great achievements in modern science, including medicine and agriculture. Research in the field of a number of scientific disciplines is strongly associated with the use of genetic technologies. One of the most intriguing examples is how genetics can lead to human enhancement. It became a hot topic for molecular geneticists and bioethicists as well. Genetic enhancement refers to the introduction the changes into a genome or epigenome intended to modify and improve nonpathological human traits. But, is it feasible or still hypothetical? How far are we from the era of genetic enhancement? Tremendous advancement in technology available for genetic research has led to the accumulation of knowledge of human genetic and epigenetic signature and the development of methodology, which are encouraging enough to make us believe that human enhancement will be a reality very soon. As a matter of fact, recent gene editing experiments carried out in human embryos have raised the question of whether interventions like the introduction of a CCR5-132 deletion, which could provide heritable resistance to HIV infection (Kang et al., 2016), ought to be considered as therapeutic editing or whether it ought to be classified as a form of "enhancement" (So et al., 2017). So, geneticists are moving extremely fast toward genetic enhancement. But, we should be aware that genetic enhancement creates significant new challenges to our regulatory capabilities. Also, prompt and thorough research in the field of bioethics is needed in order to know if Icarus' or Faustus' intentions were beneficial to mankind and whether we are able to go beyond the Sun and the alchemy of these two visionaries.

ACKNOWLEDGMENT

This work has been funded by Grants from the Ministry of Education, Science and Technological Development, Republic of Serbia (III 41004)

REFERENCES

Agar, N., 1998. Liberal eugenics. Public Aff. Q. 12 (2), 137–155.

Agar, N., 2004. Liberal Eugenics: In Defense of Human Enhancement. Blackwell Publishing, Oxford.

Akhmetov, I.I., Astranenkova, I.V., Rogozkin, V.A., 2007. Association of PPARD gene polymorphism with human physical performance. Mol. Biol. (Mosk) 41 (5), 852–857.

Akhmetov, I.I., Khakimullina, A.M., Popov, D.V., Missina, S.S., Vinogradova, O.L., Rogozkin, V.A., 2008. Polymorphism of the vascular endothelial growth factor gene (VEGF) and aerobic performance in athletes. Fiziol. Cheloveka 34 (4), 97–101.

Anderson, W.F., 1992. Human gene therapy. Science 256 (5058), 808–813.

Anisimov, V.N., 2013. Metformin: do we finally have an anti-aging drug? Cell Cycle 12 (22), 3483–3489.

Arcasoy, M.O., Karayal, A.F., Segal, H.M., Sinning, J.G., Forget, B.G., 2002. A novel mutation in the erythropoietin receptor gene is associated with familial erythrocytosis. Blood 99 (8), 3066–3069.

Baltimore, D., Baylis, F., Berg, P., Daley, G.Q., Doudna, J.A., Lander, E.S., et al., 2016. 'On human gene editing: international summit statement by the organizing committee. Sci. Dent. Tech. 32, 55–56.

Baoutina, A., Alexander, I.E., Rasko, J.E., Emslie, K.R., 2007. Potential use of gene transfer in athletic performance enhancement. Mol. Ther. 15 (10), 1751–1766.

Barton-Davis, E.R., Shoturma, D.I., Musaro, A., Rosenthal, N., Sweeney, H.L., 1998. Viral mediated expression of insulin-like growth factor I blocks the aging-related loss of skeletal muscle function. Proc. Natl. Acad. Sci. U. S. A. 95, 15603–15607.

Ben-Zaken, S., Meckel, Y., Nemet, D., Eliakim, A., 2013. Can IGF-I polymorphism affect power and endurance athletic performance? Growth Horm. IGF Res. 23 (5), 175–178.

Berger, S.L., Kouzarides, T., Shiekhattar, R., Shilatifard, A., 2009. An operational definition of epigenetics. Genes Dev. 23 (7), 781–783.

Bernatchez, P.N., Soker, S., Sirois, M.G., 1999. Vascular endothelial growth factor effect on endothelial cell proliferation, migration, and platelet-activating factor synthesis is Flk-1-dependent. J. Biol. Chem. 274 (43), 31047–31054.

Blendon, R.J., Gorski, M.T., Benson, J.M., 2016. The public and the gene-editing revolution. N. Engl. J. Med. 374, 1406–1411.

Bojovic, K., Stankovic, B., Kotur, N., Krstic-Milosevic, D., Gasic, V., Pavlovic, S., Zukic, B., Ignjatovic, D., 2017. Genetic predictors of celiac disease, lactose intolerance, and vitamin D function and presence of peptide morphins in urine of children with neurodevelopmental disorders. Nutr. Neurosci. 1–11.

Boorse, C., 1977. Health as a theoretical concept. Philos. Sci. 44 (4), 542–573.

Bostrom, N., 2007. In defence of posthuman dignity. Bioethics 19 (3), 202–214.

Brothers, K.B., Rothstein, M.A., 2015. Ethical, legal and social implications of incorporating personalized medicine into healthcare. Pers. Med. 12 (1), 43–51.

Buchanan, A., 2009. Human nature and enhancement. Bioethics 23 (3), 141–150.

Buchanan, A., Brock, D.W., Daniels, N., Wikler, D., 2000. From Chance to Choice: Genetics and Justice. Cambridge University Press, Cambridge.

Burke, M.W., Antommaria, A.H., Bennett, R., Botkin, J., Clayton, E.W., Henderson, G.E., Holm, I.A., Jarvik, G.P., Khoury, M.J., Knoppers, B.M., Press, N.A., Ross, L.F., Rothstein, M.A., Saal, H., Uhlmann, W.R., Wilfond, B., Wolf, S.M., Zimmern, R., 2013. Recommendations for returning genomic incidental findings? We need to talk!. Genet. Med. 15 (11), 854–859.

Canton, J., 2002. The impact of convergent technologies and the future of business and the economy. In: Roco, M.C., Bainbridge, W.S. (Eds.), Converging Technologies For Improving Human Performance-Nanotechnology, Biotechnology, Information Technology and Cognitive Science. National Science Foundation Report, World Technology Evaluation Center (WTEC) Inc., USA, pp. 71–72.

Caplan, A., 1992. If I Were a Rich Man Could I Buy a Pancreas?: and Other Essays on the Ethics of Health Care. Indiana University Press.

Chan, S., Harris, J., 2007. In support of human enhancement. Stud. Ethics Law Technol. 1(1).

Clarkson, P.M., Devaney, J.M., Gordish-Dressman, H., Thompson, P.D., Hubal, M.J., Urso, M., Price, T.B., Angelopoulos, T.J., Gordon, P.M., Moyna, N.M., Pescatello, L.S., Visich, P.S., Zoeller, R.F., Seip, R.L., Hoffman, E.P., 2005. ACTN3 genotype is associated with increases in muscle strength in response to resistance training in women. J. Appl. Physiol. 99 (1), 154–163.

Collins, F.S., Green, E.D., Guttmacher, A.E., Guyer, M.S., 2003. A vision for the future of genomics research. Nature 422 (6934), 835–847.

Collins, M., Posthumus, M., 2011. Type V collagen genotype and exercise-related phenotype relationships: a novel hypothesis. Exerc. Sport Sci. Rev. 39 (4), 191–198.

Collins, M., Posthumus, M., Schwellnus, M.P., 2010. The COL1A1 gene and acute soft tissue ruptures. Br. J. Sports Med. 44 (14), 1063–1064.

Committee on Human Gene Editing and National Academies of Sciences, Engineering, and Medicine, 2017. Human Genome Editing: Science, Ethics, and Governance. National Academies Press, Washington, DC.

DeGrazia, D., 2014. Moral enhancement, freedom, and what we (should) value in moral behaviour. J. Med. Ethics 40 (6), 361–368.

ENCODE Project Consortium, 2012. An integrated encyclopedia of DNA elements in the human genome. Nature 489 (7414), 57–74.

Engelbart, D.C., 1962. Augmenting Human Intellect: A Conceptual Framework. Summary Report AFOSR-3223 under Contract AF 49 (638)-1024, SRI Project 3578 for Air Force Office of Scientific Research. Stanford Research Institute, California. Retrieved on 1 March 2007.

Esposito, M., 2005. Ethical implications of pharmacological enhancement of mood and cognition. Penn Bioethics J. 1 (1), 1–4.

Eynon, N., Meckel, Y., Sagiv, M., Yamin, C., Amir, R., Sagiv, M., Goldhammer, E., Duarte, J.A., Oliveira, J., 2010. Do PPARGC1A and PPARalpha polymorphisms influence sprint or endurance phenotypes? Scand. J. Med. Sci. Sports 20 (1), e145–50.

Fattori, E., Cappelletti, M., Zampaglione, I., Mennuni, C., Calvaruso, F., Arcuri, M., Rizzuto, G., Costa, P., Perretta, G., Ciliberto, G., La Monica, N., 2005. Gene electro-transfer of an improved erythropoietin plasmid in mice and non-human primates. J. Gene Med. 7 (2), 228–236.

Fedotovskaia, O.N., Popov, D.V., Vinogradova, O.L., Akhmetov, I.I., 2012. Association of the muscle-specific creatine kinase (CKMM) gene polymorphism with physical performance of athletes. Fiziol. Cheloveka 38 (1), 105–109.

Frankfurt, H.G., 1988. Freedom of the Will and the Concept of a Person. Humana Press, pp. 127–144.

Fukuyama, F., 2002. Our Posthuman Future: Consequences of Biotechnological Revolution. Farrar, Straus and Giroux, New York.

Funk, C, Kennedy, B Podrebarac Sciupac, E 2016, 'U.S. Public Opinion on the Future Use of Gene Editing' Available at: http://www.pewinternet.org/2016/07/26/u-s-public-opinion-on-the-future-use-of-gene-editing/.

Galton, F., 1869. Hereditary Genius: An Inquiry Into its Laws and Consequences (Vol. 27). Macmillan.

Galton, F., 1894. Natural inheritance. Macmillan and Company.

Gaudard, A., Varlet-Marie, E., Bressolle, F., Audran, M., 2003. Drugs for increasing oxygen and their potential use in doping: a review. Sports Med. 33 (3), 187–212.

Ginevičienė, V., Pranculis, A., Jakaitienė, A., Milašius, K., Kučinskas, V., 2011. Genetic variation of the human ACE and ACTN3 genes and their association with functional muscle properties in Lithuanian elite athletes. Medicina (Kaunas) 47 (5), 284–290.

Gordijn, B., Chadwick, R. (Eds.), 2008. Medical Enhancement and Posthumanity. 2. Springer Science & Business Media.

Grube, G.M., Reeve, C.D.C., 1992. Plato: Republic. Hackett, Indianapolis, IN.

Guth, L.M., Roth, S.M., 2013. Genetic influence on athletic performance. Curr. Opin. Pediatr. 25 (6), 653–658.

Habermas, J., 2003. Faith and knowledge. In: Habermas, J. (Ed.), The Future of Human Nature, 1st ed., Polity, Cambridge, pp. 101–115.

Harris, J., 2011. Moral enhancement and freedom. Bioethics 25 (2), 102–111.

Harris, J., 2013. Moral progress and moral enhancement. Bioethics 27 (5), 285–290.

He, W., Goodkind, D., Kowal, P., 2015. An Aging World: 2015. International Population Reports, US Census Bureau.

Indian Council of Medical Research, 2000. Ethical Guidelines on Biomedical Research Involving Human Subjects. Indian Council of Medical Research, New Delhi.

Interdisciplinary Study Group "Gene Technology Report", 2008. Gene Therapy in Germany. An Interdisciplinary Survey. Berlin-Brandenburg Academy of Science and Humanities, Berlin.

International Human Genome Sequencing Consortium, 2004. Finishing the euchromatic sequence of the human genome. Nature 431 (7011), 931–945.

Isasi, R., Kleiderman, E., Knoppers, B.M., 2016. Editing policy to fit the genome? Science 351, 337–339.

Isasi, R., Knoppers, B.M., 2015. Oversight of human inheritable genome modification. Nat. Biotechnol. 33, 454–455.

Jinek, M., Chylinsk, I.K., Fonfara, I., Hauer, M., Doudna, J.A., Charpentier, E., 2012. A programmable dual-RNA–guided DNA endonuclease in adaptive bacterial immunity. Science 17, 816–821.

Juengst, E., 1998. What does enhancement mean. In: Parens, E. (Ed.), Enhancing human traits: Ethical and social implications. Georgetown University Press, Washington, DC, pp. 19–47.

Kang, X., He, W., Huang, Y., Yu, Q., Chen, Y., Gao, X., et al., 2016. Introducing precise genetic modifications into human 3PN embryos by CRISPR/Cas mediated genome editing. J. Assist. Reprod. Genet. 33, 581–588.

Kass, L., 2003. Beyond Therapy: Biotechnology and the Pursuit of Happiness. Harper Perennial.

Kastelein, J.J.P., Colin, J.D., Ross, C.J.D., Hayden, M., 2013. From mutation identification to therapy: discovery and origins of the first approved gene therapy in the western world. Hum. Gene Ther. 24, 472–478.

Knoppers, B.M., Isasi, R., Caulfield, T., Kleiderman, E., Bedford, P., Illes, J., et al., 2017. Human gene editing: revisiting Canadian policy. NPJ Regen. Med. 2(3).

Kwon, Y., Natori, Y., Tanokura, M., 2017. New approach to generating insights for aging research based on literature mining and knowledge integration. PLoS One 12(8).

Le Cong, F., Ran, A., Cox, D., Shuailiang, L., Barretto, R., Habib, N., Hsu, P.D., Xuebing, W., Wenyan, J., Marraffini, L.A., Zhang, F., 2013. Multiplex genome engineering using CRISPR/ Cas systems. Science 15, 819–823.

Lin, J., Wu, H., Tarr, P.T., Zhang, C.Y., Wu, Z., Boss, O., Michael, L.F., Puigserver, P., Isotani, E., Olson, E.N., Lowell, B.B., Bassel-Duby, R., Spiegelman, B.M., 2002. Transcriptional co-activator PGC-1 alpha drives the formation of slow-twitch muscle fibres. Nature 418 (6899), 797–801.

Lippin, Y., Dranitzki-Elhalel, M., Brill-Almon, E., Mei-Zahav, C., Mizrachi, S., Liberman, Y., Iaina, A., Kaplan, E., Podjarny, E., Zeira, E., Harati, M., Casadevall, N., Shani, N., Galun, E., 2005. Human erythropoietin gene therapy for patients with chronic renal failure. Blood 106 (7), 2280–2286.

MacArthur, D.G., North, K.N., 2007. ACTN3: a genetic influence on muscle function and athletic performance. Exerc. Sport Sci. Rev. 35 (1), 30–34.

Maffulli, N., Margiotti, K., Longo, U.G., Loppini, M., Fazio, V.M., Denaro, V., 2013. The genetics of sports injuries and athletic performance. Muscles Ligaments Tendons J 3 (3), 173–189.

Mann, V., Hobson, E.E., Li, B., Stewart, T.L., Grant, S.F., Robins, S.P., Aspden, R.M., Ralston, S.H., 2001. A COL1A1 Sp1 binding site polymorphism predisposes to osteoporotic fracture by affecting bone density and quality. J. Clin. Invest. 107 (7), 899–907.

Marchant, G.E., Lindor, R.A., 2013. Personalized medicine and genetic malpractice. Genet. Med. 15 (12), 921–922.

Marsit, C.J., 2015. Influence of environmental exposure on human epigenetic regulation. J. Exp. Biol. 218 (Pt 1), 71–79.

McCaughey, T., Sanfilippo, P.G., Gooden, G.E., Budden, D.M., Fan, L., Fenwick, E., Rees, G., MacGregor, C., Si, L., Chen, C., Liang, H.H., Baldwin, T., Pébay, A., Hewitt, A.W., 2016. A global social media survey of attitudes to human genome editing. Cell Stem Cell 18, 569–572.

McPherron, A.C., Lawler, A.M., Lee, S.J., 1997. Regulation of skeletal muscle mass in mice by a new TGF-beta superfamily member. Nature 387, 83–90.

Mehlman, M.J., 1999. How will we regulate genetic enhancement? Wake For. Law Rev. 34, 671–714.

Meyer, J., 1988. The fate of the mentally ill in Germany during the third Reich. Psychol. Med. 18 (3), 575–581.

Miah, A., 2001. Genetic technologies and sport: the new ethical issue. J. Philos. Sport 28, 32–52.

Mokone, G.G., Gajjar, M., September, A.V., Schwellnus, M.P., Greenberg, J., Noakes, T.D., Collins, M., 2005. The guanine-thymine dinucleotide repeat polymorphism within the tenascin-C gene is associated with achilles tendon injuries. Am. J. Sports Med. 33 (7), 1016–1021.

Mokone, G.G., Schwellnus, M.P., Noakes, T.D., Collins, M., 2006. The COL5A1 gene and Achilles tendon pathology. Scand. J. Med. Sci. Sports 16 (1), 19–26.

Montgomery, H.E., Marshall, R., Hemingway, H., Myerson, S., Clarkson, P., Dollery, C., Hayward, M., Holliman, D.E., Jubb, M., World, M., Thomas, E.L., Brynes, A.E., Saeed, N., Barnard, M., Bell, J.D., Prasad, K., Rayson, M., Talmud, P.J., Humphries, S.E., 1998. Human gene for physical performance. Nature 393 (6682), 221–222.

Murray, T.H., 2002. Reflections on the ethics of genetic enhancement. Genet. Med. 4 (6 Suppl), 27S–32S.

Niemi, A.K., Majamaa, K., 2005. Mitochondrial DNA and ACTN3 genotypes in Finnish elite endurance and sprint athletes. Eur. J. Hum. Genet. 13 (8), 965–969.

Ori, A., Toyama, B., Harris, M., Ingolia, N., Hetzer, M., Beck, M., 2015. Integrated transcriptome and proteome analyses reveal organ-specific proteome deterioration in old rats. Cell Systems 1 (3), 224–237.

Otlowski, M., Taylor, S., Bombard, Y., 2012. Genetic discrimination: international perspectives. Annu. Rev. Genomics Hum. Genet. 13, 433–454.

Pavlovic, S., Stojiljkovic, M., Tosic, N., Zukic, B., Ugrin, M., Karan-Djurasevic, T., Spasovski, V., 2017. Genomics as a basis for precision medicine. Biologia Serbica 39 (1), 46–52.

Pavlovic, S., Ugrin, M., Stojiljkovic, M., 2015. Novel therapy approaches in beta thalassemia syndromes—a role of genetic modifiers. In: Munshi, A. (Ed.), Hemoglobin Disorders. InTech, Rijeka, Croatia, pp. 75–94.

Pavlovic, S., Zukic, B., Stojiljkovic-Petrovic, M., 2014. Molecular genetic markers as a basis for personalized medicine. J. Med. Biochem. 33, 8–21.

Persson, I., Savulescu, J., 2008. The perils of cognitive enhancement and the urgent imperative to enhance the moral character of humanity. J. Appl. Philos. 25 (3), 162–177.

Porter, D.L., Levine, B.L., Kalos, M., Bagg, A., June, C.H., 2011. Chimeric antigen receptor-modified T cells in chronic lymphoid leukemia. N. Engl. J. Med. 365 (8), 725–733.

Prior, S.J., Hagberg, J.M., Paton, C.M., Douglass, L.W., Brown, M.D., McLenithan, J.C., Roth, S.M., 2006. DNA sequence variation in the promoter region of the VEGF gene impacts VEGF gene expression and maximal oxygen consumption. Am. J. Physiol. Heart Circ. Physiol. 290 (5), H1848–55.

Puthucheary, Z., Skipworth, J.R., Rawal, J., Loosemore, M., Van Someren, K., Montgomery, H.E., 2011. The ACE gene and human performance: 12 years on. Sports Med. 41 (6), 433–448.

Rakic, V 2017, 'Enhancements: how and why to become better, how and why to become good', Camb. Q. Healthc. Ethics, July 2017, vol. 26, no. 3, pp 358–363.

Resnik, D., 2000. The moral significance of the therapy-enhancement distinction in human genetics. Camb. Q. Healthc. Ethics 9 (3), 365–377.

Rezaei, M., Zarkesh-Esfahani, S.H., 2012. Optimization of production of recombinant human growth hormone in Escherichia coli. J. Res. Med. Sci. 17 (7), 681–685.

Rivera, M.A., Dionne, F.T., Simoneau, J.A., Pérusse, L., Chagnon, M., Chagnon, Y., Gagnon, J., Leon, A.S., Rao, D.C., Skinner, J.S., Wilmore, J.H., Bouchard, C., 1997. Muscle-specific creatine kinase gene polymorphism and VO2max in the HERITAGE family study. Med. Sci. Sports Exerc. 29 (10), 1311–1317.

Rogers, M.S., D'Amato, R.J., 2006. The effect of genetic diversity on angiogenesis. Exp. Cell Res. 312 (5), 561–574.

Russell, A.P., Feilchenfeldt, J., Schreiber, S., Praz, M., Crettenand, A., Gobelet, C., Meier, C.A., Bell, D.R., Kralli, A., Giacobino, J.P., Dériaz, O., 2003. Endurance training in humans leads to fiber type-specific increases in levels of peroxisome proliferator-activated receptor-gamma coactivator-1 and peroxisome proliferator-activated receptor-alpha in skeletal muscle. Diabetes 52 (12), 2874–2881.

Savulescu, J., 2006. Justice, fairness, and enhancement. Ann. N. Y. Acad. Sci. 1093 (1), 321–338.

Savulescu, J., Persson, I., 2012. Moral enhancement, freedom and the god machine. The Monist 95 (3), 399.

Shono, N., Urata, H., Saltin, B., Mizuno, M., Harada, T., Shindo, M., Tanaka, H., 2002. Effects of low intensity aerobic training on skeletal muscle capillary and blood lipoprotein profiles. J. Atheroscler. Thromb. 9 (1), 78–85.

Simkulet, W., 2016. Intention and moral enhancement. Bioethics 30 (9), 714–720.

Skodric-Trifunovic, V., Buha, I., Jovanovic, D., Vucinic, V., Stjepanovic, M., Spasovski, V., Vreca, M., Skakic, A., Gasic, V., Andjelkovic, M., Pavlovic, S., 2015. Variants in VDR and NRAMP1 genes as susceptibility factors for tuberculosis in the population of Serbia. Genetika 47 (3), 1021–1028.

So, D., Kleiderman, E., Touré, S.B., Joly, Y., 2017. Disease resistance and the definition of genetic enhancement. Front. Genet. 8–40.

Tachibana, M., Amato, P., Sparman, M., Woodward, J., Sanchis, D.M., Ma, H., Gutierrez, N.M., Tippner-Hedges, R., Kang, E., Lee, H.S., Ramsey, C., Masterson, K., Battaglia, D., Lee, D.,

Wu, D., Jensen, J., Patton, P., Gokhale, S., Stouffer, R., Mitalipov, S., 2013. Towards germline gene therapy of inherited mitochondrial diseases. Nature 493 (7434), 627–631.

Tacutu, R, Craig, T, Budovsky, A, Wuttke, D, Lehmann, G, Taranukha, D, Costa, J, Fraifeld, VE, de Magalhães, JP 2013, 'Human aging genomic resources: integrated databases and tools for the biology and genetics of aging' Nucl. Acids Res., vol. 41(Database issue), pp D1027–33.

Ugrin, M., Milacic, I., Skakic, A., Klaassen, K., Komazec, J., Pavlovic, S., Stojiljkovic, M., 2017. Molecular genetic strategy for diagnosis of congenital adrenal hyperplasia in Serbia. Genetika 49 (2), 457–467.

van Deursen, J., Heerschap, A., Oerlemans, F., Ruitenbeek, W., Jap, P., ter Laak, H., Wieringa, B., 1993. Skeletal muscles of mice deficient in muscle creatine kinase lack burst activity. Cell 74 (4), 621–631.

Wenstrup, R.J., Florer, J.B., Brunskill, E.W., Bell, S.M., Chervoneva, I., Birk, D.E., 2004. Type V collagen controls the initiation of collagen fibril assembly. J. Biol. Chem. 279 (51), 53331–53337.

Wolpe, P., 2002. Treatment, enhancement, and the ethics of neurotherapeutics. Brain Cogn. 50 (3), 387–395.

Chapter 16

Should Incidental Findings Arising From Prenatal Testing be Reported Indiscriminately to Patients?

Valentina Kaneva* and Ina Dimitrova[†]
*Faculty of Philosophy, Sofia University "St. Kliment Ohridski", Sofia, Bulgaria, [†]Faculty of Philosophy and History, Plovdiv University "Paisii Hilendarski", Plovdiv, Bulgaria

1. INTRODUCTION

In the last decade, providing the results of genetic tests to patients has become an issue of growing expert and public interest. A large amount of literature is already available on the interpretation and disclosure of the genetic information obtained through new sequencing technologies as well as on anticipating and managing incidental findings. The rapid development of genetics and genomics and the implementation of large-scale sequencing in diagnostics raise important practical questions about the handling of incidental findings in both clinical and research settings. Recently, many professional societies, national and international commissions have issued guidelines and recommendations for health-care professionals on disclosing the results, including incidental or additional findings, from genetic and genomic testing to patients or research participants in different contexts.

Incidental findings are usually defined as results that clinicians and researchers were not looking for, but happen to discover, or results that have "potential health or reproductive importance" but were "beyond the aims of the study" (Wolf et al., 2008, p. 219). They are distinct from the primary findings, which are results that clinicians and researchers are looking for when they perform a specific test. The term "incidental" has been contested, especially with the new sequencing technologies entering the field that might not be aimed at answering the initial diagnostic question (Presidential Commission for the Study of Bioethical Issues, 2013, p. 3).

Clinical Ethics at the Crossroads of Genetic and Reproductive Technologies.
https://doi.org/10.1016/B978-0-12-813764-2.00016-7

Genetic testing for monogenic conditions requires DNA analyses of a specific gene able to provide answers to a particular question of clinical relevance. Recently, genetic tests have started looking at a large panel of genes by using newer technologies such as microarrays, or DNA chips. They use a relatively targeted approach for detecting certain conditions and might also use filters to select relevant variants after sequencing. Even if the targeted approach reduces the number of incidental findings or of findings that could not be interpreted, the new sequencing techniques actually increase the likelihood of incidental or secondary findings. Whole-genome sequencing (WGS) and whole-exome sequencing (WES) offer an enormous amount of raw data and a large number of findings and unclassified variants that require a complex analysis to generate information relevant for diagnostics or therapy (Van El et al., 2013, p. S1).

Both in clinical medicine and medical research, the main issues generated by incidental findings concern the nature of the generated genetic and genomic information and its assessment in terms of clinical utility, benefits, and possible harms. These are the basic considerations determining the professional judgment on what kind of findings should be returned to patients or research participants. The main ethical issues refer primarily to the disclosure of information and the risks generated by information sharing that might do more harm than good (Parens, 2015; Johnston et al., 2017). They concern: the scope of an obligation for clinicians and researchers to return incidental and secondary findings from genetic tests and criteria for the decision what kind of results should be returned; their obligations for informing patients and providing support, respecting their preferences and willingness to know but also not to know, and respect the interests of those who might have a potential interest but are usually to protect themselves (e.g., fetuses, embryos, minors). The attention of professional bodies and policy makers is directed above all on structures and procedures already available in current practice as those of informed consent and genetic counseling with an emphasis on the education of clinicians and genetic counselors in developing of pre- and posttest genetic counseling (Skirton et al., 2014; ACMG, 2015). It is questionable whether they will be sufficient with regard to the rapid development of new sequencing technologies and changes that they potentially impose on current practices.

The question of clinical utility of genetic tests is crucial in the recent discussions on returning the results of the genetic tests. In the Recommendations of European Society of Human Genetics (ESHG), clinical utility of a test refers to the positive risk-benefit ratio and meaningful options in case of a positive result (Van El et al., 2013, p. S4). From the overview provided, it became clear that the clinical utility of WGS followed by WGA was reported for rare diseases. Whole-genome techniques might be however applied in patients with symptoms but also in presymptomatic testing and population-screening programs: preconception carrier screening; neonatal screening for rare diseases; in the context of prenatal testing or in the future also in routine prenatal screening. (Van El et al., 2013, p. S2). Clinical utility is further determined by considerations of what is medically actionable, or is related to a proven thera-peutic or preventive intervention available (Beskow and Burke, 2010). Clinical utility is distinguished from personal utility associated with reproductive decision-making or life planning. According to Beskow and Burke, personal preferences are important but not determinative of investigators' obligation. In order to decide whether to offer some kind of information they should first assess the nature and the value of the information (Beskow and Burke, 2010).

The distinction between clinical care and research is important even if it seems to be blurred in the context of genome-wide testing in diagnostic setting. A diagnostic test seeks answers to clinical questions related to the medical condition of a patient; a research test seeks to prove a hypothesis and there might be no direct benefit for the patient involved as participant in research (Matthijs et al. 2016, p. 5). The results from a diagnostic test might be however hypothesis generating. And in research context the aim could be to explore the unclassified variant whereby the information gained might lead to a clinically relevant finding. The information with uncertain clinical implications may be not useful and rather burdening for a patient or family members. However, it might be important to researchers and might lead to clinically relevant findings after further studies. The situation becomes more complex with the intertwining of clinical diagnostic testing with research activities as biobanks research and the development of national and international variant databases.

In this chapter, we will focus on the complexity of ethical issues related to reporting incidental findings from genetic tests in prenatal care and on the main problems that practitioners and patients face in clinical and research settings. We do not refer to incidental findings that arise from imaging technologies despite of their importance associated with reproductive technologies. The spe-cific problems related to direct-to-consumer (DTC) testing and those of large-scale genetic databases and biobanks for research fall beyond the scope of the article. We examine the ethical issues that concern the disclosure of informa-tion, especially information about possible incidental or secondary findings from genetic testing in prenatal context, and try to outline the most important points discussed in the recently developed guidelines and recommendations. We embrace the recommendations of several prominent organizations, which emphasize the crucial importance of the efforts to adopt comprehensive pre- and posttest counseling, in order to guarantee that patients have achieved a gen-uine understanding of the complex genetic information and are sufficiently sup-ported to make informed decisions.

2. INCIDENTAL FINDINGS IN THE PRENATAL SETTING

2.1 Challenges of Definition

Prenatal testing undergoes a rapid development, thanks to the growing precision of the imaging technologies and to the advances in genetic and genomic tech-nologies. But this development also poses unprecedented challenges to our understanding and management of incidental findings. As it was already mentioned, it has become evident that the usual way of defining incidental findings has been recently widely contested due to its empirical and conceptual shortcomings, arising mainly in the context of large-scale sequencing, which is inherently prone to generating enormous amounts of information. A major issue is the implication that such findings are "either an incident or not expected, whilst the discovery of such [genetic] variants is intrinsic to genome-wide screening technologies" (Hehir-Kwa et al., 2015, p. 1602); "many specific IFs are common and recurrent" (Parker, 2008, pp. 341–342). So, different terms as "unexpected," "unsought," "unanticipated," "additional information," or "unsolicited information" (Hastings et al., 2012), "secondary variants" (Christenhusz et al., 2013), and "unexpected diagnoses" (Srebniak et al., 2014) have been proposed. The US Presidential commission for the study of bioethical issues divides incidental findings in two groups—anticipatable and unanticipable—where the first denote findings, for which it is well known that they could occur during the procedure performed, and the second are findings that could not be expected given the current state of knowledge. The Commission also adds two additional categories: secondary finding ("finding that is actively sought… that is not the primary target") and discovery finding (e.g., the "wellness scans" or "tell me everything" tests) (Presidential commission, 2013, pp. 27–29).

This terminological diversity additionally complicates the discussions, provoking "a general plea from patient groups and clinicians that consistent terminology [to] be used" (Hehir-Kwa et al., 2015, p. 1602). The issue of standardizing terminology is crucial, since that will "aid the comparison and further development of ideas coming from scientific articles and clinical and research reports. Moreover, it is necessary to clarify current and future legal and ethical obligations, and to maintain the trust of participants and patients. Agreeing on appropriate terminology is thus no secondary issue in the incidental findings debate" (Christenhusz et al., 2013, p. 1331). As some researchers emphasize it is also important to "identify the patient preferences because the language clinicians use to explain these types of results has the potential either to promote effective patient decision-making or to cause confusion or misperceptions about the nature of the findings, undermining effective patient decision-making" (Tan et al., 2017, p. 2).

Due to the unresolved nature of this discussion, in this chapter we will still adhere to using "incidental" as the most commonly used term.

2.2 Current Practice, New Developments, and Specific Features of the Prenatal Setting

In order to grasp adequately the meaning of "incidental" in the prenatal setting, we should briefly describe the current practice and the new developments as large-scale tests as microarrays, genome and exome sequencing, which most likely would be ever more used in this setting. Additionally, with the advances of technology and the accumulation of knowledge and experience, findings that have been considered genuinely "incidental" in the past, are increasingly acquiring the status of anticipated and often identified. For example, some early publications qualify as unexpected trisomy 18, triploidy, and trisomy 21, because "professionals ... were not yet acquainted with the new technique of analyzing chromosomes derived from amniotic fluid or chorionic villi" (Zwieten et al., 2005, p. 18). A similar instance was related to the incidental identifications of sex chromosome abnormalities in the early times of prenatal testing (Christian et al., 2000; Zwieten et al., 2005, p. 18).

Incidental findings in prenatal settings are often identified during first-trimester screening, which allows the quantification of the risks associated with some genetic abnormalities. A definitive diagnosis could be accomplished by an invasive test—chorionic villus sampling or amniocentesis. During invasive testing is usually performed a full karyotyping, which involves the microscopic examination of cells (De Jong et al., 2015, p. 3). Another noninvasive technique is the ultrasound scan, which can detect major structural abnormalities. It is interesting to note that the incidental findings, which this type of technology could generate, and which can properly be called so, are the maternal findings. The fetal findings can hardly be defined as incidental, since the scan is performed precisely with the purpose to detect whatever is abnormal.

Newer alternative to the traditional karyotyping is the chromosome microarray analysis (CMA), which is offered "especially but not exclusively after detection of enlarged [nuchal translucency] or other ultrasound abnormalities... This gives a higher yield of clinically relevant abnormalities, including microdeletion and microduplication syndromes. Genome-wide CMA may also lead to findings of clinical significance not related to the ultrasound abnormalities that prompted further testing, or to variants considered benign or of unclear clinical significance" (Dondorp et al., 2015, p. 533). The latter pose serious interpretative and counseling challenges, because in the prenatal setting there are often not identifiable major phenotypic effects. A matter of controversy is also the practice to analyze the parents in order to facilitate the interpretation of the fetal findings (Vetro et al., 2012, pp. 923–924). Besides these difficulties, some researchers note with anxiety a general "tendency to use broader arrays than would be needed for clarifying karyotyping or ultrasound outcomes and an implicit shift from a prenatal diagnosis to a prenatal screening paradigm. This causes microarrays to have an ambiguous character" (De Jong et al., 2014, p. 170). For example, in 2012, the American College of Obstetrics and Gynecologists (ACOG) advised chromosomal microarray to be recommended in all cases when an invasive prenatal test has been assigned due to ultrasound abnormalities and to be considered as an alternative to fetal karyotype in every other invasive test, regardless of indication, including advanced maternal age or maternal anxiety (De Jong et al., 2015, p. 5).

Another recent breakthrough is the introduction of noninvasive prenatal testing (NIPT)—a fetal genetic testing, based on analysis of cell free DNA in maternal plasma. Its advantage is the "large decrease in the need for invasive follow-up testing, entailing an equivalent reduction of iatrogenic pregnancy losses" (De Jong et al., 2015, p. 6). There are several applications of the noninvasive prenatal genetic testing: "to estimate the chance that a fetus has Down, Edwards or Patau syndromes...; to get a definite or near definite diagnosis of other specific genetic conditions in some cases (e.g., achondroplasia and Apert syndrome)...; to determine fetal sex; to perform whole genome or exome sequencing," which has been carried out only in a research setting up to now (Nuffield council, 2017, p. 3).

Among the incidental findings that could be detected by NIPT, are genetic information about the pregnant woman and maternal malignancy. It is important to note here that such findings could be detected only if particular processing methods are used, or they could be masked if the pregnant woman has requested that in a pretest counseling (Nuffield council, 2017, p. 65). With regard to the ethical issues, it is also important to mention that the development of this field is driven predominantly by the international commercial NIPT sector (Nuffield council, 2017, p. 17), which is another reason to expect exercising pressure toward routinization of ever more wide prenatal testing. As Erik Parens warns "it would be naïve not to notice the momentum building for more sequencing in the prenatal context" (Parens, 2015, p. 19).

The application of a new large-scale diagnostic tool—prenatal diagnostic exome sequencing—has raised a controversy concerning the scope of sequencing (targeted vs comprehensive), which directly impacts the number and the type of incidental findings and variants of uncertain significance (VOUS) that are detected. "Proponents of targeted exome argue that this better fits its intent as a diagnostic, rather than exploratory tool, and would help overcome some of the challenges and concerns about the wide-scale use of [exome sequencing] and its impact on patient counseling... However, this would miss medically actionable variants for the fetus and/or other at risk family members... Conversely, a blanket adoption of comprehensive [exome sequencing] could strain already limited clinical resources and result in additional burden to families by uncertain results" (Westerfield et al., 2017, p. 77). The authors propose adopting a two-tiered system, which implies that the decision-making will be a joint effort between the clinician and the family. This will best guarantee that the individual need and preferences are respected.

The prenatal context has some unique features, which hamper the possibilities for reaching conceptual clarity and consensus whether and how to communicate incidental findings. Prenatal testing is a complex practice, involving different techniques, different parties concerned—the future child, the parents, the family, the relatives, and different professionals involved. It is important the fact that we still lack public consensus on what should be sought. All these circumstances lead to the fact that when speaking about incidental findings in the prenatal setting different taxonomies are blending, all having specific features and counseling challenges.

It should be noted that the nature of the "indication" differs substantially from pediatric and adult genetic testing. In the latter cases it is "typically undertaken to provide a diagnosis and inform medical management and treatment options for individuals with an observable phenotype" (Westerfield et al., 2014, p. 1020). In prenatal testing, due to the limitations of prenatal imaging and prenatal dysmorphology, often there is no known or incompletely defined phenotype or clinically detected abnormalities.

We should also keep in mind that many different professionals are involved in prenatal testing, all having special focus, interests, and expectations: "for instance, the obstetricians' attitude may have changed because of the increasing use of nuchal translucency measurement (NTM). The detection of Down syndrome is a much more specific goal of NTM than it is for general cytogenetic diagnosis. So obstetricians may already be focused more exclusively on finding Down's syndrome than most cytogeneticists are in the current practice of prenatal diagnosis" (Zwieten et al., 2005, p. 20).

Besides these specificities, it is important that this practice also has an unique ethical and social status, which additionally complicates the matter. It still bears a good amount of moral sensitivity and has a special standing as a public health service: officially it is "not understood as aimed at prevention, in the sense of reducing the burden of disease in society... [but] as enabling autonomous reproductive choice" (Dondorp and van Lith, 2015, p. ii). It has always been a controversial practice and now, when the tendency is to offer it for an expanding range of abnormalities and to offer it earlier during pregnancy and through noninvasive procedures, "the issue of scope [is put] on the agenda: for what range of conditions should prenatal screening be offered and on the basis of what criteria should that be decided?" (De Jong and de Wert, 2015, p. 46). And if there is a consensus that its main purpose is to enhance the individual reproductive choices, then "indication" could be any individual choice-enhancing reason for screening. This means that practically, in some cases there could be no "unexpected" findings in prenatal setting, because it has been performed in order to find "anything wrong." If we use the taxonomy, proposed by the Presidental commission, these should be called "discovery findings."

Last but not least it should be noted that the prenatal context bears some of the features, characterizing the pediatric context of incidental findings, where we have a triadic relationship (Wilfond and Carpenter, 2008)—in this case between the future child, the parents and the professionals and this introduces additional complexity, since "within the context of fetal diagnosis the individual being tested cannot be the one providing consent. Parents, who provide consent on behalf of the unborn child as the child's representatives, are also bound by the beneficence principle to act in the best interest of the fetus" (The-Hung Bui et al., 2014, p. 13).

3.   REPORTING INCIDENTAL FINDINGS IN THE PRENATAL SETTING

The decisions about disclosuring of incidental findings are directly related to some core moral and legal principles and concepts, such as autonomy, beneficence and harm, rights and duties, and touch even on some broad debates, regarding the nature of contingency, knowledge, deception, and truth-telling. At the same time, such decisions have to be made by medical practitioners in their everyday practice and are painful real-life instances of the intractability of these conceptual problems. Is it always good to know more and how does information enhance our autonomy? In what cases knowing more is harmful, because the toxicity of this knowledge could disempower and deprive us of well-being, and not knowing could be a bliss? Who should decide and whether it could be justified to leave such decisions, concerning ourselves and our loved ones, to third parties such as the medical professionals?

Such decisions should practically resolve a clash of rights and duties between the two parties involved (generally speaking, since in the real practice often more protagonists are involved)—patients and professionals, which in many instances is conceptualized as a conflict between autonomy and paternalism. From the patient perspective, we usually speak of a right to know or a right not to know and the focus of the discussion is which of them genuinely enhances patient autonomy. From the professionals' perspective, we can defend an obligation to tell or an obligation not to tell. Not disclosing the full information available could be qualified as deception of the patient, but also as the so-called therapeutic privilege—which practically is the right of the professional not to tell, if the physician considers the information more harmful than beneficial for the patient on a net balance. In this case, it could be argued that recognizing the right not to tell (and affirming the therapeutic privilege) enhances the autonomy of the physician.

In this part we shall further discuss these issues, organizing the arguments in two groups: pro disclosure and pro qualified disclosure—reporting only a subset of findings, due to the decision of the professionals or due to the pretest option in the informed consent procedure to report only those, which the patient would like to know.

Obviously, the reasoning almost always starts with considerations regarding the potential benefits and harms of reporting incidental findings. But as it was already emphasized, the prenatal context is quite a complex field and it is not easy to define unequivocally what is "beneficial" and for whom. We always have to deal with several variables: the category of the finding, depending on the technology used (imaging or genetic), the nature of the finding (actionable, not actionable, of uncertain significance), the party concerned (fetus, future child, parents, relatives), the professionals involved, and their primary focus. Contrary to this diversity is the nature of the general recommendations and guidelines, whose aim is to encompass a broad array of cases. In such a sense, we should keep in mind that the concrete appraisal and decision for (non-)disclosure of some incidental findings is and should be made considering the specific case and its unique contextual circumstances. And if we go beyond the narrower, purely medical, notion of best interests, and deem prenatal testing as also a morally and socially sensitive issue, where broader interests of groups and communities are at stake, then offering recommendations becomes an exceptionally challenging task.

3.1   Arguments Pro Disclosure

Pro disclosure recommendations usually embrace the view that this is the approach that respects and enhances patient autonomy, expressed in the capability to make free and informed choices. In other words, moral priority should be given to our right to know and the right not to know should be rejected. This is considered as a natural move, ensuing from the general shift in contemporary health care from paternalistic models toward autonomy-respecting ones, with informed consent being the procedural epitome of this transformation.

Two crucial features of incidental findings are considered enough for them to be reported. They must be pathogenic and clinically actionable. This directly implies that their knowledge could allow either curative or preventive strategies. As some researchers indicate "[t]his can justifiably be called the strongest reason, because all of the other reasons we identified in the literature, either for or against disclosure or offering caution, present themselves either in accord with or in contrast to confirmed clinical utility and the possibility of treatment or prevention" (Christenhusz et al., 2013, p. 249). In other words, the decision pro reporting tends to be easier when the findings are considered actionable and this is valid for different contexts (clinical, research, DTC), different categories (results, found with imaging or large-scale genetic technologies) and for the all parties involved— future children, mother, parents, relatives.

An interesting point in the context of prenatal testing is the ambiguous meaning of "actionable" incidental findings with regard to the fetus. If these are the results "for which it is well known that a therapeutic or preventive measure exists that can significantly benefit the health of the individual in whom it was discovered" (Westerfield et al., 2014, p. 1019), then there are cases as the conditions for which usually the pregnancy is terminated, in which we cannot characterize unequivocally "actionable" as beneficial, due to the earlier mentioned special moral challenges and societal implications of the practice. This again illustrates how obscure could be conventional terms, when used in prenatal context, which is imbued with so much moral sensitivity and unsettledness.

We should mention two groups of cases that could be considered problematic. To the first group belong the cases of pathogenic finding in the fetus leading to the discovery of clinically relevant genetic information for the parents, siblings, or relatives, and the genetic abnormality is considered amenable through preventive measures such as life-style changes or medical interventions, or it allows more adequate reproductive decisions. The question is then whether this information should be sought and reported to parents or it is better to try to avoid it. Generally, it is considered that "not reporting such prenatal diagnostic results to parents carries a risk of causing harm and is a strategy that is difficult to endorse" (The-Hung Bui, 2014, p. 13).

The second group consists of cases in which the fetus is found to have an actionable, early-onset condition, but the parents believe that their right not to know must be respected. On such cases, various recommendations and guidelines take a similar approach, recommending that parents cannot refuse to be informed, since they are actionable and the child would benefit clinically (Sénécal et al., 2015, p. 544; see also Hens et al., 2011). Such recommendations show that there are clear cases, when the right not to know is outweighed by the beneficence generated through knowledge. Often this is also the position of genetics professionals, who sometimes prioritize informing the interested parties, even when the findings are not actionable (Lemke et al., 2013; Yu et al., 2014). Perhaps the recommendations of American College of Medical Genetics and Genomics (ACMG) from 2013 are one of the most radical ones in refuting the right not to know when counterbalanced by a possible beneficence. However, as they were heavily debated and criticized (see, e.g., Wolf et al., 2013), they were reconsidered in 2016.

3.2 Arguments Against Disclosure

The main reasons against disclosure are grounded in the idea that the genetic information could be harmful. We could call it the "toxic knowledge" (Bernhardt et al., 2013), pointing out the importance of respecting the right not to know, since in this way are avoided severe psychological harm and anxiety for the patients. This anxiety could further prevent patients from making autonomous and adequate decisions, concerning their life plans and prospects, health status and life-style. This argument is often used, especially in children, because their right to an open future (Feinberg, 1992) could be compromised.

These dilemmas become relevant especially in cases when the incidental findings are clinically important, but not actionable, as some late-onset conditions, or when they are of uncertain significance. Regarding the first group, it is generally argued that disclosure should be rejected or deferred until adulthood. Usually it is recommended that children should not be tested for late-onset conditions, for which there is no treatment, since such information could compromise their right to an open future and cause significant psychological burden (Hens et al., 2011) or compromise the child's best interest through "heightened anxiety among parents, differential treatment of children based on unfavorable results, and detrimental psychosocial effects … [This] potentially violat[es] his or her right to make health decisions regarding personal genetic information" (Sénécal et al., 2015, p. 545).

Similarly, the Nuffield Report on the ethical issues of NIPT warns that, when the genetic information revealed is about "adult onset conditions, carrier status, less significant medical conditions or impairments or non-medical traits … this might be harmful to the future person. If this kind of information were generated and stored, it could undermine the ability of the future person to make their own choices about accessing their genetic information, and may result in the shutting down of some of their future life options" (Nuffield council, 2017, p. 86). The second group of cases, in which disclosure or reporting is not recommended, are the so-called variants of unknown significance (VOUS) – results that "cannot be unequivocally classified as clinically significant or benign. This could be because there is known variable expressivity or incomplete penetrance for the finding or because the finding is novel and has only very rarely or never been seen before" (Westerfield, 2014, p. 1020; see also The-Hung Bui, 2014, p. 14). Large-scale genetic technologies, yielding large amounts of information, are increasing significantly the probability to detect such findings. This has important implications for prenatal screening since the phenotypic information about the fetus is often limited.

VOUS pose another difficulty closely related to the "toxic knowledge" argument, but pointing in a slightly different conceptual direction, namely the issue of proper *communication* the complex genetic information. What does it mean to tell the patient "everything" or to convey the "truth" about an incidental finding? Is it really possible for patients "to know everything," even if they want to, since they are not professionals and could not comprehend the information in the same way as trained medical specialists? Communicating "everything" is never a straightforward matter, because what practitioners "know" is a complex heterogeneous corpus of information, tacit knowledge, skills and experience and it is always a challenge to ensure that the patient, who usually comes from an entirely different life context (sometimes also cultural context), receives the right messages and is genuinely prepared to make informed decisions.

Different studies showed that patients and parents claim that they want to know "everything," but we should bear in mind, as Erik Parens notes, that "[s]uch survey data … rest on the dubious assumption that when respondents say they want all the genomic information they can get, they have been helped to understand what they are in fact asking for … such survey results probably give a simplistic picture of what most persons really want" (Parens, 2015, p. 17). This was confirmed by recent studies regarding the women's experience with prenatal testing (see also Nuffield, 2017, p. 87) showing them to have "only a limited understanding of the indications and ramifications of the information that the tests can return… these studies clearly indicate that women have a strong preference for receiving help in parsing the distinction between empowering and disempowering information" (Johnston et al., 2017, p. 505).

The issue of communicating complex information is more cumbersome in the prenatal setting, since it has the aura of a health-care measure, whose ultimate outcome is a "healthy baby" and a "happy family." In prenatal medicine, receiving and understanding information can be seriously affected by the preceding, generally positive, set of expectations. This closely resembles therapeutic misconception, namely the mistaken assumption of research participants that research interventions will have clinical benefits (Wolf et al., 2008, p. 228; see also Appelbaum et al., 1982). It is often thought as a reason for precaution when disclosing incidental findings in the research context. The resemblance, pointed out here, boils down to the patient conviction that a given procedure (traditionally—the clinical trial, in our case—prenatal testing) entirely benefits the patient from a clinical point of view, which precludes him from paying due attention to the negative effects or burdens generated by the procedure. This is one of the reasons for a plea to develop "approaches to pretest and posttest education and counseling that empower patients to decide whether to be tested and what to do after receiving their results" (Johnston et al., 2017, p. 507).

4. HANDLING INCIDENTAL FINDINGS: INFORMED CONSENT PROCEDURES AND PRETEST COUNSELING IN GENETIC TESTING

Recently, different recommendations have been developed to address the complex issues of handling incidental findings in genetic testing. Informed consent procedures as well as pre- and posttest counseling are seen as the main instruments for protecting patients and research participants, but also for coping with the growing complexity of the genetic knowledge generated in a diagnostic setting. The challenges concern the appropriate forms of consent and counseling, the education of health professionals engaged in genetic testing in the clinical and research settings, but also structural changes in the field (genetic laboratories, genetic variants databases, and biobanks) and new actors entering the field, such as DTC test providers. In this section we will try to summarize some general recommendations that concern the incidental findings from genetic and genomic tests as well as specific recommendation that refer to the prenatal genetic testing.

The report of the U.S. Presidential Commission for the Study of Bioethical Issues (2013) provides recommendations on handling of incidental or secondary findings in the clinical, research, and DTC contexts. The ethical considerations relevant in all three contexts concern: informing the patients about the likelihood of incidental or additional findings and about the plan for disclosing and managing them before testing; development of practice guidelines for practitioners that categorize the findings; educating stakeholders; access of all individuals to adequate information, guidance and support in making informed choices. The Commission recommends additional empirical research "on types and frequency of findings that can arise from various modalities; the potential costs, benefits, and harms of identifying, disclosing and managing these findings, the recipient and practitioners preferences about discovery, disclosure, and management of incidental and secondary findings." (Presidential Commission for the Study of Bioethical Issues, 2013, p. 7).

The report requires from clinicians "skilled and insightful deliberation" combining professional judgment and contextual understanding. Besides the duties to inform patients, clinicians should engage them in shared decision-making and respect the patient's preference not to know "to the extend consistent with the clinician's fiduciary duty" (Presidential Commission for the Study of Bioethical Issues, 2013, p. 10). Even if in research settings there is no duty for researches to look for secondary finding, the researchers have obligations to the participants based on their privileged knowledge and access to private medical information.

In research context, it is argued that "the researchers have an obligation to address the possibility of discovering incidental findings not only in their protocol, but also in their consent forms and communications with those being recruited in the study and research participants" (Wolf et al., 2008, p. 227). Researchers have to establish algorithms for handling suspected incidental findings. The risks and benefits of discovering incidental findings in research require an explicit disclosure during the consent process. Further issues to be clarified include when researchers should initiate evaluation and disclosure of information obtained during the study, returning information at the request of research participants, or the reanalysis of archived data.

The ESHG requires caution and recommends a targeted approach regarding the use of sequencing or analyses of genome data, in order to avoid unsolicited findings or findings that could not be interpreted. Whether the advantages of using whole-genome techniques outweigh the disadvantages of a genetic test that may lead to unsolicited findings, should be decided by health professionals before the test and should be discussed with the patients during pretest counseling (Van El et al., 2013, p. S3). The document states that the "use of genome-wide array or WGA requires a justification in terms of necessity (the need to solve a clinical problem) and proportionality (the balance of benefits and drawbacks for the patient)" (Van El et al., 2013, p. S4).

The Canadian College of Medical Geneticists encourages caution too: "… until the benefits of reporting incidental findings are established, we do not endorse the intentional clinical analysis of disease-associated genes other than those linked to the primary indication" (Boycott et al., 2015, p. 431).

The European Commission *Recommendation on Cross Border Genetic Testing of Rare Diseases* (2015) outlines that outcomes of genetic testing could be important even if there are no treatments available in many rare disorders identified by these tests. These outcomes may allow a better understanding of the prognosis, clarify the origin of the disease, and elucidate the mode of inheritance thereby facilitating life planning and reproductive choices.

The document emphasizes the link between the rapid development of next-generation sequencing (NGS) and the appearance of national and international variant databases, as well as the opportunities for cross border genetic testing for rare diseases in Europe.

In *the Guidelines for Diagnostic Next Generation Sequencing,* the challenges of new technologies are outlined both "at the technical level and in term of data management, as well as for the interpretation of the results and for counseling" (Matthijs et al., 2016, p. 2). The role of the NGS is seen mostly in the context of rare diseases and targeted analysis of data. The handling of incidental findings is addressed in the statements 10–13 of the Guidelines and includes: laboratories to provide information on the probabilities of unsolicited findings and the local policies about dissemination of unsolicited and secondary findings, "before a NGS-based test the clinical (genetic) center to set up 'an unsolicited and secondary findings protocol' in accordance with the decision of an ethical committee"; if an opt-in or opt-out protocol is offered to the patient, pretest counseling to be offered too (Matthijs et al., 2016, pp. 3–4).

In March 2017, the Nuffield council on bioethics published its report on the ethical issues of NIPT, in which the problems of reporting incidental findings were discussed. It emphasized incidental findings as one of the ways, in which NIPT can potentially be psychologically harmful to pregnant women and their families as there are many cases in which the implications of these findings are not well understood or the possibility of detecting them is not discussed with the women before the test (Nuffield council, 2017, p. 116). In the European guidelines on prenatal diagnostic tests, the informed consent and counseling are emphasized as fostering autonomous choice. Are detailed both general and specific recommendations for different groups of women and also the main counseling topics applicable to invasive and noninvasive testing. We also recommend discussing with patients before the test whether, and how, unexpected findings will be communicated. The characteristics of the results (e.g., their accuracy) and the way they will be communicated are included in a detailed table of topics to be covered during counseling, together with several other key issues about the condition to be tested, practical aspects and psychosocial issues (Skirton et al., 2014, p. 584).

The British professional guidance on the usage of CMA in prenatal testing recommends no disclosure of any finding "not linked to potential phenotypes for the pregnancy (future child) in question or [having] no clinically actionable consequence for that child or family in the future, e.g., variants of uncertain significance" (Joint committee, 2015, p. 6).

Two key documents from the United States also advise adopting very careful approach of handling results. The ACOG's (2013) official opinion on the use of microarrays in pregnancy recommends comprehensive pre- and posttest genetic counseling from qualified personnel. The ACMG recommends joint effort of health officials, policymakers, and private companies toward accessible pretest education and counseling, during which women should be informed that there is possibility to identify maternal genomic anomalies, but this depends on the technology used. This is grounded in the principle that "[p]atient preferences for information should play a pivotal role in guiding the use of NIPS in prenatal care. This is in keeping with generally accepted genetic counseling tenets and respects that clinical utility may vary between patients" (Gregg et al., 2016, p. 3).

5. CONCLUSION

With the further development of new sequencing technologies and their implementation in different contexts—clinical, research, or DTC, the issues of interpretation and assessment of genetic information and of handling a growing number of incidental or additional findings will become more complex. The ethical analysis of the issues appertaining to genetic testing, especially in prenatal context, should focus on individual perspectives and protection, but should also incorporate societal, cultural, and political dimensions relevant to the genetic technologies that are rapidly becoming mainstream in clinical diagnostic. The ethical analysis needs a wider reflection on how genetic technologies are changing our lives and reshape the way we reproduce, including reproductive choices and the notions of motherhood and parenthood, which until recently have been considered as relatively stable. The current instruments might not be enough in the new realities that blur the distinction between clinical care and research through new types of knowledge and the use of new tools such as genetic databases and biobanks. We could only anticipate the near future by asking questions about the challenges of a new personalized medicine, of findings that might have not only diagnostic value, or of the meaning of "incidental" in regard to a variety of options determined by individual choices.

REFERENCES

American College of Medical Genetics and Genomics, 2015. ACMG policy statement: updated recommendations regarding analysis and reporting of secondary findings in clinical genome-scale sequencing. Genet. Med. 17 (1), 68–69.

American College of Obstetricians and Gynecologists and Society for Maternal-Fetal Medicine, The use of chromosomal microarray analysis in prenatal diagnosis, https://www.acog.org/-/media/Committee-Opinions/Committee-on-Genetics/co682.pdf?dmc=1&ts=20170705T0003145503, 2013, (accessed 30 September 2017).

Appelbaum, P., Roth, L., Lidz, C., 1982. The therapeutic misconception: informed consent in psychiatric research. Int. J. Law Psychiatry 2, 319–329.

Bernhardt, B., Soucier, D., Hanson, K., Savage, M., Jackson, L., Wapner, R., 2013. Women's experiences receiving abnormal prenatal chromosomal microarray testing results. Genet. Med. 15 (2), 139–145.

Beskow, L., Burke, W., 2010. Offering individual genetic research results: context matters. Sci. Transl. Med. 2(38).

Boycott, K., Hartley, T., Adam, S., Bernier, F., Chong, K., Fernandez, B., Friedman, J., Geraghty, M., Hume, S., Knoppers, B., Laberge, A., Majewski, J., Mendoza-Londono, R., Meyn, M., Michaud, J., Nelson, T., Richer, J., Sadikovic, B., Skidmore, D., Stockley, T., Taylor, S., van Karnebeek, C., Zawati, M., Lauzon, J., Armour, C., 2015. The clinical application of genome-wide sequencing for monogenic diseases in Canada: position statement of the Canadian College of Medical Geneticists. J. Med. Genet. 52, 431–437.

Bui, T.-H., Raymond, F., van den Veyver, I., 2014. Current controversies in prenatal diagnosis 2: should incidental findings arising from prenatal testing always be reported to patients? Prenat. Diagn. 34, 12–17.

Christenhusz, G., Devriendt, K., Dierickx, K., 2013. Secondary variants—in defense of a more fitting term in the incidental findings debate. Eur. J. Hum. Genet. 21, 1331–1334.

Christian, S., Koehn, D., Pillay, R., MacDougall, A., Wilson, R., 2000. Parental decisions following prenatal diagnosis of sex chromosome aneuploidy: a trend over time. Prenat. Diagn. 20, 37–40.

De Jong, A., de Wert, G., 2015. Prenatal screening: an ethical agenda for the near future. Bioethics 29 (1), 46–55.

De Jong, A., Dondorp, W., Macville, M., de Die-Smulders, C., van Lith, J., de Wert, G., 2014. Microarrays as a diagnostic tool in prenatal screening strategies: ethical reflection. Hum. Genet. 133, 163–172.

De Jong, A., Maya, I., van Lith, J., 2015. Prenatal screening: current practice, new developments, ethical challenges. Bioethics 29 (1), 1–8.

Dondorp, W., van Lith, J., 2015. Dynamics of prenatal screening: new developments challenging the ethical framework. Bioethics 29 (1), ii–iv.

Dondorp, W., Page-Christiaens, G., de Wert, G., 2015. Genomic futures of prenatal screening: ethical reflection. Clin. Genet. 89, 531–538.

European Commission, 2015. Recommendation on Cross Border Genetic Testing Of Rare Diseases in the European Union. https://ec.europa.eu/health//sites/health/files/rare_diseases/docs/2015_recommendation_crossbordergenetictesting_en.pdf. Accessed 30 September 2017.

Feinberg, J., 1992. The child's right to an open future. In: Freedom and Fulfilment: Philosophical Essays. Princeton University Press, Princeton, NJ, pp. 76–77.

Gregg, A., Skotko, B., Benkendorf, J., Monaghan, K., Baja, K., Best, R., Klugman, S., Watson, M., 2016. Noninvasive prenatal screening for fetal aneuploidy, 2016 update: a position statement of the American College of Medical Genetics and Genomics. Genet. Med. 18, 1056–1065.

Hastings, R., de Wert, G., Fowler, B., Krawczak, M., Vermeulen, E., Bakker, E., Borry, P., Dondorp, W., Nijsingh, N., Barton, D., Schmidtke, J., van El, C., Vermeesch, J., Stol, Y., Carmen Howard, H., Cornel, M., 2012. The changing landscape of genetic testing and its impact on clinical and laboratory services and research in Europe. Eur. J. Hum. Genet. 20, 911–916.

Hehir-Kwa, J., Claustres, M., Hastings, R., van Ravenswaaij-Arts, C., Christenhusz, G., Genuardi, M., Melegh, B., Cambon-Thomsen, A., Patsalis, P., Vermeesch, J., Cornel, M., Searle, B., Palotie, A., Capoluongo, E., Peterlin, B., Estivill, X., Robinson, P., 2015. Towards a European consensus for reporting incidental findings during clinical NGS testing. Eur. J. Hum. Genet. 23, 1601–1606.

Hens, K., Nys, H., Cassiman, J.-J., Dierickx, K., 2011. The return of individual research findings in paediatric genetic research. J. Med. Ethics 37, 179–183.

Johnston, J., Farrell, R., Parens, E., 2017. Supporting Women's autonomy in prenatal testing. N. Engl. J. Med. 377 (6), 505–507.

Joint Committee on Genomics in Medicine, Recommendations for the use of chromosome micro-array in pregnancy, https://www.rcpath.org/resourceLibrary/recommendations-for-the-use-of-chromosome-microarray-in-pregnancy.html, 2015, (accessed 15 September 2017).

Lemke, A., Bick, D., Dimmock, D., Simpson, P., Veith, R., 2013. Perspectives of clinical genetics professionals toward genome sequencing and incidental findings: a survey study. Clin. Genet. 84 (3), 230–236.

Matthijs, G., Souche, E., Alders, M., Corveleyn, A., Eck, S., Feenstra, I., Race, V., Sistermans, E., Sturm, M., Weiss, M., Yntema, H., Bakker, E., Scheffer, H., Bauer, P., 2016. Guidelines for diagnostic next generation sequencing. Eur. J. Hum. Genet. 24, 2–5.

Nuffield Council on Bioethics, Non-Invasive Prenatal Testing: Ethical Issues, http://nuffieldbio-ethics.org/project/non-invasive-prenatal-testing, 2017, (accessed 1 September 2017).

Parens, E., 2015. Drifting away from informed consent in the era of personalized medicine. Hast. Cent. Rep. 459 (4), 16–20.

Parker, L., 2008. The future of incidental findings: should they be viewed as benefits? J. Law Med. Ethics 36 (2), 341–351.

Presidential Commission on the Study of Bioethical Issues, 2013. Anticipate and Communicate. Ethical Management of Incidental and Secondary Findings in the Clinical, Research, and Direct-to-Consumer Contexts. Washington, DC, https://bioethicsarchive.georgetown.edu/pcsbi/sites/default/files/FINALAnticipateCommunicate_PCSBI_0.pdf. Accessed 30 September 2017.

Sénécal, K., Rahimzadeh, V., Knoppers, B., Fernandez, C., Avard, D., Sinnett, D., 2015. Statement of principles on the return of research results and incidental findings in paediatric research: a multi-site consultative process. Genome 58 (12), 541–548.

Skirton, H., Goldsmith, L., Jackson, L., Lewis, C., Chitty, L., 2014. Offering prenatal diagnostic tests: European guidelines for clinical practice. Eur. J. Hum. Genet. 22, 580–586.

Srebniak, M., Diderich, K., Govaerts, L., Joosten, M., Riedijk, S., Galjaard, R., Van Opstal, D., 2014. Types of array findings detectable in cytogenetic diagnosis: a proposal for a generic classification. Eur. J. Hum. Genet. 22, 856–858.

Tan, N., Amendola, L., O'Daniel, J., Burt, A., Horike-Pyne, M., Boshe, L., Henderson, G., Rini, C., Roche, M., Hisama, F., Burke, W., Wilfond, B., Jarvik, G., 2017. Is incidental finding the best term? A study of patients' preferences. Genet. Med. 19 (2), 176–181.

Van El, C.G., Cornel, M.C., Borry, P., Hastings, R., Fellmann, F., Hodgson, S., Howard, H., Cambon-Thomsen, A., Knoppers, B., Meijers-Heijboer, H., Scheffer, H., Tranebjaerg, L., Dondorp, W., de Wert, G., 2013. Whole-genome sequencing in health care. Recommendations of the European Society of Human Genetics. Eur. J. Hum. Genet. 21, 580–584.

Vetro, A., Bouman, K., Hastings, R., McMullan, D., Vermeesch, J., Miller, K., Sikkema-Raddatz, B., Ledbetter, D., Zuffardi, O., van Ravenswaaij Arts, C., 2012. The introduction of arrays in prenatal diagnosis: a special challenge. Hum. Mutat. 33 (6), 923–929.

Westerfield, L., Darilek, S., van den Veyver, I., 2014. Counseling challenges with variants of uncertain significance and incidental findings in prenatal genetic screening and diagnosis. J. Clin. Med. 3 (3), 1018–1032.

Westerfield, L., Braxton, A., Walkiewicz, M., 2017. Prenatal diagnostic exome sequencing: a review. Curr. Genet. Med. Rep. 5, 75–83.

Wilfond, B., Carpenter, K., 2008. Incidental findings in pediatric research. J. Law Med. Ethics 36 (2), 332–340.

Wolf, S., Lawrenz, F., Nelson, C., Kahn, J., Cho, M., Clayton, E., Fletcher, J., Georgieff, M., Hammerschmidt, D., Hudson, K., Illes, J., Kapur, V., Keane, M., Koenig, B., Leroy, B., McFarland, E., Paradise, J., Parker, L., Terry, S., Van Ness, B., Wilfond, B., 2008. Managing incidental findings in human subjects research: analysis and recommendations. J. Law Med. Ethics 36 (2), 219–248.

Wolf, S., Annas, G., Elias, S., 2013. Respecting patient autonomy in clinical genomics: new recommendations on incidental findings go astray. Science 340 (6136), 1049–1050.

Yu, J., Harrell, T., Jamal, S., Tabor, H., Bamshad, M., 2014. Attitudes of genetics professionals toward the return of incidental results from exome and whole-genome sequencing. Am. J. Hum. Genet. 95 (1), 77–84.

Zwieten, M., Willems, D., Litjens, L., Schuring-Blom, H., Leschot, N., 2005. How unexpected are unexpected findings in prenatal cytogenetic diagnosis? A literature review. Eur. J. Obstet. Gynecol. Reprod. Biol. 120, 15–21.

FURTHER READING

Appelbaum, P.S., Parens, E., Waldman, C.R., Klitzman, R., Fyer, A., Martinez, J., Nicholson Price II, W., Chung, W., 2014. Models of consent to return of incidental findings in genomic research. Hast. Cent. Rep. 44, 22–32.

Green, R., Berg, J., Grody, W., Kalia, S., Korf, B., Martin, C., McGuire, A., Nussbaum, R., O'Daniel, J., Ormond, K., Rehm, H., Watson, M., Williams, M., Biesecker, L., 2013. ACMG recommendations for reporting of incidental findings in clinical exome and genome sequencing. Genet. Med. 15, 565–574.

Parens, E., Appelbaum, P., Chung, W., 2013. Incidental findings in the era of whole genome sequencing? Hast. Cent. Rep. 43 (4), 16–19.

Chapter 17

Third Party Sharing of Genetic Information

Maureen Durnin and Michael Hoy
University of Guelph, Guelph, ON, Canada

1. INTRODUCTION

The Human Genome Project, tasked with sequencing the human DNA, was initially expected to take a total of 15 years to complete. Started in 1990, its primary mission was finished with much fanfare 2 years ahead of schedule in 2003. The speed of the project's progress and success did not go unnoticed. As is often the case, the development of policies concerning the sharing of genetic information struggled and continues to struggle to keep abreast of scientific developments. Starting in Belgium in 1992, and continuing as recently as May 2017 in Canada, countries around the world began to react to the perceived ethical and legal questions arising from how to deal with information generated by such a huge step in knowledge.[1] The outcomes decided upon by various governments range from no regulations (a wait and see attitude relying on existing ethical codes of conduct already in place) to temporary moratoria on requesting or using existing genetic tests (within monetary limits in the case of some types of insurance) to total prohibition of use of any genetic test. These reactions were primarily addressing the insurance industry but also included employers and medical ethics as applied to practitioners and institutions.

In recent years, the increased availability and reduction in cost has led to a proliferation of direct to consumer genetic tests (e.g., 23 and Me, 24genetics, Dante Labs, Gene by Gene) whose claims are not just discovery of paternity and ancestry, but also what they term "genetic health risks" as advertised on their website.[2] Many private companies now offer whole genome sequencing.

However, the brave new world prophecies anticipating the revolution in the precise predicting of diseases and development of treatments and

---

1. See Joly et al. (2010) for a thorough history of legislation regarding genetic discrimination and p. 362 specifically for reference to Belgium law. The recent law in Canada is documented at the Parliament of Canada website, http://www.parl.ca/DocumentViewer/en/42-1/bill/S-201/royal-assent (accessed on November 14, 2017).

2. See https://www.23andme.com/en-ca/dna-health-ancestry/ (accessed on November 14, 2017).

Clinical Ethics at the Crossroads of Genetic and Reproductive Technologies.
https://doi.org/10.1016/B978-0-12-813764-2.00017-9

pharmaceuticals with possibilities of targeting and curing those diseases is proving to be a far slower process. Mapping the human genome was not the end, just the beginning. The presence of genetic variations in some instances does directly cause specific outcomes. Single gene alterations were found to cause diseases such as Huntington's Chorea and Cystic Fibrosis. Other genes substantially increase the likelihood of the occurrence of disease. The presence of BRCA1 or BRCA2 genes, while not solely responsible for causing breast cancer, significantly increases the lifetime risk of contracting breast cancer.[3] For most diseases, however (i.e., so-called multifactorial diseases, which are caused by the effects of multiple gene combinations interacting with factors such as lifestyle realities and choices and environmental influences) the genetic component is much less important from a quantitative perspective. As underlined for the case of predisposition to cancer by Robson et al. (2010, p. 893), "Most inherited cancer susceptibility arises from a number of DNA sequence variants, each of which, in isolation, confers a limited increase in risk." As Rutter (2006, p. 5) points out "...we are only just learning how to pursue the long path from gene discovery to determination of the causal processes." His estimation is that this "long path" will take many years and perhaps decades.

On the other hand, many researchers in the field have much more optimistic projections. Consider McCarthy and MacArthur (2017, pp. 2–3): "The next few years will bring increasingly massive genomic data sets from patients and controls, as well as more sophisticated mechanisms for the analysis and integration of a wide variety of genomic data types. We can expect to see not only continued growth in the number of disease genes identified, but also a deepening of our understanding of the fundamental genetic architecture of human disease states, and a transformation in our ability to move from associated genes, to pathways, to biology and clinical translation."

Whether the path to accurately predicting all genetic-based disease is immediate or years in the future, it is preferable to have recourse to thoughtful considerations about the ramifications of sharing genetic information with third parties. To this end, it helps to outline and describe the motivations and concerns of some of the various groups or individuals who form the three "parties."

In this chapter, we address a wide range of practical and ethical concerns regarding the intentional and unintentional sharing of individuals' genetic information with third parties. Much has been written on many of these channels. Our goal is to summarize and highlight the key concerns for individuals

---

3. Estimates of the increased risk vary across studies. Petrucelli et al. (2015) note that the increased risk varies across pathogenic variants (see https://www.ncbi.nlm.nih.gov/sites/books/NBK1247/ (accessed November 14, 2017). In particular, they note "Breast cancer is the most common malignancy in individuals with a germline *BRCA1* or *BRCA2* pathogenic variant with a lifetime risk ranging from 46% to 87%."

who are affected by these events. Regarding the potential sharing and use of individuals' genetic test results by insurers or employers, there is a significant amount of legislation in place, at least for most developed countries. Most other potential sharing of genetic information has not been the object of legislation and is typically governed by ethical guidelines. The collection of genetic test results by research groups, often along with lifestyle choices and other health information, raises a myriad of complex concerns including whether research subjects should be privy to overall research results, whether they should be informed of any personal findings when potentially important to the future health of the individual and/or genetic family members. Some similar issues arise within the doctor-patient relationship.

We break down our analysis according to the following six situations numbered accordingly: (1) medical practitioners and institutions, (2) family and potential partners, (3) criminal investigations, (4) employers, (5) pharmaceutical investigations and clinical research, and (6) insurance industry.

2. MEDICAL PRACTITIONERS AND INSTITUTIONS

The medical community, from general practitioners to medical researchers in publicly funded institutions to pharmaceutical companies, would very much value access to as much information as possible that would lead to appropriate and specific treatment for an individual patient or, in the case of medical research, result in the possible development of cures for diseases or prophylactic measures which would deter the onset of disease. The analysis of genetic information has already led to certain prediction of some monogenetic diseases such as Huntington's Chorea. The discovery of the BRCA1 and BRCA2 genes led to the prediction of the likelihood of contracting breast cancer which can allow for measures, such as careful monitoring, that result in early detection or allow individuals more radical options such as the Angelina Jolie approach. Angelina Jolie, a well-known American actor whose mother died from breast cancer, carries the BRCA1 gene, predicted to result in an 87% chance of her contracting breast cancer.[4] As a preventative measure, Ms. Jolie opted to have both her breasts removed. In early October 2017, The Guardian newspaper in London, United Kingdom (Slawson, 2017) reported the development of a genetic test which delivers much greater accuracy in predicting the risk of contracting breast cancer for those women carrying the BRCA1 and BRCA2 gene mutations. Their quoted statistics claim that of the women who have these mutations, 50% opt to have preemptive radical mastectomies (as did Ms. Jolie). According

---

4. As reported in the Guardian Newspaper (https://www.theguardian.com/science/2016/dec/14/angelina-jolie-effect-boosted-genetic-testing-rates-study-finds-breast-ovarian-cancer) accessed November 14, 2017.

to the Guardian, the greater accuracy of the new test would provide improved certainty of risk and reduce that percentage to 36% (Slawson, 2017).

Curing or developing treatments to slow the progression of diseases and developing useful drugs to alleviate symptoms are considered lofty and laudable goals, but sharing of genetic information with medical research groups can also be problematic. Privacy concerns and fear of potential discrimination is behind the objections of umbrella groups such as the Canadian Coalition for Genetic Fairness (http://ccgf-cceg.ca/en/about-ccgf/#our-goals) and have drawn the attention of the Office of the Privacy Commissioner of Canada.[5] While these groups recognize the potential benefits of research into genetics, they are still very concerned about genetic discrimination, primarily in the areas of employment and insurance. Of course, the solution to protecting genetic privacy is to enforce regulations that will effectively protect individuals from the misuse of data collected in genetic tests, but this is problematic. We are reminded daily of the ability of hackers to access supposedly privately held information.

The transfer of sensitive health information, including genetic data, is not always anticipated by individuals. It was reported by Powles (2017) that in 2015 the National Health Service in Britain disclosed ultimately identifiable patient records. Royal Free London National Health Service A&E department (gifted) transferred 1.6 million patient records to Google's DeepMind. It was subsequently discovered that though it seemed the records were not identifiable, in fact, enough information was included, which when manipulated, would allow the identification of many of the individuals whose information was transferred without their consent. As reported by Powles (2017), it was ruled "by transferring this data and using it for app testing, the Royal Free breached four data protection principles." The NHS motivation in sharing the patient data was to support development of software that would eventually aid in facilitating public health. The transfer of this private patient information was done with the best of intentions without anticipation of financial gain but nonetheless led to a contravention of privacy laws. It serves as an example of the precariousness of stored information and its security.

As reported by Ramesh (2014) while describing how it will become possible to purchase patient information from the NHS database, which was initiated in 2014, he noted that "Drug and insurance companies will from later this year be able to buy information on patients—including mental health conditions and diseases such as cancer, as well as smoking and drinking habits—once a single English database of medical data has been created. Harvested from GP and hospital records, medical data covering the entire population will be uploaded to the repository... Advocates say that sharing data will make medical advances easier

---

5. See their "Policy statement on the use of genetic test results by life and health insurance companies" at https://www.priv.gc.ca/en/opc-news/news-and-announcements/2014/s-d_140710/ (accessed October 27, 2017).

and ultimately safe … but … privacy experts warn there will be no way for the public to work out who has their medical records or to what use their data will be put. The extracted information will contain NHS numbers, date of birth, postcode, ethnicity and gender." Again, having such detailed information as date of birth and postcode would allow for identification of specific individuals in many cases. This sharing of patient information with third parties serves as an example of the precariousness of stored information and its security.

3.   FAMILY MEMBERS, PARTNERS, AND POTENTIAL PARTNERS AS THIRD PARTY

On the surface, it seems obvious that individuals facing a possible or certain negative health outcome would want to share their genetic information with genetic family members, partners, or potential partners for many reasons: emotional support, encouraging monitoring and screening for their loved ones if there proves to be a family predisposition, avoidances of possible triggers associated with multifactorial diseases, life preparedness in the case of both monogenetic and multifactorial genetic diseases, family planning with partners in the context of inherited genetic problems. However, it is not always the case that an individual facing possible or probable negative health outcomes revealed by genetic tests will want to share the results with any third party, even genetic family members. The information may reveal information not only about health issues but also issues of paternity or ethnicity that the individual may be unwilling to share even with, or perhaps especially with, those who are closest. Wright Clayton (2003, p. 563) points out that in the United States both state and federal governments have been actively discussing who should have the right to access health information and under what conditions. She asserts, "This debate is informed appropriately by the recognition that limiting access to the medical record to the patient and the treating clinician is neither possible nor unequivocally desirable."

Wright Clayton's concerns lie in part with people's misconceptions about an inability to understand genetics and also with the physicians' two-pronged duty, which requires them to protect their patients' privacy AND to decide when exceptions to confidentiality are warranted. For example, Wright Clayton (2003, p. 567) points out that physicians are legally required to report certain infectious diseases or to warn people who have been the subjects of threats of violence uttered by patients in the physicians' care [Tarasoff v. Regents of University of California, 551 P.2d 334 (Cal. 1976)].[6] In the case of genetic information, Wright Clayton (2003, p. 567) points out that "Over the years, numerous prominent advisory bodies have said no, opining that physicians should be permitted to breach confidentiality in order to warn third parties of

6. Recorded at https://www.courtlistener.com/opinion/1175611/tarasoff-v-regents-of-university-of-california/ (accessed November 17, 2017).

genetic risks only as a last resort to avert serious harm." This certainly puts physicians on very uncertain footing. There is no exact legal or ethical definition of serious harm.[7]

Wright Clayton (2003) cites the case of a man who died of colon cancer in the 1960s. Approximately 25 years later his daughter developed the same disease. She acquired his pathology slides and discovered her father's condition resulted in her having a 50% risk of contracting the same disease. She sued the surgeon's estate because she was not informed of this risk to her health even though she was a child at the time of her father's death. The intermediate appellate court ruled that the doctor had a duty to warn the daughter [Safer v. Pack, 677 A.2d 1188 (N.J. App.), appeal denied, 683 A.2d 1163 (N.J. 1996)].[8]

As discussed in Lemmens et al. (2015), the question of age of the family member as a determining factor in performing genetic tests and sharing the results begins in utero. Prenatal or newborn testing for phenylketonuria (PKU), a genetic mutation, which, left untreated, leads to severe developmental complications is mandatory in the United States, varies by province in Canada, and has been common in prenatal and neonatal screening in many countries around the world since the mid 1960s. The American Society of Human Genetics (ASHG) has issued a Position Statement (2015) on genetic testing in children. They remark (2015, p. 9) that "Parents have wide decision-making authority, but in cases where the clinical response to a secondary finding will most likely prevent serious morbidity or mortality for the child, it can be appropriate to override a parental decision not to receive this information."

ASHG recommends that, in general, parents should be able to decline to receive secondary findings from genetic testing. However, when there is strong evidence that a secondary finding has urgent and serious implications for a child's health or welfare, and effective action can be taken to mitigate that threat, ASHG recommends that the clinician communicate those findings to parents or guardians regardless of the general preferences stated by the parents regarding secondary findings (ASHG, 2015).

In the above scenario, the patient referred to is a minor and so is under the care of the parents. However, suppose the affected individual is an adolescent. If the parent knows of a genetic reality with possible negative health effects but has chosen not to share it with family members, the adolescent may be unknowingly engaging in activities that will prove hazardous to future health or missing out on screening and monitoring that would enhance outcomes. Once again, the clinician is required to decide at what point there is a duty to breach confidentiality.

Further complications arise if the parent has purposely refused genetic testing for a myriad of reasons. For instance, it is reported (see Lemmens et al.,

7. See Nixson (2017) who summarizes the status quo in this field of inquiry.
8. Recorded at https://www.courtlistener.com/opinion/2089347/safer-v-estate-of-pack/ (accessed November 17, 2017).

2015 for a variety of sources) that only between 10% and 30% of people who possibly carry the gene for Huntington's Chorea opt to have the genetic test that will definitively predict whether they will contract the disease. Suppose the off-spring, who, in the United States at the age of 18, can legally request genetic testing, does obtain a test. Then, even if the parent is not informed of the result, if the tested person begins to act upon the result by making significant changes to lifestyle choices, undergoing additional screening or monitoring or therapy, the behavioral changes can alert the parent to unwanted information that had been previously refused and so compromises the parent's desire not to know.

The moral dilemma for physicians is complex. There is their duty to the patient who may not wish her spouse or children to learn about her future health possibilities since these may create financial, physical, or emotional burdens that could result in serious tensions within the family and even lead to separa-tion or divorce. Some patients may also wish to hide paternity or race informa-tion from their genetic family. In cases where genetic test results are incidental to medical investigations, the physician is faced with determining whether the patient actually wants to know the information or, as described in Lemmens et al. (2015), prefers to exercise the "right not to know." In the typical course of physician care for patients, it is important that physicians reveal the full truth to patients. When a physician orders medical diagnostic tests and these tests deliver results only in regards to those tests that are clearly known and under-stood by the patient, then ideally the patient would (or should) have given full informed consent to these tests and acceptance of all results that are determined. However, given that individuals are not always fully aware of the battery of tests that a physician orders, full disclosure of test results including possibly inciden-tal findings that are not anticipated by the patient may lead to the patient receiv-ing information that would not be desired.

Primary care physicians must also consider their duty to genetic family members when a genetic disease has been identified within that family. If the disease outcomes can be ameliorated by early treatment, it seems appropri-ate that these individuals should be informed (as discussed earlier in this sec-tion). In such cases where patients refuse to inform their genetic family members, the physician must consider whether, with whom, and, for children, at what age to breach patient confidentiality. This can be important when there is a possibility of averting "serious harm" by doing so. There are precedents where the medical profession has breached patient confidentiality in cases of highly communicable disease. In the interest of protecting public health, doctors are not only encouraged to share information on disease carriers, they are required to do so by law. Current opinion holds that genetic information differs from disease in that genetic information is more intrinsically the person rather than a disease the individual has contracted from some exterior source and the risk of infecting others with the disease is generated directly by the patient. Breaching confidentiality by communicating genetic information about an indi-vidual may in some circumstance reduce risk to genetic family members by allowing them to increase relevant preventive actions to avoid or ameliorate a given genetic disease. In this circumstance, however, the risk is not generated directly by the patient.

From a legal perspective, the physician duty to warn can be implemented by following a variety of guidelines. As noted in Lemmens et al. (2015), in some cases (in the United States) physician disclosure of a genetic disease to members of a patient's genetic family has been ruled to be met by warning the patient and advising the patient to reveal the information to relevant family members[9] while in another case the court has ruled in favor of a stronger duty (i.e., that the physician was responsible to directly warn relevant family members).[10] In cases where the duty to warn is required by law, a physician may not be able to consider the possibility of genetic any family member who would not want the information (i.e., would prefer to exercise the "right not to know") and of course may not know whether said family member holds such a view. Moreover, the physician may not have a doctor-patient relationship with genetic family members. Recognizing these possibilities creates legal, ethical, and social complexities when designing legal or ethical standards regarding the duty to warn in regards to genetic information.

It is apparent that physician responsibility is a hugely complex matter. There is an interesting school of thought (e.g., Foster et al., 2015) that proposes the family be considered a single unit such as family members holding a joint bank account. Under this philosophical approach, all information is shared and genetic family members become first party automatically rather than third party. Such a perspective is not generally reflected explicitly in legislation but may provide a useful starting point from which to generate a set of guidelines to help physicians decide how to handle the thorny issue of disseminating genetic information to members of a genetic family. There are, however, important limitations to the "joint bank account" approach. One must consider the effect of laws or guidelines for sharing genetic information on patient trust of physicians. Individuals may be less likely to obtain genetic information if they think it will be shared in ways in which they do not approve. In a case study from Turkey, which uses a survey instrument to elicit views of both physicians and patients about the nature of genetic information and how it should be shared, Akpinar and Ersoy (2014) point out the importance of cultural differences. In particular, in posing various questions about disclosure of genetic information, Akpinar and Ersoy (2014, pp. 1–2) remark, "The answer to these

---

9. See Lemmens et al. (2015, p. 3) "In *Pate v. Threlkel*, 661S0.2d 278 (Fl. Supr. Ct., 1995), the Supreme Court of Florida ruled that a physician has a legal 'duty to warn' the children of a person with a genetic form of cancer, but that this duty was fulfilled by warning the patient herself."
10. See Lemmens et al. (2015, p. 3), "... the Superior Court of New Jersey (Appellate Division) recognized a stronger duty, by holding that physicians sometimes have a duty to warn immediate family members themselves, when they know that they are at risk of an avoidable harm from a genetically transmissible condition (*Safer v. Estate of Pack*, 667 A.2d 1188 (N.J. Sup. Ct., 1996))."

questions will be determined by whether we take the Enlightenment-rooted individual of Anglo-Saxon culture or the family as the unit of privacy." This work demonstrates that using the perspective of who owns genetic information as a basis from which to build to generate guidelines or legislation for sharing of genetic information needs to be culture (and so country) specific.

4. CRIMINAL INVESTIGATIONS

The use of DNA testing in criminal investigations opens up a whole series of situations, which lead to sharing genetic information with third parties. Given the proliferation of police procedural dramas in films and television, it is unlikely that people are unaware of the use of forensic DNA evidence in solving crime. Police departments worldwide are touting the benefits of maintaining DNA databases. Williams and Johnson (2005) point out that criminal justice DNA databases, kept in a centralized location, would facilitate more accurate identification of possible offenders and produce more convincing and creditable trial evidence. As well, access to these databases would serve to reduce the costs to the criminal justice system and act as a deterrent to potential criminal offenders. While these reasons, which support keeping a DNA database, seem at first to be logical and useful, there are objections that arise based on concerns for privacy in both the obtaining and the storing of DNA samples. Williams and Johnson (2005) point out that in many cases individuals suspected in criminal activity often have their DNA samples retained even if they are not subsequently convicted of the crime. They comment (p. 551), ... "in stressing the exceptional information richness of genetic material it can also be argued that using DNA in forensic contexts where samples are taken without consent raises new kinds of questions about privacy and the protection of individuals from whom such samples have been taken, interrogated, and stored." The data banks kept by police departments are claimed to be secure and accessed and used only in police investigations. There have been some concerns about the adequacy of the laboratories examining samples as well as the security of the database itself.

While Williams and Johnson's research deals with the DNA database in the United Kingdom, the largest DNA database in the world is in China where over 40 million people have had their DNA collected. The samples are claimed to be for use in criminal investigations but the ESHG (The European Society of Human Genetics) (2017) points out that "Police have taken samples from...dissidents, migrants workers, and ethnic minorities," making it obvious that the process is not confined to only criminal activity.

Davis (2017) reports on the following phenomenon in an article with the alarming headline: "DNA in the Dock: how flawed techniques send innocent people to prison." The article describes a man who was initially convicted of murder using DNA samples retained from a break-in investigation where he was the victim 31 years before. DNA evidence has given the impression of being "bulletproof" to a population immersed in police procedural drama on television and in the movies. Now that minute traces of DNA can be detected and analyzed, more questions need to be asked. There is, for instance, the question of timing. The man in question was accused of murder using decades old DNA samples. He was a cab driver and his DNA could definitely be found in his cab, but also on many people who took his cab over the years. The transfer of DNA from one place to another or one person to another is also cause for concern when DNA is used in criminal investigations. Davis (2017) points out that in one well-known case in Britain, DNA evidence from semen was discovered in clothing several months later even after it had been laundered. Moreover, the same DNA was detected in clothing that had never been in contact with the individual whose DNA it was or with the contaminated clothing except from using the same washing machine. Consider if the clothing had been washed in a public laundromat. It is conceivable that the DNA could be transferred to clothing of innocent individuals in subsequent uses of the same machine.

The above examples highlight the importance of using more than DNA evidence to convict someone of a crime and especially if a DNA database is large. As known from Bayes' Law, the updated belief of a person being truly guilty as a result of an independent piece of evidence such as presence of DNA is still quite low if cross contamination is possible and the number of individuals in the database is large.

5. EMPLOYMENT

It is obvious that employers seek to employ the best possible candidates for positions. Their criteria include not only educational and experience requirements but also candidates who would bring other qualities to the job such as having low absenteeism rates or, when relevant such as for employers in the Unitd States, who self-insure, requiring lower insurance claims due to good health. To this end, having genetic tests that predict diseases may be seen to be a significant asset to employers. However, with very few exceptions, genetic testing has reached neither the point where the onset of disease can be established with a high level of accuracy as seen in the case of monogenetic diseases nor even the ability to significantly improve the prediction of the likelihood of contracting an illness in the case of multifactorial genetic diseases.

Rutter, (2006, p. 223) underlines that "genetic researchers need to take on board the evidence that many genetic effects are contingent upon one or other of several different forms of gene-environment co-action." That is, even members of the same genetic family, carrying the same DNA, will not experience expression of the problem-causing gene in the same way or at all. Using genetic information in decision-making about whether to hire potential job candidates is both ineffectual and discriminatory.

To protect against unfair hiring practices, many countries have enacted laws that prohibit the use of genetic information by employers. Examples include The Genetic Information Non-discrimination Act (GINA) May 2008 in the

United States which assures that genetic information cannot be used in deciding employment or to deny health care.[11] It is noteworthy, however, that a bill recently introduced by the House (https://www.congress.gov/bill/115th-congress/house-bill/1313/all-info) would allow employers to use genetic information (and other information such as is available through so-called wearables or fitness trackers) to provide discounts of up to 30% on their health insurance premiums. Nonparticipants would not be eligible and so would effectively face paying up to 30% more for their health insurance costs.[12] This bill would undo some of the equity enhancements of GINA.[13]

Canada's recently passed Genetic Antidiscrimination Law (May 2017) also protects against employment discrimination based on genetic test results as do many others. For Europe, the Oviedo Convention on Human rights and Biomedicine 1997 formed the guidelines for most European states to enact laws protecting against the use of predictive genetic tests for employment purposes. More specifically, article 12 (Oviedo Convention, 1997, p. 4) states that "Tests which are predictive of genetic diseases or which serve either to identify the subject as a carrier of a gene responsible for a disease or to detect a genetic predisposition or susceptibility to a disease may be performed only for health purposes or for scientific research linked to health purposes, and subject to appropriate genetic counseling."

Otlowski et al. (2012, p. 436) notes "Genetic information has potential value to employers in identifying individuals who may be susceptible to specific workplace hazards or who could pose risks if they develop genetic disease in the future." Taking the perspective of protecting workers with heightened sensitivity to toxins in the workplace, this seems a valuable use of genetic information from both the employer and employee perspective. However, it is also possible that by eliminating individuals with heightened sensitivity to certain toxins or other workplace hazards, employers may be less cautious in controlling such elements in the workplace. This would create greater risks for the (admittedly less susceptible) remaining workers. Such a practice would, of course, also limit employment opportunities for those with heightened sensitivity to workplace hazards although this may be the lesser of two evils. Nonetheless, the end result is increased unfavorable discrimination directed at those with certain genes.

---

11. As noted by Zhang (2017), there are, however, some rather major loopholes in that it does not protect against things like schooling, granting mortgages, and access to housing. Moreover, while GINA protects against discrimination in access to health insurance, it does not include protection against discrimination in access to life insurance or long-term disability insurance.

12. See Belluz (2017) for a discussion on the various types of health information that could be used.

13. This bill is still in committee. Skopos Labs gives the bill only a 12% chance of passing. (See https://www.govtrack.us/congress/bills/115/hr1313 and https://www.skoposlabs.com/, both sites accessed on November 17, 2017.)

6. PHARMACEUTICAL INVESTIGATIONS AND CLINICAL RESEARCH

Large pharmaceutical companies have the means, the manpower, and the facilities for the development and production of new drug therapies as well as drugs and tests that will positively contribute to personalized medicine allowing the medical community to offer more effective treatments for their patients on an individual basis. In their run down on various vendors in the global genetic testing market, Technavio (July 16, 2015), lists Abbott Laboratories' short description of their company. "In an effort to help doctors determine how to best treat their patients, we have pioneered innovative ways to screen, diagnose, and monitor a vast range of health conditions with greater speed, accuracy, and efficiency." Technavio (2015) also points out that "Major segments like pharmacogenomics testing and predictive testing are expected to make significant contributions to the revenue of the overall market in the next few years." With all these financial resources available, there is a proliferation of opportunities in pharmacogenics and publicly funded genetic based research feeding on the significant and encouraging discoveries begun with the Human Genome Project.

In an unusual consideration of who is, in fact, the third party in the sharing of genetic test information, it is interesting to consider the case of clinical trials. Traditionally, clinical trials used for research purposes have been conducted without directly communicating the results to the participants.[14] Results were perhaps (and only perhaps) published in scientific journals and the participants were, of course, free to find and read the results and attempt to discern their meaning. Knoppers et al. (2006) point out that the revised Declaration of Helsinki (2013) insists "negative as well as positive (global) research results should be published or otherwise publically available." The onus remains on the participant to discover where the results are shared and what they contain. In these cases, where normally the individual is the first party and the physician is the second, it turns out that the participant/individual becomes the third party since the researcher is the second party and the research itself becomes the first party.[15] That is; if in the usual understanding of physician/patient relationships, the patient is the first party and the physician is the second party, in research situations, the research results become the first party, the holder of information gleaned from tests. The researcher is the second party, the party who arranged for the tests to be performed and who reacts to the results and the participant

---

14. See Knoppers et al. (2006, p. 1170) who point out: "This approach is problematic in that whereas on the one hand, the return of clinical trial results (Table 2) relevant to health has long been the norm, on the other hand, fundamental research (Table 2) results are by their very nature not individually identifiable, understandable or significant."

15. As noted by a reviewer, this relationship creates a conflict between the dual role of physicians, as agents in both physician-patient relationship, and subject-investigator relationship. According to the principle of double loyalty, the physician-patient relationship should, almost always, prevail. See articles 3 and 8 of the Helsinki Declaration (2013).

becomes the third party, with whom the results of tests may or may not be shared depending on the ethical stance of the researcher. Knoppers et al. (2006) also add that the European Federation of International Epidemiologists Association have updated their position on sharing results from research to add "It is advisable to publish the main results in a form that reaches the participants in the study..."[16] Even here the participant remains the third party, only having indirect access to results.

However, concurrent with the advances in the field of genomics (and perhaps because of them) there have also been changes to the approach to biomedical research. Traditional clinical research had as a goal the discovery of previously uncertain realities. The research faced problems with finding participants, disseminating information, finding well-trained technicians and, of course, funding. The move to what is being termed "translational research" requires the progression from fundamental research to development of diagnostic tests, treatments, and pharmaceuticals, which would "translate" to improved health services (see Sung et al., 2003). It is expected that these changes will also result in greater funding since the possibilities for immediate returns from investment are vastly improved. Because of the changing direction of research, there is also a change in the availability and expectations of participants. Individuals who feel they will be given specific information about their health from participation in research are more likely to take part in research. The participant returns to a first party position when the research team contracts to return test results directly to the individual. The challenge for the research team becomes how to share the information successfully and in an ethical manner. In the case of genetic research, results may have negative impact on genetic family members who played no part in giving informed consent.

The development of a protocol for sharing genetic information resulting from research faces many logistical as well as ethical and legal hurdles. There is the presumption that each participant has given informed consent to be part of a study. This is the first hurdle. As noted by Mehlman (2017a,b, p. 1115), the "complexity of genomic science exacerbates the difficulty of fulfilling one of the core requirements, which is to obtain informed consent from competent adults before they serve as research subjects." Assuring informed consent for genomic research demands that a competent instructor remains available to respond to participants' further and repeated questioning. It also means that an individual needs to be informed on the possibility that any information received may be very important to members of his genetic family. Presuming

---

16. See also articles 24 and 26 of the Helsinki Declaration (2013). The former (article 24) states that "In advance of a clinical trial, sponsors, researchers and host country governments should make provisions for post-trial access for all participants who still need an intervention identified as beneficial in the trial. This information must also be disclosed to participants during the informed consent process." Article 26 includes that "All medical research subjects should be given the option of being informed about the general outcome and results of the study."

informed consent has been given and the research undertaken, the next obstacle becomes, Who is responsible for returning pertinent information to the individual participant? It is obvious that the person who delivers the results to individual participants must be in a position to fully explain the meaning of these results. A genetic counselor or the treating physician are natural candidates for this role. Other suggestions include (Knoppers et al., 2006, p. 1175) "a physician whose competence permits a full explanation of the significance of the results," the genetic researcher, or perhaps national research granting agencies (see Knoppers et al., 2006, p. 1176).[17] After the results are shared by whoever is deemed appropriate, it is important to provide on-going genetic counseling. In Canada, "Both researchers and the institutional review board now have to ensure the availability of such counseling when appropriate." (again see Knoppers et al., 2006, p. 1176).

At this point in the advancement of genomic investigations in research, there is no global protocol for delivering research results to participants and their genetic families. McGuire et al. (2008) make strong recommendations, encouraging the development of a formal research protocol which includes specifics on how data will be returned to participants, further training for primary care physicians so they will be capable of clearly and accurately delivering results and ongoing updates to the participants' health records as the research progresses.[18] As for sharing results with those they term "close genetic relatives," the stage of assuring informed consent should include making it understood by the participant that resulting data may be of concern to family members. Participants should be encouraged by the research team to share pertinent information with genetic family members and offers of help in doing so should be included. Reporting relevant results to affected family members should be undertaken using the same guidelines followed in clinical discovery of genetic problems that affect families. Sharing information with third parties should be decided on a case-by-case basis and depend on the possibility of helping the third party avoid negative outcomes.[19]

On the other side of this argument, Wright Clayton and McGuire (2012, p. 475) point out not only the increasing costs involved in returning results to individuals but also the fact that "the more encompassing guidelines and practices are with regard to return of results, the more sweeping the potential ethical and legal obligation." Added to these concerns for the researcher is the question of "how long the duty to disclose should extend." The suggested length of time is usually until the end of the funding. Thus far there are no global ethical or legal guidelines in place for return of individual genetic information to research

---

17. Knoppers et al. (2006, p. 1175) notes that "Canada's three granting councils, in their Policy Statement, seem to recommend that the genetic researcher be the one to report results back to the individuals."

18. See their recommendations 1.1, 1.2, and 1.3 (McGuire et al., 2008, pp. 6–7).

19. See their recommendations 2.1, 2.2, and 2.3 (McGuire et al., 2008, p. 7).

participants. There are recommendations by working groups and published opinions by academics in the fields of medicine and law but no absolute consensus has been reached nor legislation enacted.

7. INSURANCE

It is not surprising that insurance companies selling life, health, or long-term care insurance would like to access the results of genetic tests taken by their customers. Just like any medical information (e.g., blood cholesterol levels), such results may have predictive power for future mortality or morbidity risks and so provide insurers with valuable underwriting information. However, many people feel information about their DNA sequence is more personal than most other types of health information and many others feel it unfairly discriminatory for insurers to use genetic information to assess higher prices or even restrict access to insurance. Article 4 of the UNESCO International Declaration on Human Genetic Data (2004, pp. 7–8) stipulates that genetic data have a special status for reasons including that such information holds cultural significance for peoples or groups." As Wolpe (1997, pp. 216–218) points out in his comments on Dorothy Nelkin and Susan Lindee's The DNA Mystique, "We have biologized the idea of the soul, the essence of each person, that which can recreate us and which encapsulates all that we are," and further that "Genes are, in some sense, more truly us than we ourselves are; they are us condensed, boiled down, reduced to essence." It can therefore be argued that it is appropriate to treat genetic data differently from other medical information such as one's blood pressure, for special treatment within privacy laws.

Citizen advocacy groups, such as the Canadian Coalition for Genetic Fairness (http://ccgf-cceg.ca/en/home/) favor laws that restrict insurers from using genetic test results in their pricing formulas.[20] There is broad citizen support for such restrictions. For example, as reported by Bueckart (2004), a 2003 poll by Pollara-Earnscliffe found that 91% of Canadian respondents agreed that insurance companies should not be allowed to use genetic test results in pricing contracts.

The issue of genetic discrimination in insurance markets has received a great deal of attention by scholars in many fields of inquiry. The question of whether insurers should be allowed access to genetic test results of potential clients remains hotly debated. A debate and desire to influence public opinion has been waged by privacy groups, who favor banning insurers from accessing genetic test results of consumers with the insurance industry. Below are outlined the recent history of legislative and other responses to this concern. This is followed

---

20. A questionnaire administered in Canada to asymptomatic people at risk for Huntington's disease in 2006 found that 86% of respondents feared genetic discrimination in some sphere or other with 29.2% reporting experiences of discrimination in insurance (see Bombard et al., 2009).

by a discussion of the arguments in favor and against restricting insurers' use of genetic information of consumers.

Most Western European countries have laws or voluntary bans agreed to by the insurance industry that severely restrict or completely ban the use of genetic tests for ratemaking purposes. In some cases, such as in the United Kingdom and the Netherlands, insurers may not ask for or use genetic test results for purchases below a specific limit of coverage (£500,000 in the United Kingdom and $150,000 USD equivalent in the Netherlands, enacted in 1997 with the limit updated every 3 years to the cost of living). In 2008, the United States passed the Genetic Information Nondiscrimination Act (GINA), which covers health insurance and employment. Canada passed a genetic discrimination law, Bill S-201, in May 2017. It prevents insurance companies from using results of any genetic tests to determine coverage or pricing. Joly et al. (2010) provides a comprehensive review of international legislation.

The insurance industry concern about such bans is their belief that not allowing insurers access to the same information that their consumers have will mean that higher risk types will purchase more insurance than moderate or low risk types and this will increase the rate of claims per policy issued, a phenomenon known as adverse selection. This, it is often claimed, will lead to higher prices and/or reduced profits. In a letter from the American Academy of Actuaries to law makers in the House of Representatives, it was even claimed that such a ban "would have a direct impact on premium rates, ultimately raising the cost of insurance to everyone."[21] The possibility that even the highest risk types would pay a higher price under a ban than in the absence of a ban seems unlikely. In fact, the extent to which the average price of insurance would rise in the presence of a ban is questionable. This is a key factor in determining whether genetic nondiscrimination laws are in fact detrimental to the effective functioning of insurance markets.

There is a significant amount of research in economics that points to potentially serious problems arising from adverse selection.[22] If the fraction of high-risk types within a population is sufficiently high and insurers are either not able or not allowed by law to identify them and assess a higher price to them, then the market will lead to low-risk types ending up with relatively lower coverage of insurance. This is the result of insurers having to charge the same price to all so that low-risk types are not willing to pay for insurance. Insurers respond by charging an even higher price to cover the higher frequency of claims of the high-risk types. In principle (or theory), this can even lead to only the highest risk types in the population ending up buying any insurance—a phenomenon known as the adverse selection death spiral.[23] Such an outcome, however, is

---

21. See Durnin et al. (2012, pp. 134–135) for details.
22. Early studies include Hoy (1982) and Crocker and Snow (1986). For a comprehensive review of this economics literature see Dionne and Rothschild (2014).
23. For good descriptions of this possibility see Hendren (2014) and Butler (2002).

an extreme case and currently there are relatively few individuals in the population who have actuarially significant information about higher expected claims due to genetic test results as has been noted in earlier sections of this paper.

Some careful work has been done by Macdonald (2009, 2012) demonstrating the lack of an abundance of actuarially significant information from genetic tests relating especially to mortality risk. Therefore the simplistic notion of a market collapse or an adverse selection death spiral would not plausibly follow from banning the use of genetic information in insurance pricing. More nuanced actuarial and economic analysis is required to assess the desirability of genetic antidiscrimination laws. Fundamental examination about how society should judge alternative allocation rules is raised by this question.

First consider a very pragmatic perspective, which addresses the impact on insurance prices and coverages under a ban compared to no ban. We will consider the implications for the life insurance market but similar implications would arise for other forms of insurance such as health and long-term care insurance. Under a ban the various risk types of insureds will be pooled together and face a single price that is more favorable to high-risk types but less favorable to low-risk types compared to the outcome without a ban. It is reasonable to expect some adverse selection will arise. That is, individuals with genetic test results that indicate they are at higher risk for premature death will find the price more favorable given their high-risk level than would individuals without such information. There will be at least some additional insurance purchased (on average) by higher risk types and this will lead to a somewhat higher price for the others than would happen if there were no ban in place and insurers separated the risk types into separate classes and charged risk type specific prices. Especially in the case where the differential (genetic) risk is entirely due to inherited genes, many people would say that risk-rated prices in the absence of a regulatory ban represents unfair price discrimination.[24] On the other hand, lower risks are less likely to purchase insurance or (on average) will purchase a lower level of insurance, which is indeed a disadvantage of a ban. But one must also consider that higher risks end up with more coverage since the pooling price is more favorable than the price they face without a ban. Interestingly, Thomas (2008, 2017) demonstrates that, on balance, it is quite possible that more losses will be covered overall. Since one can reasonably argue that the purpose of insurance to both individuals and society as a whole is to offer compensation for losses, Thomas concludes that a little bit of adverse selection is a good thing. Adopting a utilitarian approach, Hoy (2006) concurs that at least

---

24. Insurers generally take the opposing view that higher risk types impose higher costs on insurers and so *should* be assessed a higher price *to avoid* price discrimination. Hoy and Lambert (2000) show that even taking such a premise that prices should reflect expected costs of insurance does not necessarily mean that price discrimination is worsened by charging a single price for insurance when, as is typical, the information is not a perfect predictor of risk class.

for cases where there is a "small enough" fraction of high-risk types in the population, the result of a ban that creates just "a little adverse selection" is higher aggregate utilitarian welfare.

Aside from allocative issues related to pricing and coverage of insurance, there are additional pragmatic reasons to favor a ban on insurers using genetic test results to price insurance differentially. Consider the case where many people are unaware of their genotype relating to a disease such as colon cancer for which prevention or monitoring may be effective health strategies. Given the possibility of being placed into a high-risk class for future insurance purchases, some people may be deterred from obtaining otherwise useful information from a genetic test; i.e., they worry more about reclassification risk than about potentially improving health-care decisions. Aside from creating possible harm for individuals themselves, Filipova-Neumann and Hoy (2014) show that foregoing genetic tests may lead to higher costs for public health providers. There are also significant concerns about the impact on family members. An individual who is beyond the age of wanting to purchase future life insurance coverage may feel no concern about her own reclassification risk and so on that basis would accept genetic tests that may prove helpful to health-care decisions. If insurers can use that information should any of the person's children apply for life (or other) insurance, however, then she may be reticent to obtain genetic tests given the implications for reclassification risk imposed on her children.

A detailed exploration of some of the issues in this section can be found in Durnin and Hoy (2012) and Durnin et al. (2012). It is argued there that, given the current modest amount of genetic information of actuarial significance for both mortality and morbidity risks, the pros of a ban on insurers' use of information clearly outweigh the cons. Therefore antigenetic discrimination laws or voluntary moratoria by the insurance industry are appropriate at least for the near and medium term future. The promise of affordable and highly specific genetic information discovered through whole genome sequencing, for example, may eventually change the insurance market environment. It is not out of the realm of possibility that eventually enough individuals will possess a sufficient amount of actuarially relevant genetic information that the effective functioning of insurance markets may be compromised by a ban. That is, it is possible that efficiency concerns from more intense adverse selection will dwarf equity considerations and a ban on insurers' use of genetic information for risk-rating purposes will not promote the social good. It is, however, also possible that this scenario will never happen. By the time such a large amount of genetic information that holds highly predictive value about the diversity of mortality and/or morbidity risk exists, it may also be the case that many genetic diseases will have effective treatments which counteract that differential effect on actuarially based insurance prices. If so, then a ban on insurers' use of genetic information will continue to promote the social good. Long-term predictions, however, are admittedly difficult to make in this area.

8. CONCLUSIONS

The issue of third party sharing of genetic information is multifaceted and complex. This chapter presents six situations that vary from genetic family members as third parties in the usual patient-doctor relationship to researchers (both private and public), employers, insurers, and criminal investigators as third parties. In these cases, third party interests are identified along with their reasons for wanting or not wanting access to the information available. All scenarios raise critical concerns about privacy and discrimination as well as potential effectiveness of genetic information and research.

In the cases of employment and insurance, arguments have mostly fallen on the side of the rights of privacy for individuals and substantial legal protection, which prohibits the sharing of genetic information with these third parties, has been enacted in many countries. In the fields of research and other medical uses of genetic information, there is still much discussion on how to handle the rights of individuals to privacy and the rights of interested third parties. This is especially important when the third parties are members of the first party's genetic family. Given that many individuals prefer to exercise their right not to know such information, it is a very difficult task to decide when the potential of "serious harm" to third parties should dominate the right of privacy of first parties. Given that the physician or clinician is typically the individual who must decide whether to honor privacy rights or the legitimate interests of genetic family members, it is not surprising that no universally accepted guidelines have been developed since decisions need to be made on a case-by-case basis.

The following general principle from the World Medical Association Declaration of Helsinki Ethical Principles for Medical Research Involving Human Subjects (World Medical Association, 2013, p. 2191) represents a solid foundation for many of the concerns expressed in this chapter about third party sharing of genetic information. "10. Physicians must consider the ethical, legal and regulatory norms and standards for research involving human subjects in their own countries as well as applicable international norms and standards. No national or international ethical, legal or regulatory requirement should reduce or eliminate any of the protections for research subjects set forth in this Declaration." Their mandate is restricted to physicians, but they recommend that their guidelines be extended to all those who conduct research on human subjects. This is further strengthened by Article 10 of the UNESCO Universal Declaration on the Human Genome and Human Rights (2004), which states "No research or research applications concerning the human genome, in particular in the fields of biology, genetics and medicine, should prevail over respect for the human rights, fundamental freedoms and human dignity of individuals or, where applicable, or groups of people."

In both the case of existing legislation for insurance and employment purposes and ethical guidelines for clinical research and the doctor-patient relationships, ongoing review is required as the field of medical genetics progresses rapidly.

REFERENCES

Akpinar, A., Ersoy, N., 2014. Attitudes of physicians and patients towards disclosure of genetic information to spouse and first-degree relatives: a case study from Turkey. BMC Med. Ethics 15 (39), 1–10.

Belluz, J., 2017. A new bill would allow employers to see your genetic information. Voxmedia, https://www.vox.com/policy-and-politics/2017/3/13/14907250/hr1313-bill-genetic-information.

Bombard, Y., et al., 2009. Perceptions of genetic discrimination among people at risk for Huntington's disease: a cross sectional survey. BMJ 338.

Bueckart, D., 2004. 'Hands off my genes, Canadians say', The Globe and Mail. 14 March. https://beta.theglobeandmail.com/news/national/hands-off-my-genes-canadians-say/article1128996/?ref=http://www.theglobeandmail.com.

Butler, J., 2002. Policy change and private health insurance: did the cheapest policy do the trick? Aust. Health Rev. 25 (6), 33–41.

Crocker, K., Snow, A., 1986. The efficiency effect of categorical discrimination in the insurance industry. J. Political Econ. 94 (2), 321–344.

Davis, N., 2017. DNA in the dock: how flawed techniques send innocent people to prison. Guardian. https://www.theguardian.com/science/2017/oct/02/dna-in-the-dock-how-flawed-techniques-send-innocent-people-to-prison.

Dionne, G., Rothschild, C., 2014. The economic effects of risk classification bans. Geneva Risk Insurance Rev. 39 (2), 184–221.

Durnin, M., Hoy, M., Ruse, M., 2012. Genetic testing and insurance: the complexity of adverse selection. Ethical Perspect. 19 (1), 123–154.

Durnin, M., Hoy, M., 2012. The potential economic impact of a ban on the use of genetic information for life and health insurance, office of the privacy commissioner of Canada. https://www.priv.gc.ca/en/opc-actions-and-decisions/research/explore-privacy-research/2012/gi_hoy_201203/.

ESHG (The European Society of Human Genetics) 2017, China's mass collection of human DNA without informed consent is contrary to the right to privacy, Press release, 13 June, 2017 url: https://www.eshg.org/13.0.html (accessed October 27, 2017)

Filipova-Neumann, L., Hoy, M., 2014. Managing genetic tests, surveillance, and preventive medicine under a public health insurance system. J. Health Econ. 34, 31–41.

Foster, C., Herring, J., Boyd, M., 2015. Testing the limits of the 'joint account' model of genetic information: a legal thought experiment. J. Med. Ethics 41, 379–382.

Hendren, N., 2014. Unravelling vs unravelling: a memo on competitive equilibriums and trade in insurance markets. Geneva Risk Insurance Rev. 39 (2), 176–183.

Hoy, M., 1982. Categorizing risks in the insurance industry. Q. J. Econ. 97 (2), 321–336.

Hoy, M., Lambert, P., 2000. Genetic screening and price discrimination in insurance markets. Geneva Pap. Risk Insurance Theory 25 (2), 245–269.

Hoy, M., 2006. Risk classification and social welfare. Geneva Pap. Risk Insurance Issu. Pract. 31 (2), 245–269.

Joly, Y., Braker, M., Huynh, M.L., 2010. Genetic discrimination in private insurance: global perspectives. N. Genet. Soc. 29 (4), 351–368.

Knoppers, M.B., Joly, Y., Simard, J., Durocher, F., 2006. The emergence of an ethical duty to disclose genetic research results: international perspectives. Eur. J. Hum. Genet. 14, 1170–1178.

Lemmens, T., Luther, L., Hoy, M., 2015. Genetic Information Access, a Legal Perspective: A Duty to Know or a Right Not to Know, and a Duty or Option to Warn? eLS. John Wiley & Sons, Ltd., Chichester. http://www.els.net.

Macdonald, A.S., 2009. Genetic Factors in Life Insurance: Actuarial Basis. Encyclopedia of Life Sciences (ELS), John Wiley & Sons, Ltd., Chichester, pp. 1–5.

Macdonald, A.S., 2012. The actuarial relevance of genetic information in the life and health insurance context. Office of the Privacy Commissioner of Canada, https://www.priv.gc.ca/en/opc-actions-and-decisions/research/explore-privacy-research/2011/gi_macdonald_201107/.

McCarthy, M.I., MacArthur, D.G., 2017. Human disease genomics: from variants to biology. Genome Biol. 18 (20), 1–3.

McGuire, L.A., Caulfield, T., Cho, K.M., 2008. Research ethics and the challenge of whole-genome sequencing. Natl. Inst. Health 9 (2), 152–156.

Mehlman, M.J., 2017a. Genomics and The Law. The Oxford Handbook of Health Law, New York, pp. 1113–1133.

Mehlman, M.J., 2017b. Genomics and the law. In: Cohen, I.G., Hoffman, A.K., Sage, W.M. (Eds.), The Oxford Handbook of Health Law. Oxford University Press, New York, pp. 1111–1133.

Nixson, W., 2017. Has the right to breach patient confidentiality created a common law duty to warn genetic relatives. QUT Law Rev. 17 (1), 147–159.

Otlowski, M., Taylor, S., Bombard, Y., 2012. Genetic discrimination: international perspecitves. Annu. Rev. Genomics Hum. Genet. 13, 433–454.

Oviedo Convention: Oviedo convention: a european legal framework at the intersection of human rights and health law, 1997, https://rm.coe.int/168007cf98 (accessed November 13, 2017).

Petrucelli, M., Daly, M.B., Pal, T., 2015. BRCA1- and BRCA2-Associated Hereditary Breast and Ovarian Cancer, GeneReviews [Internet]. https://www.ncbi.nlm.nih.gov/sites/books/NBK1247/.

Powles, J., 2017. Why are we giving away our most sensitive health data to Google? The Guardian, 5 July, viewed 5 July 2017, https://www.theguardian.com/commentisfree/2017/jul/05/sensitive-health-information-deepmind-google.

Robson, E.M., Storm, D.C., Weitzel, J., Wollins, S.D., Offit, K., 2010. American society of clinical oncology policy statement update: genetic and genomic testing for cancer susceptibility. J. Clin. Oncol. 28 (5), 893–901.

Rutter, M., 2006. Genes and Behavior: Nature-Nurture Interplay Explained. Blackwell Publishing, Oxford, UK.

Slawson, N., 2017. Test for breast cancer risk could reduce pre-emptive mastectomies. The Guardian, 8 Oct, https://amp.theguardian.com/society/2017/oct/08/test-for-breast-cancer-risk-could-reduce-pre-emptive-mastectomies. (accessed 14 November 2017).

Sung, S.N., Crowley Jr., F.W., Genel, M., Salber, P., Sandy, L., Sherwood, M.L., Johnson, B.S., Catanese, V., Tilson, H., et al., 2003. Central challenges facing the national clinical research enterprise. JAMA 289 (10), 1278–1287.

Technavio, 2015. Top companies in the global testing market. https://www.technavio.com/blog/top-companies-in-the-global-genetic-testing-market.

Thomas, R.G., 2008. Loss coverage as a public policy objective for risk classification schemes. J. Risk Insur. 75 (4), 997–1018.

Thomas, R.G., 2017. Loss Coverage: Why Insurance Works Better with Some Adverse Selection. Cambridge University Press, Cambridge, UK.

UNESCO 2004, International Declaration on Human Genetic Data, United Nations Educational, Scientific and Cultural Organization; Division of Ethics of Science and Technology Social and Human Sciences Sector, Paris, France.

Williams, R., Johnson, P., 2005. Inclusiveness, effectiveness and intrusiveness: issues in the developing uses of DNA profiling in support of criminal investigations. J. Law Med. Ethics 33 (3), 545–558.

Wolpe, P.R., 1997. If I am only my genes, what am I? Genetic essentialism and a Jewish response. Kennedy Inst. Ethics J. 7 (3), 213–230.

World Medical Association, 2013. World medical association declaration of Helsinki ethical principles for medical research involving human subjects. JAMA 310 (20), 2190–2194.

Wright Clayton, E., 2003. Ethical, legal, and social implications of genomic medicine. N. Engl. J. Med. 349 (6), 562–569.

Wright Clayton, E., McGuire, A.L., 2012. The legal risks of returning results of genomics research. Genet. Med. 14 (4), 473–477.

Zhang, S., 2017. The loopholes in the law prohibiting genetic discrimination. In: The Atlantic Daily. https://www.theatlantic.com/health/archive/2017/03/genetic-discrimination-law-gina/519216/.

FURTHER READING

Botkin, R.J., Belmont, W.J., Berg, S.J., Berkman, E.B., et al., 2015. Points to consider: ethical, legal, and psychosocial implications of genetic testing in children and adolescents. Am. J. Human Genet. 97 (1), 6–21.

Burke, W., Antommaria, H.M.A., Bennett, R., Botkin, J., et al., 2013. Recommendations for returning genomic incidental findings? We need to talk!. Genet. Med. 15 (11), 854–859.

Clarke, A., Richards, M., Kerzin-Storrar, L., Halliday, J., Young, A.M., et al., 2005. Genetic professionals' reports of nondisclosure of genetic risk information within families. Eur. J. Hum. Genet. 13, 556–562.

Fabsitz, R.R., McGuire, A., Sharp, R.R., Puggal, M., et al., 2010. Ethical and practical guidelines for reporting genetic research results to study participants. J. Am. Heart Assoc. 3, 574–580.

GeneWatch, U.K., 2008. Genetics and 'predictive medicine': selling pills, ignoring causes. May, viewed 22 June 2017, https://www.quackwatch.org/01QuackeryRelatedTopics/Tests/gpm.html.

Gillis, C., 2015. D.I.Y. DNA: genetic testing at home is risky business. *Macleans,* 23 March, viewed 22 June 2017, http://www.macleans.ca/society/health/d-i-y-dna-genetic-testing-at-home/.

Hogg, P., 2015. How genome sequencing is aiding drug research and development. ProClinical Life Sciences Recruitment Blog, 6 July, https://blog.proclinical.com/how-genome-sequencing-is-aiding-drug-research-and-development.

Lewis, C.M., Whitwell, C.L.S., Forbes, A., Sanderson, C., Mathew, G.C., 2007. Estimating risks of common complex diseases across genetic and environmental factors: the example of Crohn disease. J. Med. Genet. 44, 689–694.

Linde, V. D. D 2015, 'Valeant Pharmaceuticals pricing controversy attracts scrutiny', *Macleans,* 23 march, viewed 22, http://www.pressreader.com/canada/calgary-h'Humanerald/20151016/282067685775652/TextView.

Offit, K., Groeger, E., Turner, S., Wadsworth, A.E., Weiser, A.M., 2004. The "duty to warn" a patient's family members about hereditary disease risks. JAMA 292 (12), 1469–1473.

Pelias, M.K., 2004. Research in human genetics: the tension between doing no harm and personal autonomy. Clin. Genet. 67, 1–5.

Rasmi, A 2017, 'DNA discrimination', Maisonneuve, 16 March, viewed 2 August 2017, https://maisonneuve.org/article/2017/03/16/dna-discrimination/.

Sankar, P., Moran, S., Merz, F.J., Jones, L.N., 2003. Patient perspectives on medical confidentiality. J. Gen. Intern. Med. 18, 659–669.

Virus Works with Gene to cause Crohn's-Like Illness, 2010. ScienceDaily, 25 June, http://www.sciencedaily.com/releases/2010/06.

Wilson, J.B., Nicholls, G.S., 2015. The humane genome project, and recent advances in personalized genomics. Risk Manage. Healthc. Policy 8, 9–20.

Wright Clayton, E., 2008. Incidental findings in genetics research using archived DNA. J. Law Med. Ethics 36 (2), 286–291.

Index

Note: Page numbers followed by *f* indicate figures, *t* indicate tables, *b* indicate boxes, and *np* indicate footnotes.